Studies in Computational Intelligence

Volume 741

Series editor

Janusz Kacprzyk, Polish Academy of Sciences, Warsaw, Poland
e-mail: kacprzyk@ibspan.waw.pl

The series "Studies in Computational Intelligence" (SCI) publishes new developments and advances in the various areas of computational intelligence—quickly and with a high quality. The intent is to cover the theory, applications, and design methods of computational intelligence, as embedded in the fields of engineering, computer science, physics and life sciences, as well as the methodologies behind them. The series contains monographs, lecture notes and edited volumes in computational intelligence spanning the areas of neural networks, connectionist systems, genetic algorithms, evolutionary computation, artificial intelligence, cellular automata, self-organizing systems, soft computing, fuzzy systems, and hybrid intelligent systems. Of particular value to both the contributors and the readership are the short publication timeframe and the world-wide distribution, which enable both wide and rapid dissemination of research output.

More information about this series at http://www.springer.com/series/7092

Ivan Zelinka · Pandian Vasant
Vo Hoang Duy · Tran Trong Dao
Editors

Innovative Computing, Optimization and Its Applications

Modelling and Simulations

Editors
Ivan Zelinka
Department of Computer Science
Faculty of Electrical Engineering and Computer Science, VŠB-TU Ostrava
Ostrava-Poruba
Czech Republic

Pandian Vasant
Faculty of Science and Information Technology
Universiti Teknologi PETRONAS
Teronoh, Perak
Malaysia

Vo Hoang Duy
Ton Duc Thang University
Ho Chi Minh
Vietnam

Tran Trong Dao
Ton Duc Thang University
Ho Chi Minh
Vietnam

ISSN 1860-949X ISSN 1860-9503 (electronic)
Studies in Computational Intelligence
ISBN 978-3-319-88357-1 ISBN 978-3-319-66984-7 (eBook)
https://doi.org/10.1007/978-3-319-66984-7

Printed on acid-free paper

This Springer imprint is published by Springer Nature
The registered company is Springer International Publishing AG
The registered company address is: Gewerbestrasse 11, 6330 Cham, Switzerland

COMPSE 2016 was organized by The European Alliance for Innovation (EAI). This conference could not have been organized without the strong support and help from the staff members of Golden Sands Resort and the organizing committee of COMPSE 2016. We would like to sincerely thank Prof. Imrich Chlamtac (University of Trento and Create-NET), Barbara Fertalova (EAI), and Lucia Zatkova (EAI) for their great help and support in organizing the conference. We also appreciate the fruitful guidance and help from Prof. Gerhard Wilhelm Weber (Middle East Technical University, Turkey), Prof. Rustem Popa (Dunarea de Jos" University in Galati, Romania), Prof. Goran Klepac (Raiffeisen Bank Austria D.D., Croatia), Prof. Leopoldo Barron (Tecnológico de Monterrey, Mexico), Prof. Ivan Zelinka (VSB-TU Ostrava, Czech Republic), Dr. Jose Antonio Marmolejo (Universidad Anahuac Mexico Norte, Mexico), Mr. Kalaivanthan (Golden Sands Resort, Malaysia) and Dr. Vo Hoang Duy (Ton Duc Thang University, Vietnam).

Foreword

The first *International Conference on the Computer Science and Engineering* (*COMPSE*), ***COMPSE 2016***, was held during November 11–12, 2016, in Penang, Malaysia. The objective of this conference was to bring together experts and scientists of the research areas ***Computer Science*** and ***Optimization*** from all over the world to exchange their experiences and knowledge on contemporary and state-of-the-art research achievements in those fields. These two areas are a **Key Technology** of the world of today and tomorrow, together with ***Operational Research*** and ***Big Data Analytics***. This conference gave a valuable opportunity for the international research community to closely interact based on newest and innovative findings, discoveries, and innovations between old and new colleagues and friends, and to discuss, formulate and initiate and new and high-quality research projects.

For this firsst edition of *COMPSE*, the *Program Committee* received over 100 submissions from 25 countries and each paper was reviewed by four international reviewers. The prominent technical committee selected 44 papers for final presentation at *COMPSE 2016* in Penang, Malaysia. Excellent and high-quality works were selected and reviewed by *International Program Committee* in order to publish extended versions of them in the ***Studies in Computational Intelligence Book Series*** with ***Springer Verlag***.

This entire academic process led to the subsequently list of the chapters and their authors:

Chapter "A Performance Analysis of DF Model in the Energy Harvesting Half-duplex and Full-Duplex Relay Networks", by Tam Nguyen Kieu, Nhu Nguyen Hong, Long Nguyen Ngoc, Thanh-Duc Le, Thuan Do Dinh, Jaroslav Zdralek and Miroslav Voznak;

Chapter "A Model of Swarm Intelligence Based Optimization Framework Adjustable According to Problems", by Utku Köse and Pandian Vasant;

Chapter "Domain Model Definition for Domain-Specific Rule Generation Using Variability Model", by Neel Mani, Markus Helfert and Pandian Vasant;

Chapter "A Set-Partitioning-Based Model to Save Cost on the Import Processes", by Jania Astrid Saucedo Martínez, José Antonio Marmolejo Saucedo and Pandian Vasant;

Chapter "Combining Genetic Algorithm with Variable Neighborhood Search for MAX-SAT", by Noureddine Bouhmala and Kjell Ivar Øverg°ard;

Chapter "Enzyme Classification on DUD-E Database Using Logistic Regression Ensemble (Lorens) ", by Heri Kuswanto, Jainap N. Melasasi and Hayato Ohwada;

Chapter "Consolidation of Host-Based Mobility Management Protocols in Wireless Mesh Network", by Wei Siang Hoh, Bi-Lynn Ong R Badlishah Ahmad and Hasnah Ahmad;

Chapter "Application of Parallel Computing Technologies for Numerical Simulation of Air Transport in the Human Nasal Cavity", by Alibek Issakhov and Aizhan Abylkassymova;

Chapter "Genetic Algorithms-Based Techniques for Solving Dynamic Optimization Problems with Unknown Active Variables and Boundaries", by Abdelmonem Fouad, Daryl L. Essam and Ruhul A. Sarker;

Chapter "Text Segmentation Methods: A Critical Review", by Irina Pak and Phoey Lee Teh;

Chapter "On-Line Power Systems Security Assessment Using Data Stream Random Forest Algorithm Modification", by Aleksei Zhukov, Nikita Tomin, Denis Sidorov, Victor Kurbatsky and Daniil Panasetsky;

Chapter "Enhanced Security of Internet Banking Authentication with Extended Honey Encryption (XHE) Scheme", by Tan Soo Fun and Azman Samsudin;

Chapter "An Enhanced Possibilistic Programming Model with Fuzzy Random Confidence-Interval for Multi-Objective Problem", by Nureize Arbaiy, Noor Azah Samsudin, Aida Mustapa, Junzo Watada and Pei-Chun Lin;

Chapter "A Crowdsourcing Approach for Volunteering System", by Nurul-hasanah Mazlan, Sharifah Sakinah Syed Ahmad and Massila Kamalrudin;

Chapter "One Dimensional Vehicle Tracking Analysis in Vehicular Ad hoc Networks", by Ranjeet Singh Tomar and Mayank Satya Prakash Sharma;

Chapter "Parallel Coordinates Visualization Tool on the Air Pollution Data for Northern Malaysia ", by J. Joshua Thomas, Raaj Lokanathan and Justtina Anantha Jothi;

Chapter "Application of Artificial Bee Colony Algorithm for Model Parameter Identification", by by Olympia Roeva;

Chapter "A Novel Weighting Scheme Applied to Improve the Text Document Clustering Techniques", by Laith Mohammad Abualigah, Ahamad Tajudin Khader and Essam Said Hanandeh;

Chapter "A Methodological Framework to Emulate The Visual Of Malaysian Shadow Play With Computer-Generated Imagery", by Kheng-Kia Khor;

To all the ***authors*** of this valuable book, I extend my cordial appreciation and thanks for having shared their devotion and expertise with the academic family and mankind. Furthermore, I convey my gratitude to the four ***editors*** of this compendium, *Prof. Dr. Ivan Zelinka*, *Prof. Dr. Pandian Vasant*, *Prof. Dr. Vo Hoang Duy*, and *Prof. Dr. Tran Trong Dao*, for their hard work and vision, of having gathered such a remarkable variety of rich contributions. Finally, I am very thankful to the publishing house of ***Springer Verlag***, for having ensured and made reality a premium work of applied importance and of future impact for tomorrow's world and the next generations.

Now, I wish us all a lot of joy when reading this interesting book, and I hope that a great benefit will be gained from it personally and societally.

May 2017

Prof. Dr. Gerhard-Wilhelm Weber
Middle East Technical University, Ankara, Turkey

Gerhard-Wilhelm Weber is a Professor at Institute of Applied Mathematics, METU, Ankara, Turkey. His research is on optimization and optimal control (continuous, discrete, and stochastic), Operational Research, financial mathematics, economics, on earth, bio, neuro and human sciences, dynamics, data mining, statistical learning, inverse problems, and development. He is involved in the organization of scientific life internationally. He received both his Diploma and Doctorate in mathematics and economics/business administration at Aachen University of Technology (RWTH Aachen), and his Habilitation at Darmstadt University of Technology (TU Darmstadt). He held Chair Professorships by proxy at the University of Cologne, Germany, and Chemnitz University of Technology, Germany, before he worked at Cologne Bioinformatics Center and then, in 2003, went to Ankara. At IAM, METU, he is in the Programs of Financial Mathematics and Scientific Computing. He is a member of five further graduate schools, departments, and institutes of METU, and has different affiliations at University of Siegen (Germany), University of Aveiro (Portugal), Federation University (Ballarat, Australia), and University of North Sumatra (Medan, Indonesia), as well as at EURO and IFORS.

Preface

The first edition of the International Conference on the Computer Science and Engineering (COMPSE), COMPSE 2016 was held during November 11–12, 2016, at **Golden Sands Resort** in Penang, Malaysia. The objective of the international conference is to bring the experts and scientist in the research areas of Computer Science and Optimization from all over the world to share their knowledge and experiences on the current research achievements in these fields. This conference provides a golden opportunity for global research community to interact and share their novel findings and research discoveries among their colleagues and friends.

For this edition, the Program Committee received over 100 submissions from 25 countries and each paper was reviewed by at least four expert reviewers from across the globe. The prominent technical committee has selected the best 44 papers for final presentation at the conference venue of Golden Sands Resort in Penang, Malaysia. The organizing would like to sincerely thank all the authors and the reviewers for their wonderful job for this conference. The best and high-quality papers has been selected and reviewed by International Program Committee in order to publish the extended version of the paper in the Studies Computational Intelligence Book Series with SPRINGER.

The following is the synopsis of the best selected chapters for the Studies Computational Intelligence Book Series with SPRINGER.

Chapter "A Performance Analysis of DF Model in the Energy Harvesting Half-duplex and Full-Duplex Relay Networks", by Tam Nguyen Kieu, Nhu Nguyen Hong, Long Nguyen Ngoc, Thanh-Duc Le, Thuan Do Dinh, Jaroslav Zdralek and Miroslav Voznak;

Chapter "A Model of Swarm Intelligence Based Optimization Framework Adjustable According to Problems", by Utku Köse and Pandian Vasant;

Chapter "Domain Model Definition for Domain-Specific Rule Generation Using Variability Model", by Neel Mani, Markus Helfert and Pandian Vasant;

Chapter "A Set-Partitioning-Based Model to Save Cost on the Import Processes", by Jania Astrid Saucedo Martínez, José Antonio Marmolejo Saucedo and Pandian Vasant;

Chapter "Combining Genetic Algorithm with Variable Neighborhood Search for MAX-SAT", by Noureddine Bouhmala and Kjell Ivar Øverg°ard;

Chapter "Enzyme Classification on DUD-E Database Using Logistic Regression Ensemble (Lorens) ", by Heri Kuswanto, Jainap N. Melasasi and Hayato Ohwada;

Chapter "Consolidation of Host-Based Mobility Management Protocols in Wireless Mesh Network", by Wei Siang Hoh, Bi-Lynn Ong R Badlishah Ahmad and Hasnah Ahmad;

Chapter "Application of Parallel Computing Technologies for Numerical Simulation of Air Transport in the Human Nasal Cavity", by Alibek Issakhov and Aizhan Abylkassymova;

Chapter "Genetic Algorithms-Based Techniques for Solving Dynamic Optimization Problems with Unknown Active Variables and Boundaries", by Abdelmonem Fouad, Daryl L. Essam and Ruhul A. Sarker;

Chapter "Text Segmentation Methods: A Critical Review", by Irina Pak and Phoey Lee Teh;

Chapter "On-Line Power Systems Security Assessment Using Data Stream Random Forest Algorithm Modification", by Aleksei Zhukov, Nikita Tomin, Denis Sidorov, Victor Kurbatsky and Daniil Panasetsky;

Chapter "Enhanced Security of Internet Banking Authentication with Extended Honey Encryption (XHE) Scheme", by Tan Soo Fun and Azman Samsudin;

Chapter "An Enhanced Possibilistic Programming Model with Fuzzy Random Confidence-Interval for Multi-Objective Problem", by Nureize Arbaiy, Noor Azah Samsudin, Aida Mustapa, Junzo Watada and Pei-Chun Lin;

Chapter "A Crowdsourcing Approach for Volunteering System", by Nurul-hasanah Mazlan, Sharifah Sakinah Syed Ahmad and Massila Kamalrudin;

Chapter "One Dimensional Vehicle Tracking Analysis in Vehicular Ad hoc Networks", by Ranjeet Singh Tomar and Mayank Satya Prakash Sharma;

Chapter "Parallel Coordinates Visualization Tool on the Air Pollution Data for Northern Malaysia ", by J. Joshua Thomas, Raaj Lokanathan and Justtina Anantha Jothi;

Chapter "Application of Artificial Bee Colony Algorithm for Model Parameter Identification", by Olympia Roeva;

Chapter "A Novel Weighting Scheme Applied to Improve the Text Document Clustering Techniques", by Laith Mohammad Abualigah, Ahamad Tajudin Khader and Essam Said Hanandeh;

Chapter "A Methodological Framework to Emulate the Visual of Malaysian Shadow Play With Computer-Generated Imagery", by Kheng-Kia Khor;

We hope this book series will be a wonderful reference for the global researchers in the field of computational intelligence and its applications in real-world complex problems. The innovative methodologies adopted in this book will be very fruitful to solve complicated and large-scale problems in industrial applications. Finally, we would like to sincerely thank Mr. Ramamoorthy Rajangam and Ms. Victoria Meyer for their great help and support.

Ostrava-Poruba, Czech Republic — Prof. Dr. Ivan Zelinka
Teronoh, Malaysia — Dr. Pandian Vasant
Ho Chi Minh, Vietnam — Dr. Vo Hoang Duy
Ho Chi Minh, Vietnam — Dr. Tran Trong Dao
November 2016

Acknowledgements

The editors would like to sincerely thank the participants, presenters, authors, and reviewers of the International Conference on the Computer Science and Engineering (COMPSE), COMPSE 2016, Penang, Malaysia. The following reviewers have tremendously contributed in providing an excellent review report on the chapters submitted to the Studies Computational Intelligence Book Series of SPRINGER. Many thanks for their marvelous help and support. Their valuable contributions are highly appreciated and acknowledged.

Igor Litvinchev (Mexico)
Tam Nguyen Kieu (Vietnam)
Utku Kose (Turkey)
Neel Mani (Ireland)
Jose Antonio Marmolejo (Mexico)
Heri Kuswanto (Indonesia)
Abdelmonem Fouad (Australia)
Denis Sidorov (Russia)
Ranjeet Singh Tomar (India)
J. Joshua Thomas (Malaysia)
Miroslav Voznak (Czech Republic)
Kheng-Kia Khor (Malaysia)

Finally, the editors would like to thank Mr. Ramamoorthy Rajangam and Ms. Viktoria Meyer for their tireless effort and hard work for coordination of this book project.

Contents

A Performance Analysis of DF Model in the Energy Harvesting Half-Duplex and Full-Duplex Relay Networks 1
Kieu-Tam Nguyen, Hong-Nhu Nguyen, Ngoc-Long Nguyen, Thanh-Duc Le, Jaroslav Zdralek and Miroslav Voznak

A Model of Swarm Intelligence Based Optimization Framework Adjustable According to Problems 21
Utku Kose and Pandian Vasant

Domain Model Definition for Domain-Specific Rule Generation Using Variability Model 39
Neel Mani, Markus Helfert, Claus Pahl, Shastri L Nimmagadda and Pandian Vasant

A Set-Partitioning-Based Model to Save Cost on the Import Processes 57
Jania Astrid Saucedo-Martínez, José Antonio Marmolejo-Saucedo and Pandian Vasant

Combining Genetic Algorithm with Variable Neighborhood Search for MAX-SAT 73
Noureddine Bouhmala and Kjell Ivar Øvergård

Enzyme Classification on DUD-E Database Using Logistic Regression Ensemble (Lorens) 93
Heri Kuswanto, Jainap N. Melasasi and Hayato Ohwada

Consolidation of Host-Based Mobility Management Protocols with Wireless Mesh Network 111
Wei Siang Hoh, Bi-Lynn Ong, R. Badlishah Ahmad and Hasnah Ahmad

Application of Parallel Computing Technologies for Numerical Simulation of Air Transport in the Human Nasal Cavity 131
Alibek Issakhov and Aizhan Abylkassymova

Genetic Algorithms-Based Techniques for Solving Dynamic Optimization Problems with Unknown Active Variables and Boundaries . . . 151
AbdelMonaem F. M. AbdAllah, Daryl L. Essam and Ruhul A. Sarker

Text Segmentation Techniques: A Critical Review . . . 167
Irina Pak and Phoey Lee Teh

On-Line Power Systems Security Assessment Using Data Stream Random Forest Algorithm Modification . . . 183
Aleksei Zhukov, Nikita Tomin, Denis Sidorov, Victor Kurbatsky and Daniil Panasetsky

Enhanced Security of Internet Banking Authentication with EXtended Honey Encryption (XHE) Scheme . . . 201
Soo Fun Tan and Azman Samsudin

An Enhanced Possibilistic Programming Model with Fuzzy Random Confidence-Interval for Multi-objective Problem . . . 217
Nureize Arbaiy, Noor Azah Samsudin, Aida Mustapa, Junzo Watada and Pei-Chun Lin

A Crowdsourcing Approach for Volunteering System Using Delphi Method . . . 237
Nurulhasanah Mazlan, Sharifah Sakinah Syed Ahmad and Massila Kamalrudin

One Dimensional Vehicle Tracking Analysis in Vehicular Ad hoc Networks . . . 255
Ranjeet Singh Tomar and Mayank Satya Prakash Sharma

Parallel Coordinates Visualization Tool on the Air Pollution Data for Northern Malaysia . . . 271
J. Joshua Thomas, Raaj Lokanathan and Justtina Anantha Jothi

Application of Artificial Bee Colony Algorithm for Model Parameter Identification . . . 285
Olympia Roeva

A Novel Weighting Scheme Applied to Improve the Text Document Clustering Techniques . . . 305
Laith Mohammad Abualigah, Ahamad Tajudin Khader and Essam Said Hanandeh

A Methodological Framework to Emulate the Visual of Malaysian Shadow Play With Computer-Generated Imagery . . . 321
Kheng-Kia Khor

A Performance Analysis of DF Model in the Energy Harvesting Half-Duplex and Full-Duplex Relay Networks

Kieu-Tam Nguyen, Hong-Nhu Nguyen, Ngoc-Long Nguyen, Thanh-Duc Le, Jaroslav Zdralek and Miroslav Voznak

Abstract Energy harvesting (EH) structure based on the ambient wireless communication technique, has lastly carried out the development approach to widen the existing time of the wireless networks. The energy collecting composition for the half duplex and full duplex wireless networks is examined. By employing the time switching-based relaying (TSR) networks and the Decode-and-Forward (DF) structure, we imply the approximated section of the operation probability and then estimate the throughput of the half-duplex and full-duplex wireless networks. We can see an essential outcome clearly based on the position of relaying, the achievement rate, the noise at the source and the relay as well as the energy transition component in TSR, effect on their outage possibility and throughput. Ultimately, the performances of the full-duplex (FD) and half-duplex (HD) wireless network architectures is different, the results are collected to demonstrate that, for a normal maximal networks, sometimes, FD one are more optimal than HD model and contrariwise, for instance, when the distance between the relay and the transmitter isn't great, The outage possibility of the HD model is better than FD one. Or when rises η, the HD throughput becomes better etc. With the same method, we also get the results illustrate that DF of two antennas is better than that of one antenna. Consequently, based on the practical condition, we determine what model is better.

K.-T. Nguyen (✉) · H.-N. Nguyen · T.-D. Le · J. Zdralek · M. Voznak
VSB-Technical University of Ostrava, Ostrava, Czech Republic
e-mail: nguyenkieutam@tdt.edu.vn

H.-N. Nguyen
e-mail: nguyenhongnhu@hotec.edu.vn

T.-D. Le
e-mail: lethanhduc@hotec.edu.vn

J. Zdralek
e-mail: Jaroslav.Zdralek@vsb.cz

M. Voznak
e-mail: miroslav.voznak@vsb.cz

N.-L. Nguyen
Ton Duc Thang University, 19 Nguyen Huu Tho St., 7th Dist, Ho Chi Minh City, Vietnam
e-mail: nguyenngoclong@tdt.edu.vn

I. Zelinka et al. (eds.), *Innovative Computing, Optimization and Its Applications*, Studies in Computational Intelligence 741,
https://doi.org/10.1007/978-3-319-66984-7_1

Keywords Energy harvesting · Full-duplex · Half-duplex
Decode and forward · Throughput

1 Introduction

The EH structure is developing as a promising method which suitable to the radio cooperation as well as the sensor systems to operate in the ambient environment has limited technology and economic. The application of the energy collectors, replace the normal power sources, in wireless cooperative communication is also considered. For the displaying purpose, the basic three-node Gaussian relay channel with the DF relaying network, in which the source and relay nodes transmit the information with the power drawn from the EH sources is investigated in [1–4].

Moreover, the technology of radio frequency energy collecting is a worth -expecting model to sustain the activities of the wireless networks. In experience, in a radio network, a second user could be installed with the RF energy harvesting function. Such network where the second user could effect the channel access to transmit information or to collect the RF energy when the selected channel is available or being used by the first user, respectively, is also studied in [5–7].

In practice, the authors in [8, 9] had taken the energy constrained relay node placement problem in an energy-harvesting network in which the energy harvesting ability of the candidate locations. On the other hand, the transmittance of power and information from the resource node to the relay node is executed by two method, i) the TSR and ii) the power splitting-based relaying (PSR) is also considered in [10, 11], too. The authors in [12, 13] had considered the employment of collective energy, in place of traditional power sources, in wireless communication systems. The classic three-node Gaussian relay channel with decode-and-forward (DF) relaying, in which the source and relay nodes transfer with power extracted from energy-harvesting resources is studied.

Furthermore, a wireless-energized cooperation communication system comprising of one hybrid access-point (AP), one resource, and one relay is investigated. In contrary to frequent cooperation systems, the resource and relay in the used network do not have power supply. They must be based on the energy collected from the source signals for their communication transmittance as in [14–17].

On the other hand, a energy collecting cognitive radio (CR) network acting in slotted scheme, where the secondary user (SU) does not supply power and is used energy harvested from the surroundings are examined. The SU can only employ either energy collecting, spectrum sensing or information transfer at a time because of hardware restriction such that a time gap is partitioned into three non-overlapping parts, as the authors revealed in [18–20].

So, in this paper, we consider the outage probability and throughput of the full-duplex (FD) relaying network with a new capacity of energy collecting and information transfer. Based on the analytical expressions, their outage probability and throughput are studied and according as the practical condition, we can employ what model to achieve much profit.

In this article, we examine a radio HD relaying and FD relaying system applying TSR protocol to perform the simultaneous wireless power and information transfer (SWPIT) scheme so as to optimize the outage probability and to maximize their throughput.

The main content of the paper is summarized as follows. By comparing two techniques of FD and HD relaying, our results show that based on factors we used, we could valuate the impact of the relays to the determination of optimal outage probability and maximizing throughput for an appropriately optimized network.

The rest of the article is organized as follows. In Sect. 2, the system's model of the EH enabled HD and FD relaying network are described. Section 3 demonstrates the analyses of the outage probability and throughput. Section 4 presents the result of simulations. Finally, the conclusion is drawn in Sect. 5.

2 Network Model

As observed in Fig. 1, we study a DF relaying cooperative system, where the data is transferred from the source node, called, S, to the destination node, called, D, through an immediate relay node, R. Every node has two antennas, one for transmitting the communication, the rest for receiving information. Assuming that between S and D, there is not existence of the contact link, due to the far distance. Therefore, that DF relay helped the connect between them. At the beginning, the DF relay harvests energy from the resource data. And then, that collected power is used to convey the information to the target. d_0, d_1 and d_2 are denoted for the distance between $S \rightarrow R$ and $R \rightarrow D$.

Supposing that due to the interference neutralize structure is unperfected, the residual self-interference (RSI) channel at R is named by f. For this reason, always, a certain amount of self-interference existed.

With the TSR model in Fig. 2, the communication procedure is split into two stages. At first, the energy is transmitted and returned between the source and the

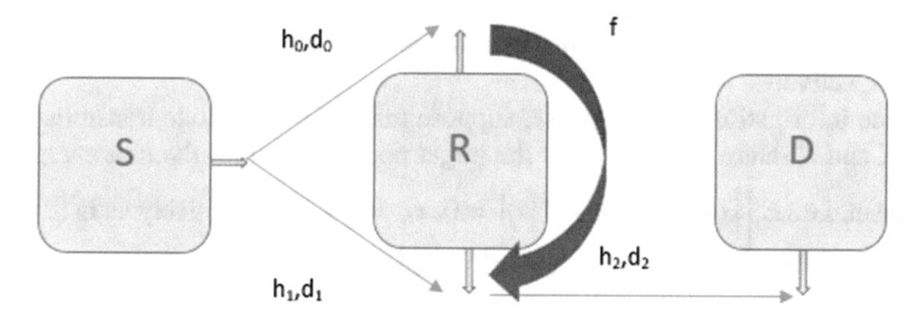

Fig. 1 The Frame diagram of TSR system

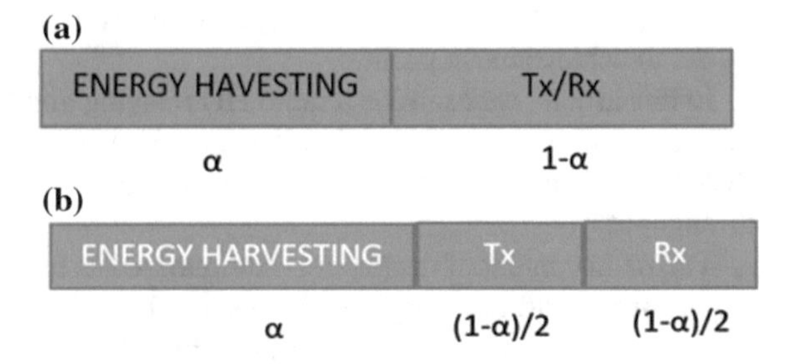

Fig. 2 Illustration of the parameters of TSR protocol: **a** The full-duplex model. **b** The half-duplex model

relay in a cycle of $\alpha T, (0 < \alpha < 1)$ and in the second stage, in case of full-duplex model, $(1 - \alpha) T$ is used to transmit the data,then in case of the half-duplex are $(1 - \alpha) T/2$ for the data receiver and $(1 - \alpha) T/2$ for data transmit. In which α is time switching coefficient and T is time for the transmitted data examination.

2.1 *The Full Duplex Relaying Model*

In this section, two cases enabled at R is investigated.

A. Only One Antenna Is Utilized to Harvest Power, A rest Is Utilized to Transmit Communication: Following the energy harvesting cycle, the received data at the relay as

$$y_{RFD} = \sqrt{\frac{P_S}{d_1^m}} h_1 x_S + n_R. \tag{1}$$

where P_S is the source transmittance energy, n_R is the addition white Gaussian noise at R with zero-mean and variance of σ_R^2.

As state by the wireless received energy, the collected power at the relay is given by [11].

$$E_h = \eta \alpha T \frac{P_S |h_1|^2}{d_1^m}. \tag{2}$$

where m is the exponential path loss, η is the power conversion factor, h_0, h_1, h_2 are the channel gain factors between the resource-relay line and the relay-destination line, respectively.

In the information transfer phase, suppose that the source node transmits signal x_S to R and R return the data x_R to the target node. They have the unit energy and zeromean, i.e., $E\left[\left|x_j\right|^2\right] = 1$ and $E\left[x_j\right] = 0$, x_j, $j = S, R$, respectively as in [21]. So, the received data at the relay under a self-noise resource is given as

$$y_{RFD} = \sqrt{\frac{P_S}{d_1^m}} x_S h_1 + f x_R + n_R. \tag{3}$$

It is widely known that the energy collected, then supported the performance for the next stage, is given by

$$P_{RFD} = \frac{E_h}{(1-\alpha)T} = \rho P_S \frac{|h_1|^2}{d_1^m}. \tag{4}$$

where ρ is denoted as $\rho = \frac{\alpha\eta}{1-\alpha}$.
Then, we get signal at the receiver as

$$y_{DFD} = \frac{h_2}{\sqrt{d_2^m}} x_R + n_D. \tag{5}$$

where n_D is the addition white Gaussian noise at destination node with a zero-mean and variance of $\sigma_D^2 = \sigma_R^2 = \sigma^2$, for simplicity. Following [10], we have

$$x_{RFD} = \sqrt{\frac{P_R}{P_S}} x_R(i-\tau). \tag{6}$$

Replacing Eqs. (4) and (6) into Eq. (5), we get the received signal at the destination node as

$$y_{DFD} = \frac{h_2}{\sqrt{d_2^m}} \left(\sqrt{\frac{P_R}{P_S}} x_R(i-\tau) \right) + n_D. \tag{7}$$

Through some simple calculations, we get the new formula as

$$\gamma_{DF}^{FDO} = \min\left(\frac{1}{\rho|f|^2}, \frac{\rho P_S |h_1|^2 |h_2|^2}{d_1^m d_2^m \sigma^2} \right). \tag{8}$$

Supposing that the channel gains $|h_0|^2, |h_1|^2, |h_2|^2$ are independent and identically distributed (i.i.d.) exponent.

B. Both Antennas Are Used to Collect Energy, But Only One Is Utilized to Transfer The Information: The same as previous section, terms d_0, and h_0 denote the distance from the source to the relay as well as the fading channel factor of the remainder antenna at relay. To simplify the issue, we assume $d_0 = d_1$, and $E\left\{ n_R n_R^\dagger \right\} = \sigma^2 I$ with I is unit matrix. the received signal at the relay can be taken as

$$y_R = \sqrt{\frac{P_S}{d_1^m}} \begin{pmatrix} h_0 \\ h_1 \end{pmatrix} x_S + n_R. \tag{9}$$

and

$$E_h = \eta\alpha T \frac{P_s(|h_0|^2 + |h_1|^2)}{d_1^m}. \tag{10}$$

So, the signal at R can be rewritten as

$$y_R = \sqrt{\frac{P_S}{d_1^m}} x_S(h_0 + h_1) + f x_R + n_R. \tag{11}$$

and

$$P_R = \frac{E_h}{(1-\alpha)T} = \rho P_S \frac{(|h_0|^2 + |h_1|^2)}{d_1^m}. \tag{12}$$

base on Eq. (7),

Of course, a new formula can be gotten

$$\gamma_{DF}^T = \min\left\{ \frac{|h_1|^2}{\rho|f|^2(|h_1|^2 + |h_0|^2)}, \frac{\rho P_S(|h_1|^2 + |h_0|^2)|h_2|^2}{d_1^m d_2^m \sigma^2} \right\}. \tag{13}$$

2.2 The Half-Duplex Relaying

Equivalent to the full-duplex model, in the half-duplex, we get:

$$y_{RHD} = \sqrt{\frac{P_S}{d_1^m}} h_1 x_S + n_R. \tag{14}$$

and

$$P_{RHD} = 2\rho P_S \frac{|h_1|^2}{d_1^m}. \tag{15}$$

After that, the received data at the target such as

$$y_{DHD} = \frac{h_2}{\sqrt{d_2^m}} x_R + n_D. \tag{16}$$

As same as the above calculate, we get

$$\gamma_{DF}^{HD} = \min\left(\frac{P_S|h_1|^2}{d_1^m\sigma^2}, \frac{2\rho P_S|h_1|^2|h_2|^2}{d_1^m d_2^m\sigma^2}\right). \tag{17}$$

3 The Outage Probability and Throughput Analysis

At here, we examine the outage probability and the throughput of the half-duplex and full-duplex relaying networks with the energy collecting and information transmittance. Based on that analytic conditions, the outage probability and throughput of two models are distinguished perfectly.

3.1 The Outage Probability Analysis

The outage probability of the relaying system in the delay-limited scheme is computed as

$$P_{out} = \Pr\left(\gamma \le Z\right). \tag{18}$$

where R is the target rate and $Z = 2^R - 1$.

Proposition 1 *The outage probability of the energy harvesting one way enabled full-duplex relay with DF protocol is derived as:*

$$P_{out}^{DFO} = 1 - \left(1 - e^{\frac{1}{\rho\lambda_r Z}}\right)\left(2\sqrt{\frac{\sigma^2 d_1^m d_2^m Z}{\rho\lambda_s\lambda_d P_S}}\right)K_1\left(2\sqrt{\frac{\sigma^2 d_1^m d_2^m Z}{\rho\lambda_s\lambda_d P_S}}\right). \tag{19}$$

where λ_s, λ_d, λ_r are the mean value of the exponential random variables h_1, h_2, f, respectively and $K_1(x)$ is Bessel function denoted as (8.423.1) in [22].

Proof Rely on the cumulative distribution, function of x is computed by

$$F_x(b) = \Pr\left(x \le b\right) = 1 - 2\sqrt{b/\lambda_s\lambda_d}K_1\left(2\sqrt{b/\lambda_s\lambda_d}\right). \tag{20}$$

and y can be modeled with the probability distribution function $f_y(c) = \left(1/\lambda_r\right) e^{(-c/\lambda_r)}$.

We define $x = |h_1|^2|h_2|^2$ and $y = |f|^2$, from Eqs. (8) and (18), we get

$$P_{out}^{DFO} = Pr\left(\min\left(\frac{1}{\rho y}, \frac{\rho P_S x}{d_1^m d_2^m\sigma^2}\right) \le Z\right). \tag{21}$$

Then

$$P_{out}^{FDF} = 1 - Pr\left(\frac{1}{\rho y} \geq Z\right) \Pr\left(\frac{\rho P_S x}{d_1^m d_2^m \sigma^2} \geq Z\right). \tag{22}$$

where $Pr\left(\frac{1}{\rho y} \geq Z\right) = Pr\left(\frac{1}{\rho Z} \geq y\right) = \int_0^{\frac{1}{\rho Z}} f_y(t)dt = \frac{1}{\lambda_r} \int_0^{\frac{1}{\rho Z}} e^{-\frac{t}{\lambda_r}} dt$ and
$\Pr\left(\frac{\rho P_S x}{d_1^m d_2^m \sigma^2} \geq Z\right) = 1 - \Pr\left(x \leq \frac{Z d_1^m d_2^m \sigma^2}{\rho P_S}\right) = 1 - F_x\left(\frac{Z d_1^m d_2^m \sigma^2}{\rho P_S}\right)$.

Based on Eq. (20), we have our wanted result after some calculations by hand, then the proposition 1 is achieved through some simple applications.

Proposition 2 *The outage probability of the two-antennas protocol is given as*

$$\begin{aligned} P_{out}^{DFT} &= 1 - (1 - \rho\lambda_r Z(1 - e^{\frac{-1}{\rho\lambda_r Z}}))\times \\ &(\frac{2d_1^m d_2^m ZN}{\lambda_s \lambda_d \rho P_S}) \times K_2\left(2\sqrt{\frac{d_1^m d_2^m ZN}{\lambda_s \lambda_d \rho P_S}}\right). \end{aligned} \tag{23}$$

where $K_2(x)$ is Bessel function denoted as (8.423.1) in [22]

Proof We denote $M = \frac{X+Y}{Y}$ and $N = X + Y$ with $X = |h_0|^2$ and $Y = |h_1|^2$, where M and N are built independent random variables. Now, we can illustrate the joint distribution function of X and Y as

$$f_{X,Y}(x, y) = \frac{1}{\lambda_S^2} e^{-\frac{x+y}{\lambda_S}}. \tag{24}$$

alternatively, we see that $X = MN$ and $Y = N(1 - M)$.

Employing Jacobian transformer (M, N) on to (X, Y),

$$f_{M,N}(m, n) = \frac{n}{\lambda_S^2} e^{-\frac{n}{\lambda_S}}. \tag{25}$$

distinctly that M and N are independence, and M is set to follow the uniform distribution with *pdf* $f_M(m) = 1, 0 \leq M < 1$ and N is set up following the gamma distribution with *pdf*

$$f_N(n) = \frac{n}{\lambda_S^2} e^{-\frac{n}{\lambda_S}}. \tag{26}$$

At last, let $W = N|h_2|^2$ and $T = \frac{M}{|f|^2}$ We easily get $F_W(w) = 2\frac{w}{\lambda_S \lambda_D} K_2(2\sqrt{\frac{w}{\lambda_S \lambda_D}})$, $F_T(t) = \lambda_r t(1 - e^{-\frac{1}{\lambda_r t}})$.

As the statistical property of W and T, we obtain

$$\begin{aligned} P_{out}^{DFT} &= Pr\left\{\min\left\{\frac{T}{\rho}, \frac{\rho P_S W}{d_1^m d_2^m \sigma^2}\right\} \leq Z\right\} \\ &= 1 - Pr(T \geq \rho Z)\Pr(W \geq \frac{d_1^m d_2^m \sigma^2 Z}{\rho P_S}). \end{aligned} \tag{27}$$

a hoped result is achieved after some uncomplicated manipulations.

Proposition 3 *The outage probability of HD DF relaying networks is taken by*

$$P_{out}^{HDF} = 1 - e^{\frac{\sigma^2 d_1^m z}{\lambda_s P_S} - \frac{d_1^m}{2\lambda_d \rho}} - \frac{1}{\lambda_d} \int_0^{\frac{d_2^m}{2\rho}} e^{\frac{\sigma^2 d_1^m d_2^m z}{2\rho \lambda_s P_S t}} e^{-\frac{t}{\lambda_d}} dt. \tag{28}$$

Proof From Eqs. (17) and (18), we get

$$P_{out}^{HDF} = \Pr\left(\frac{P_S |h_1|^2}{d_1^m \sigma^2} \min\left(1, \frac{2\rho |h_2|^2}{d_2^m}\right) \leq Z\right). \tag{29}$$

For the time being, from conditioned on $t = \min\left(1, \frac{2\rho|h_2|^2}{d_2^m}\right)$, we have $P_{out}^{HDF} = 1 - e^{\frac{\sigma^2 d_1^m Z}{\lambda_s P_S t}}$. Averaging over t supplies the expected result.

3.2 The Throughput Analysis

In the Proposition 1, the outage probability of two models, since the relay collects the energy from the source data and utilizes that energy to decrypt and transmit the information to the target is a function of the power collecting time α and varies when α rises from 0 to 1, in the delay-limited transfer protocol, the transmitter is sent at a fix rate R bits/sec/Hz and $(1 - \alpha)T$ is the effective communication time. So, the throughput of the FD model is given as

$$\begin{cases} \tau_{DF}^{O} = \left(1 - P_{out}^{DFO}\right) R(1 - \alpha) \\ \tau_{DF}^{T} = \left(1 - P_{out}^{DFT}\right) R(1 - \alpha). \end{cases} \tag{30}$$

also, the throughput of the HD system is given as

$$\tau_{HD} = \left(1 - P_{out}^{HDF}\right) R \frac{(1 - \alpha)}{2}. \tag{31}$$

Unfortunately, it is difficult to calculate the maximal throughput mathematically but we can achieve such value by the simulation approach as demonstrated in the next part.

4 Numerical Results

In this part, by using Matlab, we utilize the extracted critical results to investigate the outage probability as well as throughput of two DF models. We fix the resource transmittance rate $R = 3$ (bps/Hz) (except for Figs. 9, 10, 15 and 16), $\alpha = 0.3$ (except for Figs. 17 and 18) and so the outage SINR threshold is taken by $Z = 2^R - 1$. The energy conversion factor is fixed to be $\eta = 0.4$ (except for Figs. 5 and 6), the path loss exponent factor is fixed to be $m = 3$. For uncomplicated, we set the distance $d_1 = d_2 = 1$ (except for Figs. 3, 4, 11 and 12). Also, we set $\lambda_s = \lambda_d = 1; \lambda_r = 0.1$ and $P_S(dB) = 20, \sigma^2 = 0.1$ (except for Figs. 7, 8, 13 and 14). These values are chosen because they make curves nearly asymptote together.

Figure 3 shows the outage probability contrasted with the relay position. In the estimation, we evaluate $d_1 + d_2 = 2$, and d_1 changes from 0 to 2. As shown in Fig. 3, for both models, the best outage probability is gotten. So, the relay should be installed near to the resource or targeted to optimize the outage probability of the FD model. The suggested FD model gets the worst outage probability process at midway. Furthermore, Fig. 3 associate the half-duplex model gets the outage probability activities is better than that of the FD model when $d_1 \leq 0.6$, it means the relay is deployed close to the source. Otherwise, when $d_1 \geq 0.6$ the outage operation of FD model is better than that of the HD model. Their curves have the crossover at $d_1 = 0.6$. This thing is uncomplicated to explain when d_1 increases, d_2 decreases and based on Eqs. (19) and (28), we get this result.

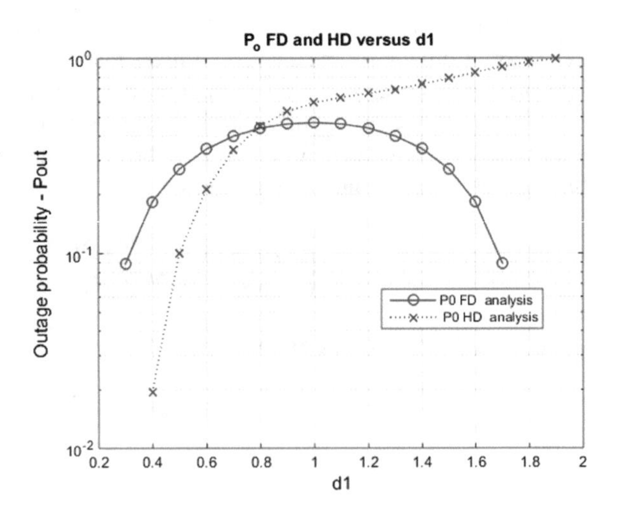

Fig. 3 The outage probability of FD and HD model versus d_1

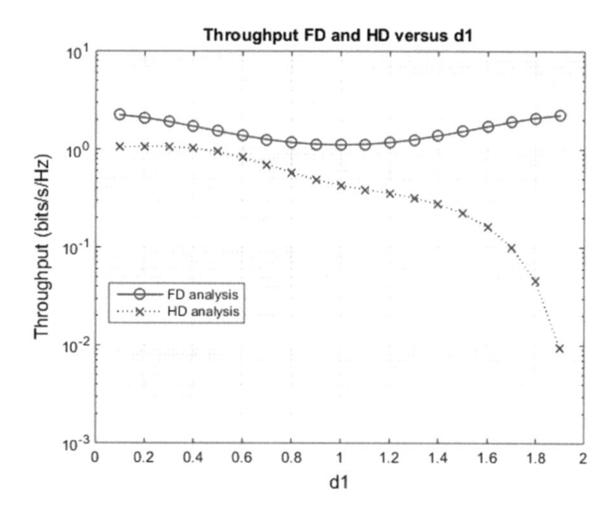

Fig. 4 The throughput of FD and HD model versus d_1

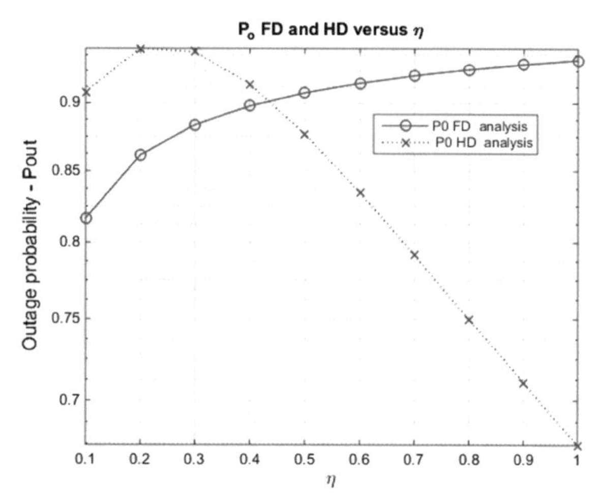

Fig. 5 The outage probability of FD and HD model versus η

As observed in Fig. 4, the throughput of full-duplex outperforms than that of the half-duplex one, although its outage probability is worse than that. We have this thing: When d_1 increases, P_{RHD} and P_{RFD} decrease, P_{out}^{HD} and P_{out}^{FD} decrease but not enough to make τ_{HD} better than τ_{FD} described in Eqs. (30) and (31).

As described in Figs. 5 and 6 that η factor had altered the outage probability and throughput of both models. In Fig. 5, When $\eta \leq 0.2$, as higher the η, as worse the HD outage probability gets. And this model achieves the worst value as $\eta \in (0.2, 0.3)$. If η prevails over this space, the higher the η is, the better the HD outage probability turns out. Meanwhile, as much η, as worse the FD outage probability reaches. This problem is not hard to explain, when η is big, much power is used to transfer signal to destination, so the HD outage probability gets the better presentation, on the antagonistic, for FD model, high power at the relay makes the noises create impact on its

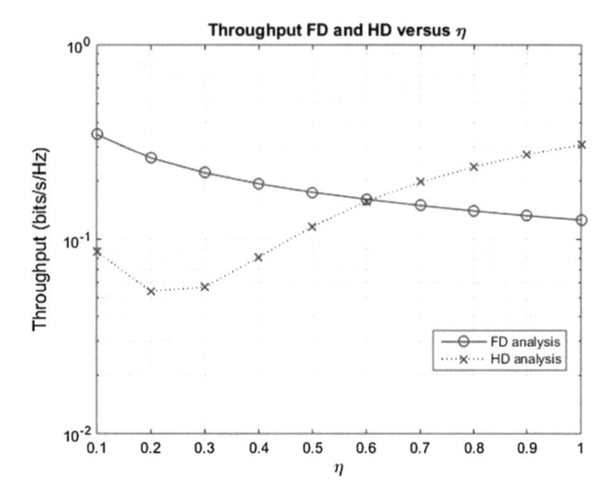

Fig. 6 The throughput of FD and HD model versus η

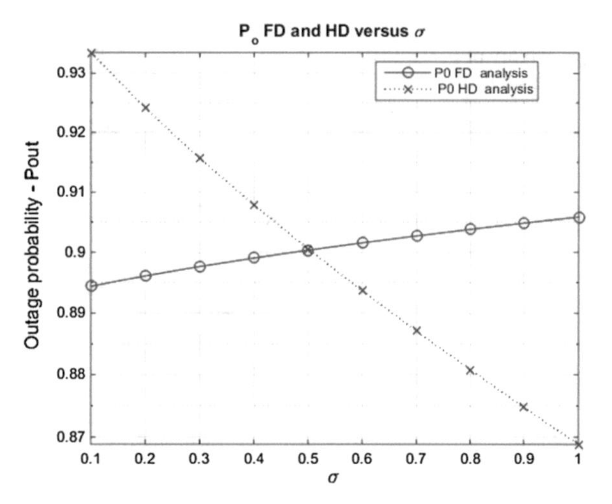

Fig. 7 The outage probability of FD and HD model versus σ

activity, hence its outage probability turns worse. This matter is illustrated in Fig. 6, as much η, as better the HD throughput gets, but for FD model, the circumstance is opposite. It is observable that two models have the cross point at $\eta = 0.6$.

Figures 7 and 8 expose that the loop back noise has the effect on their action. In this case, when $\sigma \leq 0.5$, the FD outage probability is better than the one of the HD model but when $\sigma \geq 0.5$, everything operate differently. In both cases, the FD throughput is always better than that of the HD model. This also prove that the noise at the relay has huge effect to their operation.

Besides, Figs. 9 and 10 show the plots of outage probability and throughput of two models versus the achievement rate R. As we can see in Fig. 9, as higher R, the outage probability of FD model is downward, meanwhile the HD outage probability

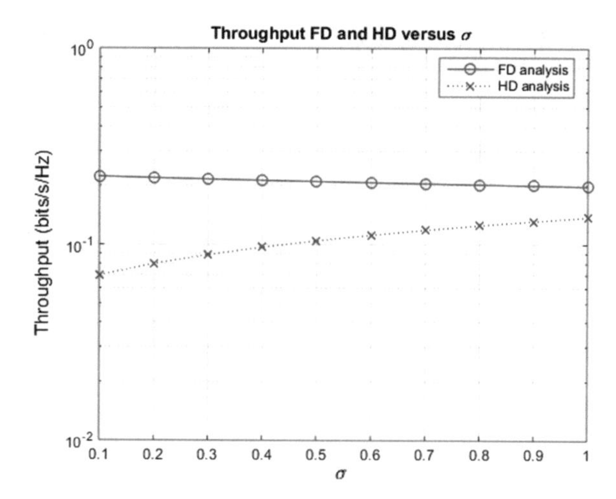

Fig. 8 The throughput of FD and HD model versus σ

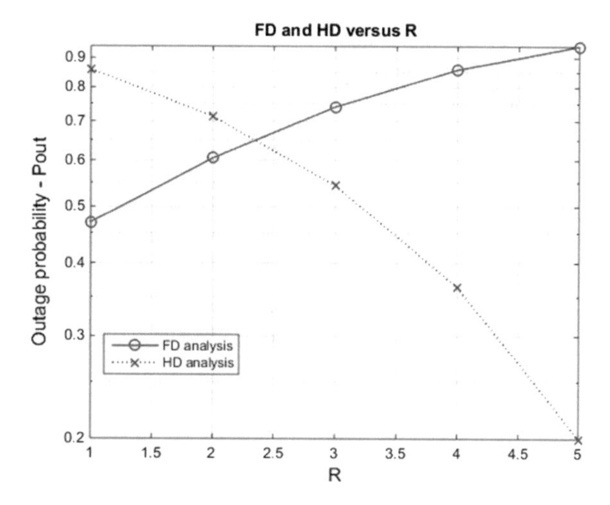

Fig. 9 The outage probability of FD and HD model versus R

is opposite. When $R \leq 2$, the FD outage probability is better than the HD one, but every thing is contrariwise as $R \geq 2$. Eqs. (19), (28), (30) and (31) verify this thing.

On the other hand, Fig. 10 is also suitable for above issue, it means as better outage probability gets, as better the throughput obtains. We have the crossover of their throughput at $R = 3$.

It can be seen from Fig. 11, the outage probability of one models decreases when relay position is as far away the source, and it gets the worst value at the midway between S and D. The outage probability of one antenna is better than that of two antennas since $d_1 \leq 0.25$ or ≥ 1.75, but since $d_1 \geq 0.25$ or ≤ 1.75, every thing is on the contrary. It is easily to explain because as far rethe source or the destination, less the transfer power is, the two antennas model can harvest more energy to do without difficulty the communication transmittance. But, since the transmit energy is large, it

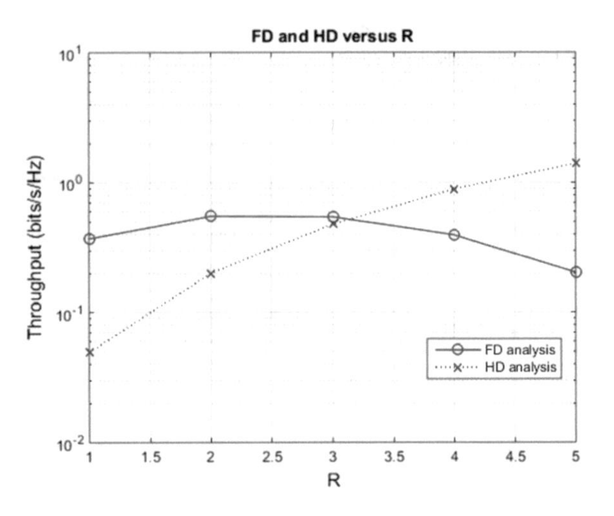

Fig. 10 The throughput of FD and HD model versus R

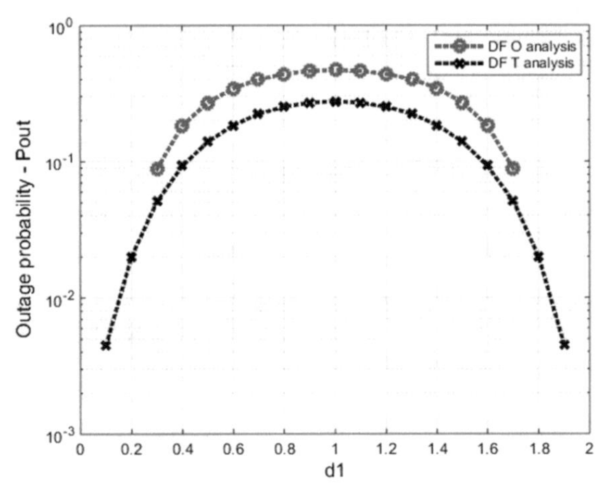

Fig. 11 Outage probability of DFT and DFO model versus d_1

means that the relay is close to the resource or the destination, the unwanted amount of power is collected, which is really harmful because it creates strong loop back interference, which reduces the systems action.

This is fit to Fig. 12, described where the outage probability is better than, their throughput is also better than, too. As observed, when the delay is close to the resource or the destination, one antenna model performance is better than that of two antennas one, since the delay as far the resource or the destination, the two antennas model outperforms than the other.

Figures 13 and 14 recognize that the outage probability and throughput of two antennas model are better than clearly that of one antenna model. As big σ, as bad their performance. They get the worst value when noise is at the greatest point. This is

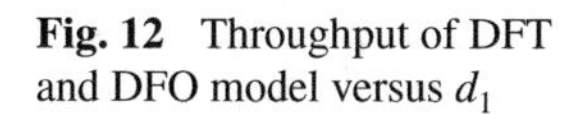

Fig. 12 Throughput of DFT and DFO model versus d_1

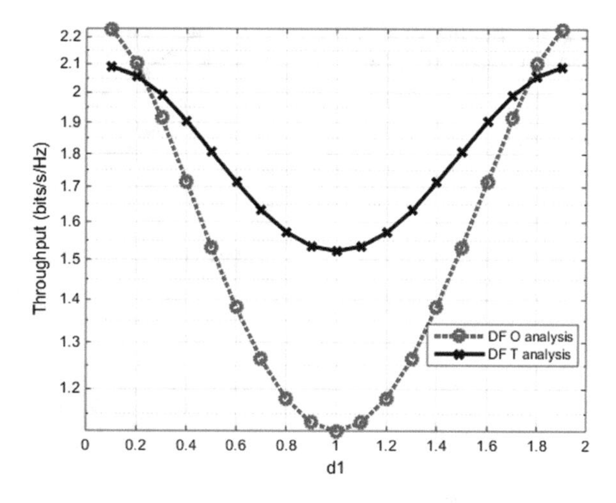

Fig. 13 Outage probability of DFT and DFO model versus σ

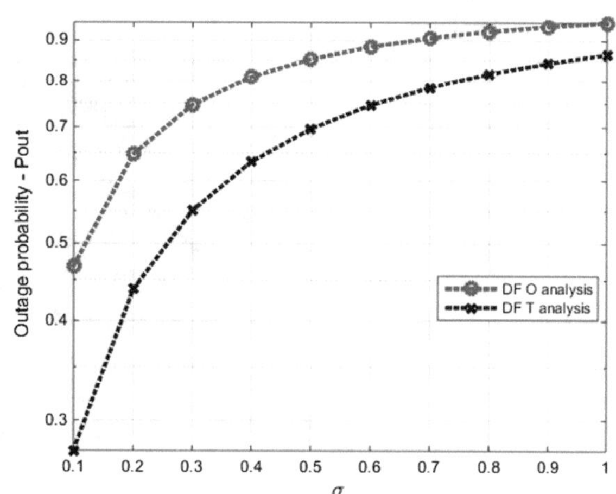

easily to understand because as much noise, as much harmful power is made, which causes interference to the transmission signal.

Besides, Fig. 15 studies the impact of the transmittance rate R on the outage probability of two models. When R raises, the outage probability of two antennas model lessens, it gets the worst performance as great R. It is noted that two antennas model is normally better than that of one antenna model.

Moreover, Fig. 16 lead us the fact that visually, as small the transfer rate, as small the throughput; and they get the best value at $R = 3$ when the transmittance rate ≥ 3, the outage probability raises dramatically, which again decreases the throughput. Easily, we note that the throughput of the two antennas scheme is better than that of the one antenna scheme. This could be explained easily because of the loop back interference in two antennas model and the more transmittance energy create it to

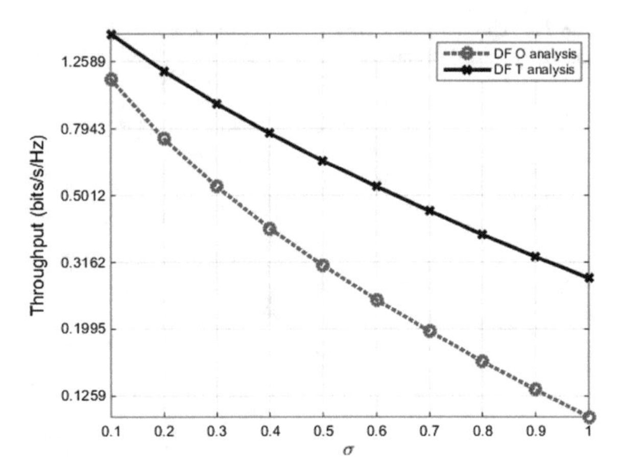

Fig. 14 Throughput of DFT and DFO model versus σ

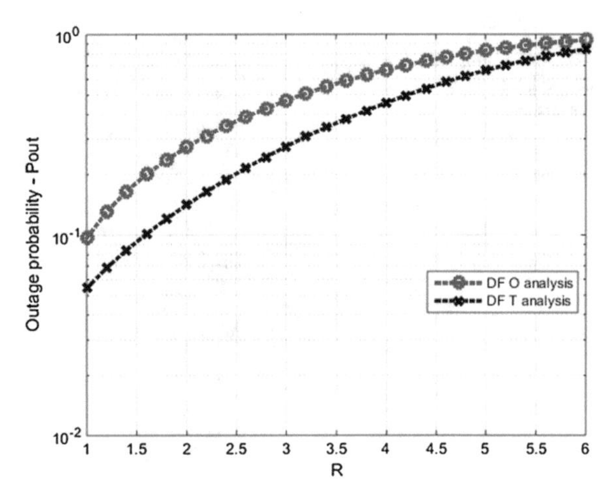

Fig. 15 Outage probability of DFT and DFO model versus R

attain the better outage probability and throughput performance than that of one antenna scheme for any fixed transmission rate.

As observed in Fig. 17 shows that the curves change depended on the time fraction α influencing on the performance of systems. When $\alpha = 0.2$, for one antenna model, as big α is, as bad its outage probability gets. It gets the best score as $\alpha = 0.2$. However, for two antennas models, as much α. As good its outage probability gets. It is also no difficult to understand, the noise they get to convert energy enough great, which helps it to overcome the influence of the power divided initially.

And as we can see in Fig. 18, the throughput of two antennas model is matching as analyzed in mentioned Fig. 17. The throughput of one antenna is the same, too. It gets the best curve when $\alpha = 0.2$. The two antennas model is with $\alpha = 0.3$. As a result, it is noted that the two antennas model outperforms than that of one antenna model.

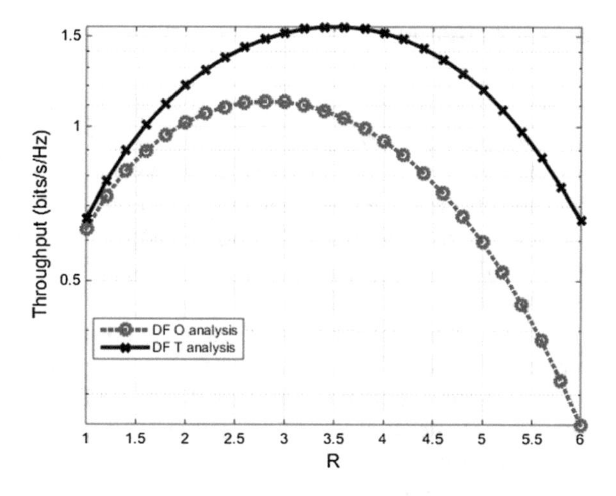

Fig. 16 Throughput of DFT and DFO model versus R

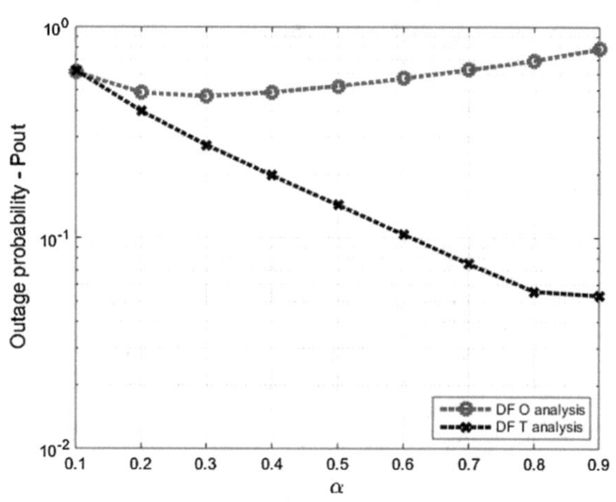

Fig. 17 Outage probability of DFT and DFO model versus α

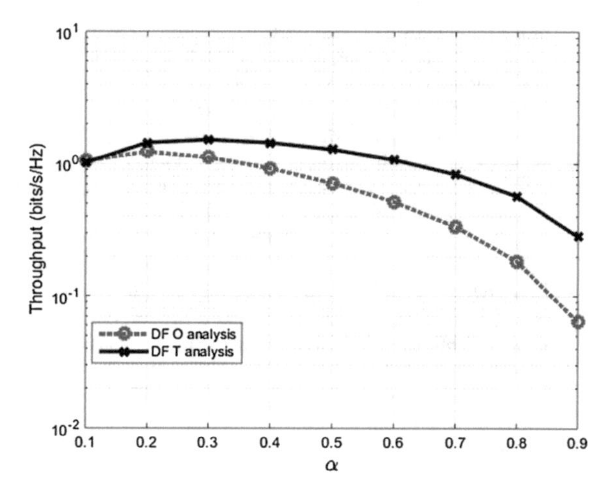

Fig. 18 Throughput of DFT and DFO model versus α

5 Conclusion

In this paper, we make comparison between the half-duplex and full-duplex relaying networks in the radio energy harvesting and data exchange protocol, in which a power constrained relaying node take the delivery of the energy from the received RF data and utilize that energy to decrypt and transmit the data to the target. Thus, their operation is also compared. To determine the achieved system, the analytic appearances for the outage probability and throughput in TSR protocol can be executed and established by the MATLAB modelling.

As expected, over the simulations, it is known that the full-duplex model outperforms than that of the half-duplex one in some cases. And in another case, the HD relaying system could be better than that of the FD one. Also, we get the DF of two antennas is better than that of one antenna. So, based on our practical circumstance, we can determine which model to be used to achieve the best benefit. The numerical analysis in this article has also provided practical vision to the impact of various system parameters to the performance of wireless energy collecting and data processing using DF relay nodes.

References

1. Chuan, H., Rui, Z., & Shuguang, C. (2012). Delay-constrained Gaussian relay channel with energy harvesting nodes. In *IEEE International Conference on in Communications ICC* (pp. 2433–2438).
2. Dang Khoa, N., & Ochi, H. (2014). On the impact of transceiver impairments to cognitive DF relay networks. In *IEEE Asia Pacific Conference on in Circuits and Systems (APCCAS)* (pp. 125–128).
3. Nasir, A. A., Xiangyun, Z., Durrani, S., & Kennedy, R. A. (2014). Throughput and ergodic capacity of wireless energy harvesting based DF relaying network. In *IEEE International Conference on in Communications (ICC)* (pp. 4066–4071).
4. Zihao, W., Zhiyong, C., Ling, L., Zixia, H., Bin, X., & Hui, L. (2014). Outage analysis of cognitive relay networks with energy harvesting and information transfer. In *IEEE International Conference on in Communications (ICC)* (pp. 4348–4353).
5. Ke, X., Pingyi, F., & Ben Letaief, K. (2015). Time-switching based SWPIT for network-coded two-way relay transmission with data rate fairness. In *IEEE International Conference on in Acoustics, Speech and Signal Processing (ICASSP)* (pp. 5535–5539).
6. Krikidis, I., Timotheou, S., Nikolaou, S., Gan, Z., Ng, D. W. K., & Schober, R. (2014). Simultaneous wireless information and power transfer in modern communication systems. *IEEE Communications Magazine*, *52*, 104–110.
7. Liu, Y., & Wang, X. (2015). Information and Energy Cooperation in OFDM Relaying: Protocols and Optimization. *IEEE Transactions on Vehicular Technology PP*, 1–1.
8. Mousavifar, S. A., Yuanwei, L., Leung, C., Elkashlan, M., & Duong, T. Q. (2014). Wireless energy harvesting and spectrum sharing in cognitive radio. In *2014 IEEE 80th Vehicular Technology Conference (VTC Fall)* (pp. 1–5).
9. Mugen, P., Yuan, L., Quek, T. Q. S., & Chonggang, W. (2014). Device-to-device underlaid cellular networks under rician fading channels. *IEEE Transactions on Wireless Communications*, *13*, 4247–4259.
10. Riihonen, T., Werner, S., & Wichman, R. (2011). Hybrid full-duplex/half duplex relaying with transmit power adaptation. *IEEE Transactions Wireless Communications*, *10*, 3074–3085.

11. Huang, C., Zhang, R., & Cui, S. (2013). Throughput maximization for the gaussian relay channel with energy harvesting constraints. *IEEE Journal on Selected Areas in Communications*, *31*(8), 14691479.
12. Tutuncuoglu, K., Varan, B., & Yener, A. (2013). Energy harvesting two-way half-duplex relay channel with decode-and-forward relaying: Optimum power policies. In *2013 18th International Conference on Digital Signal Processing (DSP)* (pp. 1–6).
13. Zhengguo, S., Goeckel, D. L., Leung, K. K., & Zhiguo, D. (2009). A stochastic geometry approach to transmission capacity in wireless cooperative networks. In *2009 IEEE 20th International Symposium on Personal, Indoor and Mobile Radio Communications* (pp. 622–626).
14. He, C., Yonghui, L., Luiz Rebelatto, J., Uchoa-Filho, B. F., & Vucetic, B. (2015). Harvest-then-cooperate: Wireless-powered cooperative communications. *IEEE Transactions on Signal Processing*, *63*, 1700–1711.
15. Hong, X., Zheng, C., Zhiguo, D., & Nallanathan, A. (2014). Harvest-and-jam: Improving security for wireless energy harvesting cooperative networks. In *Global Communications Conference (GLOBECOM)* (Vol. 2014, pp. 3145–3150), IEEE.
16. Huan, Y., Yunzhou, L., Xibin, X., & Jing, W. (2014). Energy harvesting relay-assisted cellular networks based on stochastic geometry approach. In *International Conference on Intelligent Green Building and Smart Grid (IGBSG)* (Vol. 2014, pp. 1–6).
17. Jia, X., Xing, Z., Jiewu, W., Zhenhai, Z., & Wenbo, W. (2014). Performance analysis of cognitive relay networks with imperfect channel knowledge over Nakagami-m fading channels. In *Wireless Communications and Networking Conference (WCNC)* (Vol. 2014, pp. 839–844). IEEE.
18. Sibomana, L., Zepernick, H. J., & Hung, T. (2014). Wireless information and power transfer in an underlay cognitive radio network. In *2014 8th International Conference on Signal Processing and Communication Systems (ICSPCS)* (pp. 1–7).
19. Sixing, Y., Erqing, Z., Zhaowei, Q., Liang, Y., & Shufang, L. (2014). Optimal cooperation strategy in cognitive radio systems with energy harvesting. *IEEE Transactions on Wireless Communications*, *13*, 4693–4707.
20. Sixing, Y., Zhaowei, Q., & Shufang, L. (2015). Achievable throughput optimization in energy harvesting cognitive radio systems. *IEEE Journal on Selected Areas in Communications*, *33*, 407–422.
21. Kieu, T. N., Do, D.-T., Xuan, X. N., Nhat, T. N., & Duy, H. H. (2016). Wireless information and power transfer for full duplex relaying networks: Performance analysis. In: *AETA 2015: Recent Advances in Electrical Engineering and Related Sciences. Cham: Springer International Publishing, 2016, ch. Wireless Information and Power Transfer for Full Duplex Relaying Networks: Performance Analysis* (p. 5362). https://doi.org/10.1007/978-3-319-27247-4-5.
22. David, H. A. (1970). *Order Statistics*. New York, NY, USA: Wiley.

A Model of Swarm Intelligence Based Optimization Framework Adjustable According to Problems

Utku Kose and Pandian Vasant

Abstract Swarm Intelligence has been a popular sub-field of Artificial Intelligence as including intelligent approaches, methods, and techniques for optimization problems. When the associated literature is examined, it can be seen that there are lots of different Swarm Intelligence techniques and researchers are still in an unstoppable interest in designing and developing newer ones. At this point, it is possible to imagine a general Swarm Intelligence system combining all developed techniques to enable researchers, who just want to solve their optimization problems with necessary features and functions. In this context, objective of this study is to introduce essentials of a modular framework, which is currently being developed for such purposes. The system introduced here consists of feature and function sets of different Swarm Intelligence techniques and employs an adaptive, adjustable infrastructure, which is also improved by additions of new techniques and results regarding to performed optimization tasks.

Keywords Swarm intelligence · Intelligent optimization
Artificial intelligence · Swarm intelligence framework

1 Introduction

Artificial Intelligence is known as one of the strongest scientific fields of the future. In time, many different Artificial Intelligence oriented approaches, methods, and techniques have been introduced and nowadays, this scientific field takes interest of almost all fields of the modern life. One cannot deny that this situation is because of

U. Kose (✉)
Computer Sciences Application and Research Center, Usak University, Usak, Turkey
e-mail: utku.kose@usak.edu.tr

P. Vasant
Faculty of Science and Information Technology, Universiti Teknologi Petronas, Perak, Malaysia
e-mail: vasantglobal@gmail.com

I. Zelinka et al. (eds.), *Innovative Computing, Optimization and Its Applications*, Studies in Computational Intelligence 741,
https://doi.org/10.1007/978-3-319-66984-7_2

its multi-disciplinary nature and wide solution scope covering real-world based problems of different fields. This situation has made Artificial Intelligence a wider and wider research platform in time and a need for introducing new sub-fields under it has appeared with more interests on specific problems types. Optimization is known as one of these problem types and Swarm Intelligence is a sub-field of Artificial Intelligence as taking great number of researchers' interests. In the literature, it is possible to see that Swarm Intelligence is widely used in a multidisciplinary manner [8, 9, 46, 50, 51].

Swarm Intelligence is generally focused on using nature-inspired mechanisms to design multi-agent based optimization algorithms. The main idea is to employ some agents—particles to use a collective intelligence for searching optimum solutions within a continuous or discrete solution space. At this point, the collective intelligence is designed as inspirations from behaviors of living organisms or natural dynamics. Because of that Swarm Intelligence algorithms have different features and functions to perform optimization processes. Different kinds of Swarm Intelligence algorithms are remarkable improvement signs of the associated literature because there is always a competitive state among all developed techniques to achieve more effective and more efficient solution ways for optimization. But on the other hand, researchers from especially different fields may find it difficult to adapt each of Swarm Intelligence techniques to their problems. Rather than doing this, it can be better to have a general framework of Swarm Intelligence based optimization and use it through the problems. Such framework can be developed by designing a modular application system in which different features and solution mechanisms of different Swarm Intelligence techniques are stored to run for optimization problems given as input to the framework. Furthermore, it will be also a remarkable improvement to keep obtained optimization solutions for similar problems in the past and adapt them in newly encountered problems if desired.

In the context of the explanations above, objective of this study is to introduce essentials of a modular Swarm Intelligence framework, which is currently being developed for such purposes. The system introduced here consists of feature and function sets of different Swarm Intelligence techniques and employs an adaptive, adjustable infrastructure, which is also improved by additions of new techniques and results regarding to performed optimization tasks. Generally, this framework is result of imaginations about a model of Swarm Intelligence system combining all developed techniques to enable researchers, who just want to solve their optimization problems with necessary features and functions. In detail, the framework enables users to define both infrastructure and problem input by using modular elements, which are connected in an algorithmic flow to achieve an optimum solution process for the considered problem. The authors believe that this framework will improve running of optimization tasks in different fields by combining the most recent technological background of the Swarm Intelligence and intelligent optimization literature.

According to the subject of the study, remaining content of the paper is organized as follows: The next section is devoted to background of the study considered here. In detail, readers are enabled in this section to have enough idea about what is intelligent optimization and what is Swarm Intelligence taking place on the background of the introduced framework. The section also includes some brief explanation about general structures—essentials of typical Swarm Intelligence techniques—algorithms. Following that, the third section gives remarkable details about the framework. In this way, general modular structure of the framework is explained briefly. After the third section, details regarding to working mechanism of the framework in case of a problem input are explained under the fourth section. Finally, the paper ends with some explanations on conclusions and future work.

2 Background

Before focusing on details of the Swarm Intelligence based optimization framework model, it is better to give brief information regarding to concepts lying on the background. So, this section has focused on the concept of intelligent optimization, Swarm Intelligence, and must-know subjects about Swarm Intelligence based techniques—algorithms.

2.1 The Concept of Intelligent Optimization

Until the current state, many different approaches, methods, and techniques have been employed for solving optimization problems. In this sense, classical optimization has been based on mathematical rules and formed traditional solution ways of optimization. However, because of more advanced optimization problems including more variables, more complex mathematical models, and even more detailed constraints, alternative techniques have been designed to be employed in optimization tasks. As a remarkable example of such techniques, Hill Climbing [10, 49] employs a more heuristic approach to find optimum according to plan, classical techniques. Because of that, it is possible to classify Hill Climbing as a technique just before the era of Artificial Intelligence based solution ways. When this technique was not strong enough for more advanced problems appeared, it has been the time for the Artificial Intelligence.

The concept of intelligent optimization stands generally for Artificial Intelligence based optimization processes (or at least the applications done over this approach). Intelligent optimization is related to using collective agents (it is possible to call them briefly as particles), which are moving intelligently-like to find the optimum within some iterative algorithmic steps [23, 32]. In detail, these algorithmic steps include some specific phases of adjusting parameters of the particles or solution environment as making the algorithmic steps sound more intelligently according to crisp algorithmic steps. Combination of such steps make the whole technique following something heuristic to find its way towards optimum of the considered problem. It is also important to mention that intelligent optimization on combinatorial optimization problems can benefit from graphs in order to form a general solution space with candidate solution members for the problem, as different from global optimization solution space structures. Here, we can understand that although optimization processes are associated with global, combinatorial or even dynamic optimization problems from different fields, it is still possible to use intelligent optimization [36, 38, 39, 52].

By considering former approaches, methods, and techniques, essential characteristics of the intelligent optimization can be listed briefly as follows (Fig. 1):

- Heuristic ways of searching.
- Collective behavior philosophy on the background.
- Role based solution ways.
- More complex algorithmic flows.
- Inspiration often from living organisms, nature and/or natural dynamics.

Today, intelligent optimization is associated with the Swarm Intelligence. Except from different features and working mechanisms in detail of developed algorithms, Swarm Intelligence generally explains the intelligent optimization in a philosophy associated with employment of particle groups behaving collectively to obtain a unique, intelligent solution way.

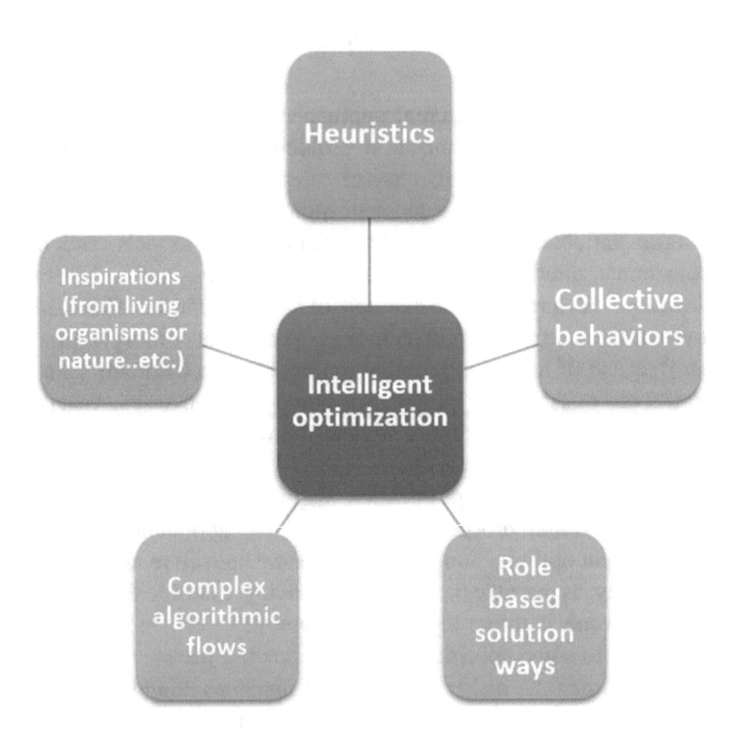

Fig. 1 Essential characteristics of the intelligent optimization

2.2 Swarm Intelligence

As firstly suggested by Beni and Wang [6] in the context of cellular robotic systems, Swarm Intelligence deals with the behaviors of natural or artificial objects behaving collectively towards an objective and without being connected to a central authority [31, 34, 35]. Considering as the sub-field of Artificial Intelligence, Swarm Intelligence is associated with the techniques—algorithms inspiring from natural swarms to perform optimization oriented operations. A typical system examined under Swarm Intelligence can be defined as a group of many members behaving collectively by interacting also with a complex environment to perform—complete common tasks for ensuring common benefit(s) [4, 32, 43]. Some examples of swarms from nature as inspiration for Swarm Intelligence oriented studies are birds, fishes, ants, bees, and any other living organisms searching for some food sources or interacting with the nature in common objectives [21, 44]. Additionally, we can also include some dynamics in the nature under the Swarm Intelligence oriented studies. Because although such dynamics are related to lifeless or/and abstract concepts, routine events in them often inspire the field of Swarm Intelligence for new techniques—algorithms to be developed.

By considering the related literature, some essential features and functions that Swarm Intelligence based systems should have can be expressed briefly as follows [32]:

- A swarm with members having similar/same characteristics.
- Interactions among swarm members and also with the environment.
- Intelligent moves—behaviors that cannot be understood specifically but meaningful on the general picture of the swarm.
- No central authority controlling swarm members.
- Spontaneously appeared common rules within the swarm.

Except from the explanations, it is also possible to examine also technical details of Swarm Intelligence based techniques—algorithms to understand more about working mechanisms of such solution options. Previously mentioned details have enabled the author(s) to design general structure of the framework model introduced in this study while the technical details provided within next sub-title have shaped main working mechanisms of the framework—model.

2.3 Essentials of Swarm Intelligence Based Techniques—Algorithms

In the associated literature, there are many different Swarm Intelligence based techniques—algorithms that are widely used within optimization tasks. From a general view, it is possible to indicate that these techniques—algorithms are mostly for global optimization or combinatorial optimization but there are also specifically improved ones that can be used for i.e. dynamic optimization [7, 11, 39]. Although techniques provided to the Swarm Intelligence literature are inspired from the nature, there are some specific approaches that are widely employed to ensure the mechanism of searching for the optimum under some typical swarm oriented moves. At this point, it is too important for one to understand main dynamics of the related approaches to develop a general framework, which can cover all the introduced techniques—algorithms and be able to include newer ones that have not introduced yet. Some popular Swarm Intelligence algorithms in the literature are as follows:

- Particle Swarm Optimization (PSO) (global) [18, 29].
- Ant Colony Optimization (ACO) (combinatorial) [15–17].
- Intelligent Water Drops Algorithm (IWDs) (combinatorial) [47].
- Artificial Bee Colony (ABC) (global) [27].
- Cuckoo Search (CS) (global) [60].
- Firefly Algorithm (FA) (global) [58].
- Genetic Algorithm (GA) (global) [25, 33, 40].
- Clonal Selection Algorithm (CSA) (global) [3, 12].

Readers are referred to [55] for a wider list of the algorithms.

If already introduced Swarm Intelligence based techniques—algorithms by 2017 are analyzed in detail, remarkable essentials of the techniques—algorithms can be expressed briefly as follows [32]:

- N number of particles, which are able to search within the solution space of optimization problems [Here, the particle is a general name and also used by PSO. But particles are called with different concepts (i.e. individual for GA, bee for ABC, ant for ACO, cuckoo for CS) in different techniques——algorithms.].
- For each particle, specific parameters inspired from the natural dynamics to which the technique—algorithm is connected.
- Roles designed for the particles to achieve a better interaction environment towards a successful optimization.
- Iteration based algorithmic steps with equations designed for moves—intelligent behaviors of the particles.
- In some cases, Evolutionary Computation [19, 20, 41] oriented approaches for eliminating worse particles and improving better ones to achieve the optimum objectives easily and as soon as possible.
- Random walk [24, 59] based or specifically designed methods to move the particles towards optimum point(s) of the problem.
- Some sets of rules for ensuring necessary interaction among particles to share experiences—information about obtained outputs so far for the objective function(s) of the problem and also movements towards better (or the best) particle(s) found better outputs—values for the considered optimization problem.
- Sometimes, special sets of rules to keep past values of particles are used for benefiting from past experiences through optimization process.

- In case of multi-objective optimization in addition to single-objective optimization, specifically designed solution approaches to keep past values of the particles to form pareto optimums (for more information: [1, 13, 14, 30].

As it can be understood, it is possible to form a general algorithmic structure of intelligent optimization system combining the features and functions of all techniques—algorithms under common, modular code structures. More details regarding to that are explained under the following section.

3 A Model of Swarm Intelligence Based Optimization Framework

The model designed for the optimization framework considered here is briefly for improving optimization processes by employing features—functions of different Swarm Intelligence based techniques—algorithms and get desired optimization outputs if it is even possible to form a unique algorithmic flow during the optimization process. Designing such framework is not an easy task and requires good programming approaches to deal with all details leading to adaptive, adjustable optimization characteristics of the system. In this context, the authors have employed a well-known programming approach to achieve that under a modular structure. It is important to express that the framework is currently in development processes (with final stages) but this section is devoted to details regarding to model background of the framework, which was carefully designed.

3.1 Programming Approach

The history of computer programming includes many innovations because of developments occurred in software and hardware technologies. Needs for faster computation times and more practical algorithmic solution ways have always attracted researchers—developers to think about something new. Increasing diversity of programming environments and changed features of human—computer interaction have also caused to that. As a result of the developments, a programming approach called as Object Oriented Programming has appeared by achieving good connection between real-world and programming concepts in a logical manner.

Object Oriented Programming (OOP) has been introduced in late 1960s as an alternative for improving complexity and size problems in classical, procedural programming [26, 42] and gained popularity more and more so far. Because of its focus on using object—class based hierarchical organization in operation elements (i.e. variables) and employing features—functions to keep specific values or operational functions in them and finally alternative approaches like creating inheritance between objects, running elements with same name but different

Table 1 Main class structures in the model of the framework

Class name	Feature—mechanism
Algorithm	Infrastructure for stored Swarm Intelligence algorithms
CombOpt	Specific definitions—features—functions related to combinatorial optimization
Core	Management of the algorithmic flow and optimization processes
Evaluation	General evaluation definitions for determining how to evaluate solutions
Evolution	Evolutionary computing based solution features and mechanisms
GlobalOpt	Specific definitions—features—functions related to global optimization
Experience	Experience infrastructure for the already solved optimization problems
Loop	General definition for loops during the optimization processes
MathModel	Mathematical model definition of the problem and the related details
Move	Definitions regarding to movement of particles in the solution space
MultiObj	Definitions regarding to multi-objective optimization
Param	Specific parameter definitions that can be used during optimization processes
Parser	Parsing mechanism for mathematical model definitions of the problem
Particle	Particle definitions and general usage in the optimization process
Resources	Management and use of resources (i.e. CPU, main memory) of the computer
SingleObj	Definitions regarding to single-objective optimization
Visual	Support for visual outputs regarding to optimization processes performed

functions (polymorphism) [26, 28, 42, 48] has made it simpler and faster to form algorithms for problems.

OOP is a great way to create Swarm Intelligence based systems easily because the particle philosophy in Swarm Intelligence often directs researchers to design operation elements with features and functions as similar to the mechanisms within OOP. So, an OOP approach was applied while developing the framework.

Main class structures and their features—mechanisms in the model of the framework are explained briefly (in an alphabetical order) under the Table 1. This class structure has been designed according to all adjustable and controllable features—functions of an optimization processes considered in the context of Swarm Intelligence. Some of the classes are hidden on the background while some ones are associated with the modules on the general model structure of the framework.

3.2 General Model Structure

Considering the class structure explained under the previous sub-title, it is possible to indicate that model structure of the framework should be examined under some specific connections between the Core class and other classes. Briefly the class of Core is the main management center of the framework and it has some connection legs to other classes in order to achieve the related adjustable, adaptive Swarm Intelligence based optimization model. From a general view, model structure over the Core class seems like a 'spider' having several legs. On the other hand, parallel connections between these legs (with 'connectors') make the model more similar to an also 'spider web' (Fig. 2).

It is important to indicate that the model itself has been built over some kind of intelligent mechanisms (which are explained under the next sub-title) to ensure the adjustable, adaptive optimization processes. Some details regarding to the model structure can be expressed briefly as follows:

- The Core class has many functions to manage flow among the modules, manage input(s), output(s), and direct the solution processes generally.
- Model structure has a total of 436 flag variables in the type of boolean to achieve a simple but effective enough flow.
- On each leg of the general model, there is a connector in order to decide if the solution flow will continue by passing another module on another leg or turning back to the Core in order to decide what to perform next.
- The model is able to evaluate the current optimization problem over some criterions. As connected with the Evaluation class, it is possible to evaluate both processes and the found solution candidates to improve solution way thanks to alternative Swarm Intelligence based solution steps.
- In addition to the evaluation mechanism mentioned above, the model also uses the Experience class to check if there is any similar mathematical model that the system has solved in the past. So, mathematical structures parsed before the Parser class are stored as default by the model—system. On the other hand, the model can also save optimization flows of the past successful problems over chaotic time series. Depending on user choose, the time series can be related to particle moves, optimum value changes, or optimum path changes (for especially combinatorial optimization processes) in the past operations.

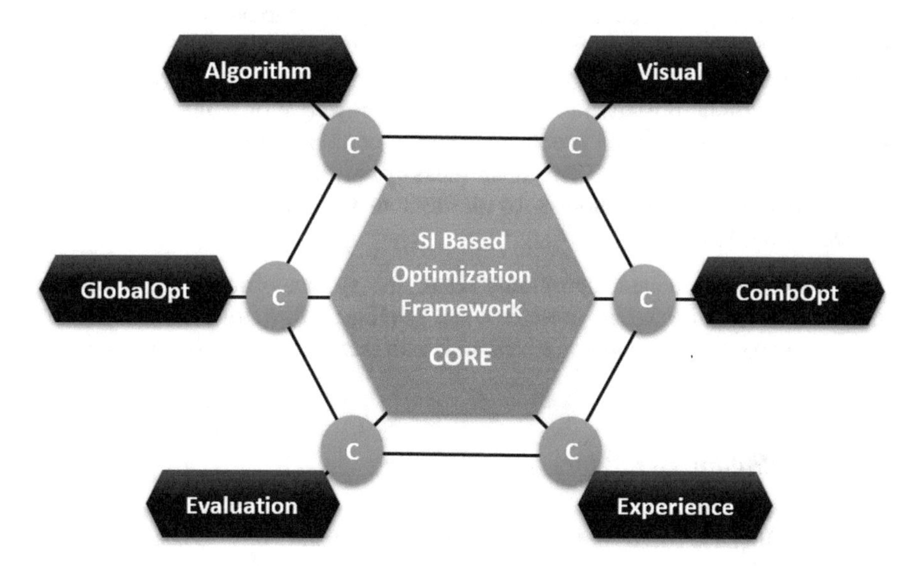

Fig. 2 General model structure of the framework

- Considering the class structures mentioned in Table 1, the Core class has six connections to six different—'visible' modules. These modules are based on their corresponding classes. At this point, the other remaining modules——classes are connected with the related six modules—classes: 'Algorithm', 'CombOpt', 'GlobalOpt', 'Evaluation', 'Experience', and 'Visual' on the background. From a general view, the framework has an easily understandable model structure in this way. More detailed schema regarding to the whole model structure is shown in Fig. 2.

As seen from the Fig. 2, the model is able to deal with different types of optimization problems and run any necessary algorithmic steps by using visible modules and the related classes on the front-side or background.

4 General Working Mechanism of the Framework

General working mechanism of the Swarm Intelligence based optimization framework is briefly based on some algorithmic steps corresponding to some intelligent behaviors of finding best steps during optimization. Differences of solution steps in different optimization problems depend on the following conditions:

- Predefined parameters and conditions for the related optimization problem (i.e. predefined objective value, initial values for some parameters of specific algorithms, values about solution space, constraints, candidate solution elements for combinatorial optimization cases),
- Experience of the system in the past optimization tasks (i.e. similar/same mathematical model, similar/same time series regarding to optimization process),
- Choosing whether a specific algorithm or adaptive process will be used for the optimization problem,
- Number of total objective functions as single-objective or multi-objective,
- Type of the optimization problem (global or combinatorial optimization),
- Available resources (i.e. CPU, main memory) of the computer used,
- All possible adaptive chooses by the system for particle moves—behaviors, evaluation, loops…etc.

Among the mentioned algorithmic steps, the system—framework sometimes follow intelligent behaviors to achieve an intelligent flow towards the problem solution. Some details on that mechanism is provided under the following sub-title.

4.1 Intelligent Behaviors

Swarm Intelligence based framework consists of a general Fuzzy Logic [45, 56, 57] based reasoning system supported with some probability to determine which move will be performed next at the specific break points of the algorithmic flow. The related break points are:

- Evaluation phases over objective function(s) output(s), time series…etc.,
- When it is detected that the flow of the optimization process seems getting worse and needs to be adjusted (this is achieved via time series analysis, error calculations, density on optimum points…etc.),
- In case of many variables, if the distribution of the particles is heterogeneous,
- When a specific number of iteration—loop determined by the user is reached,
- When a specific number of parameters like total eliminated particles, number of produced individuals—particles, number of particles (with a role) in a specific group is obtained. Such criterions are determined by the user.

Running the related break points are controlled by some rules defined under a Mamdani model [45, 56] based Fuzzy Inference System. In addition to these break point rules, flow of the algorithmic steps is determined with also some other fuzzy rules specifically defined for some features of the system. There are a total of 384 fuzzy rules in the latest form of the system. Some short ones of these rules are provided under the Table 2.

In addition to the obtained values at the output(s), some probability values are used to improve value of the output(s) and allow the Core module to run a choosing mechanism in order to determine which move will be performed next. Probability values for a fuzzy rule is calculated according to Eq. 1.

$$pr_i = r_i + \frac{r_{totaluse}}{r_{generaluse}} \tag{1}$$

In the Eq. 1, choosing probability of the ith rule is determined by summing a random value (0, 1) with its use rate calculated by dividing total use in the current

Table 2 Some short fuzzy rules included in the framework

Input(s)	Output(s)
IF (aimed iteration number is MEDIUM) AND (heterogeneous state is HIGH)	(Particle move change is HIGH) AND (evaluation rate is MEDIUM)
IF (error value is LOW) OR (total eliminated particle is LOW)	(Evaluation phase active is VERY HIGH)
IF (particles with role 1 is HIGH) AND (density on optimum is MEDIUM)	(Evaluation rate is HIGH)
IF (number of variables is HIGH) AND (number of locals is HIGH)	(Number of particles is HIGH) AND (experience use is VERY HIGH)
IF (completed iteration is HIGH) AND (time series similarity is MEDIUM)	(Adjust particle positions is HIGH)
IF (random walk choose is VERY HIGH) AND (number of variables is VERY LOW)	(Adjust algorithm is VERY HIGH) AND (evaluation rate is VERY HIGH)
IF (lev flight rate is HIGH) AND (heterogeneous state is MEDIUM)	(Adjust algorithm is VERY LOW) AND (update parameters randomly is HIGH)
…etc.	…etc.

problem by general use so far. Here, random value is used for improving choosing probability with a chance factor even a fuzzy rule has not chosen yet. On the other hand, choosing probability of a fuzzy rule is improved by increasing the last probability with a random value (0, 1) after it is used in the problem.

It is important that having choosing probability value for each fuzzy rule is not always a good option for the system to determine the next move. So, in case of more than one dominant fuzzy rule, a 'roulette wheel' mechanism [40, 54] inspired from Genetic Algorithms is run under the Core module.

Another intelligent behavior employed within the framework is related to analyzing past time series as under the Experience module. In this context, the first mechanism of the system—framework is to compare flow of some particles in the current solution process with past time series stored under the infrastructure. If there is no similar/same time series, future states of the chaotic time series for current particles is forecasted by using an Artificial Neural Networks [53, 61] based forecasting structure. If forecasted states seem similar/same with any of the ones in past time series, then algorithmic flow is directed towards the states of that past time series. Additionally, it is also possible to evaluate any of the fuzzy rules to determine next move, after understanding the current situations according to forecasted future states over the current time series of the particles.

4.2 Parser

One of the most remarkable functions of the developed Swarm Intelligence optimization framework is its parsing infrastructure. As associated with the Parser class, it is possible for the Core module to analyze mathematical model of the optimization problem given as input and direct the algorithmic solution way according to its essential details. Briefly, the parser of the Core requires the user to define the mathematical model of the problem via some definition rules. After defining a mathematical model, the following details—aspects of the model are parsed directly:

- Objective function(s),
- Variables and info on dimension,
- Constant(s),
- Constraint(s),
- Optimum values (if necessary),
- Optimization type (global or combinatorial),
- Solution space limitations,
- Problem type if it is specific for the literature,
- Any additional parameters.

4.3 Algorithm Infrastructure

The framework employs algorithmic codes regarding to some well-known Swarm Intelligence techniques—algorithms under the Algorithm module. This module is able to employ a whole algorithm or make it possible to use some aspects of an algorithm according to decisions by the Core module. As general, the Algorithm module is also responsible for parsing Swarm Intelligence based techniques—algorithms to include their specific parts in the system. A parsing process applied to an algorithm given to the framework results to extraction of the following factors:

- Optimization type (global, combinatorial or both of them),
- Single and multi-objective solution steps—calculations,
- Equations employed in the algorithm,
- Particles (with parameters, specific group—role types),
- Evolutional mechanisms (if possible),
- Specific evaluation mechanisms,
- Particle move characteristics on random walk or other moves (i.e. Levy),
- Specific connections among particles,

4.4 Visual Support

The module called as Visual is for some visual support for the processes regarding to optimization. In detail, it is possible to run a dynamic solution space simulation showing moves of particles—solution objects, or changes in error along the algorithmic flow. If users desire, it is also possible to track specific particles over time series or solution space behaviors over some specific graphics. Thanks to the Visual module—class, it is even possible to have statistical graphics regarding to the whole performed optimization processes, at the end.

4.5 An Example Scenario

Typical working mechanism of the Swarm Intelligence based framework can be explained briefly over the example pseudo-code scenario below (Some abbreviations are: **UC**: user chooses; **RE**: run experience module; **EP**: evaluation phase; **MP**: move particles; **PM**: parse the mathematical model; **NA**: no algorithm chosen; **EC**: run evolutionary computing mechanisms; **FR**: run fuzzy rules; **S**: step):

S1. Input ← Problem/mathematical model/specific problem/UC
S2. PM and take the algorithm on UC. If NA, follow the loop. In all cases FR:
S3. RE and benefit from experience or if NA, choose move type and MP else just MP according to UC (algorithm). In all cases FR.

S4. Objective function(s) → Output and EP
S5. FR and EC if necessary.
S6. Run visual elements according to UC
S7. If the desired optimization and/or UC stopping criterion(s) is not met yet, return to the S3 (in all cases FR just before turning back to S3).
S8. Save the flow—results under the experience module.

5 Discussion

Considering the developed Swarm Intelligence based framework, it is possible to express the following points:

- As mentioned before, the model within this framework is an alternative way of ensuring unique solution ways for all kinds of optimization problems.
- With its algorithmic infrastructure on the background, it is possible to improve the model with even the most recent Swarm Intelligence based techniques—algorithms.
- It is important that the model is able to deal with all kinds of optimization problems. But of course, it depends on the trained infrastructure, which is a sign for the strength of the system—model.
- Employing OOP is an important way for improving speed of the model on the optimization problems. Connection between OOP and Swarm Intelligence is an important factor that should be examined by the readers to understand more about Swarm Intelligence and programming it.
- It is clear that features and functions of the model has been designed according to recent literature on Swarm Intelligence. At this point, it may be important to think about future state of the Swarm Intelligence literature. Because new approaches in developing Swarm Intelligence based algorithms can result to newer additions to the infrastructure of the model.
- It is important to apply such framework for many different kinds of optimization problems. In this way, effectiveness and success of this framework—model will be understood in detail.
- In the literature, it is possible to see effective, hybrid approaches including both Swarm Intelligence oriented techniques and different Artificial Intelligence or traditional techniques to deal with real-world based optimization problems [2, 5, 22, 37]. So, infrastructure of the framework may be analyzed in detail with hybrid approaches and especially traditional techniques, in order to understand its solution scope and employ any possible improvements if possible.

6 Future Research Directions

It is possible to express ongoing or planned future research directions associated with the introduced framework—model here as follows:

- In detail, the system of the framework is being developed as based on Object Oriented Programming and will have different versions aiming different platforms like desktop, Web, and even mobile.
- Future works will also include reports on experiences regarding to use of the framework and any possible software based further developments, which have not indicated in this study, yet.
- Considering the software based developments, it is possible to point future research works planned on solving different kinds of optimization problems by using the framework.
- For the framework, it is also possible to evaluate its running time performances and usability situations by employing different kinds of evaluation methods. The planned future works also include such approaches.
- Except from its technical application ways, the framework can be used also for educational purposes. Such works on educational evaluation of the framework will be done in the future.

7 Conclusion

In this paper, the authors have introduced a model of Swarm Intelligence based optimization framework that can be used for global and combinatorial optimization problems by running adaptive and adjustable system features and functions. Because the system has a modular structure, it is possible to use essential parts of well-known Swarm Intelligence techniques—algorithms and all add new modules if necessary. Here, the main objective is to provide a comprehensive framework that can be used easily for problem solving process without considering details of which algorithm to use or how to adjust its parameters, any other values…etc. It is also important that the system of this model has some abilities like searching for the most effective way of solving difficult parts of an optimization problem if it seems leading to an undesired solution. Along such processes, it is possible for the framework to create its own unique optimization algorithm for the problem.

As it is also indicted that the framework introduced here is in development process and will be available as soon as possible. Currently, the mentioned works and plans under the previous section are in progress in order to improve the framework and its infrastructure. On the other hand, the authors believe that this work is an important reference for the researchers interested in Swarm Intelligence and developing such application framework.

Acknowledgements The work presented in this paper has been presented in the EAI International Conference on Computer Science and Engineering (COMPSE) 2016. The authors are also grateful to the independent reviewers for their comments for improving the work.

References

1. Abbass, H. A., Sarker, R., & Newton, C. (2001). PDE: A pareto-frontier differential evolution approach for multi-objective optimization problems. In *2001. Proceedings of the 2001 Congress on Evolutionary Computation* (Vol. 2, pp. 971–978). IEEE.
2. Abualigah, L. M., Khader, A. T., Al-Betar, M. A., & Hanandeh, E. S. (2017). A new hybridization strategy for krill herd algorithm and harmony search algorithm applied to improve the data clustering. In *First EAI International Conference on Computer Science and Engineering (COMPSE) 2016*. EAI.
3. Aickelin, U., Dasgupta, D., & Gu, F. (2014). Artificial immune systems. In *Search methodologies* (pp. 187–211). USA: Springer.
4. Andreasik, J. (2009). The knowledge generation about an enterprise in the KBS-AE (Knowledge-based system-acts of explanation). In *New challenges in computational collective intelligence* (pp. 85–94). Heidelberg: Springer.
5. Behnamian, J. (2015). Combined electromagnetism-like algorithm with tabu search to scheduling. In *Handbook of research on artificial intelligence techniques and algorithms* (pp. 478–508). IGI Global.
6. Beni, G., & Wang, J. (1993). Swarm intelligence in cellular robotic systems. In *Robots and biological systems: Towards a new bionics?* (pp. 703–712). Heidelberg: Springer.
7. Branke, J., Kaußler, T., Smidt, C., & Schmeck, H. (2000). A multi-population approach to dynamic optimization problems. In *Evol design and manufacture* (pp. 299–307). London: Springer.
8. Brownlee, J. (2011). Clever algorithms: Nature-inspired programming recipes. Retrieved June 1, 2017, from http://www.cleveralgorithms.com/nature-inspired/index.html.
9. Cai, T. (2015). Application of soft computing techniques for renewable energy network design and optimization. In *Handbook of research on artificial intelligence techniques and algorithms* (pp. 204–225). IGI Global.
10. Chalup, S., & Maire, F. (1999). A study on hill climbing algorithms for neural network training. In *Proceedings of the 1999 Congress on Evolutionary Computation, 1999. CEC 99*. (Vol. 3, pp. 2014–2021). IEEE.
11. Chiang, A. C. (2000). *Elements of dynamic optimization*. Illinois: Waveland Press Inc.
12. De Castro, L. N., & Timmis, J. I. (2003). Artificial immune systems as a novel soft computing paradigm. *Soft Computing-A Fusion of Foundations, Method. and Applications* 7(8), 526–544.
13. Deb, K. (2014). Multi-objective optimization. In *Search methodologies* (pp. 403–449). USA: Springer.
14. Deb, K., Sindhya, K., & Hakanen, J. (2016). Multi-objective optimization. In *Decision sciences: theory and practice* (pp. 145–184). CRC Press.
15. Dorigo, M. (1992) *Optimization, learning and natural algorithms* (In Italian). Ph.D. Thesis, Politecnico di Milano, Italy.
16. Dorigo, M., Maniezzo, V., & Colorni, A. (1991). *Positive feedback as a search strategy*. Technical Report 91–016, Dipartimento di Elettronica, Politecnico di Milano, Italy.
17. Dorigo, M., Maniezzo, V., & Colorni A (1996) Ant system: Optimization by a colony of cooperating agents. *IEEE Transactions on Systems, Man, and Cybernetics, Part B (Cybernetics)*, *26*(1), 29–41.
18. Eberhart, R., & Kennedy, J. (1995). A new optimizer using particle swarm theory.In *1995 Proceedings of the Sixth International Symposium on Micro Machine and Human Science, MHS'95* (pp. 39–43). IEEE.

19. Fogel, D. B. (2000). What is evolutionary computation? *IEEE Spectrum, 37*(2), 26–32.
20. Fogel, D. B (2006). *Evolutionary computation: Toward a new philosophy of machine intelligence*. Wiley.
21. Garnier, S., Gautrais, J., & Theraulaz, G. (2007). The biological principles of swarm intelligence. *Swarm Intelligence, 1*(1), 3–31.
22. Ghanem, W., & Jantan, A. (2017). Hybridizing bat algorithm with modified pitch-adjustment operator for numerical optimization problems. In *2016 First EAI International Conference on Computer Science and Engineering (COMPSE)*. EAI.
23. Guler, G., & Kose, U (2016) Intelligent optimization for logistics. In *Scientific Conference*. 2016 (pp. 131–137). Slovakia. https://doi.org/10.18638/scieconf.2016.4.1.380.
24. Hassanien, A. E., & Emary, E. (2016). *Swarm intelligence: Principles, advances, and applications*. CRC Press.
25. Holland, J. H. (2012). Genetic algorithms. *Scholarpedia 7*(12), 1482. Çevrimiçi. http://www.scholarpedia.org/article/Genetic_algorithms. Accessed 26 March 2017.
26. Kalemis, D. (2013). *The fundamental concepts of object-oriented programming*. CreateSpace Independent Publishing.
27. Karaboga, D., & Basturk, B. (2007). A powerful and efficient algorithm for numerical function optimization: artificial bee colony (ABC) algorithm. *Journal of Global Optimization, 39*(3), 459–471.
28. Karacay, T. (2016). *Object Programming with Java* (In Turkish). Seckin Press.
29. Kennedy, J. (2011). Particle swarm optimization. In *Encyclopedia of machine learning* (pp. 760–766). USA: Springer.
30. Konak, A., Coit, D. W., & Smith, A. E. (2006). Multi-objective optimization using genetic algorithms: A tutorial. *Reliability Engineering and System Safety, 91*(9), 992–1007.
31. Kose, U. (2015). Present state of swarm intelligence and future directions. In *Encyclopedia of information science and technology* (3rd ed., pp. 239–252). Hershey: IGI Global.
32. Kose, U. (2017). *Development of artificial intelligence based optimization algorithms* (In Turkish). Ph.D. Thesis. Selcuk University, Turkey.
33. Kramer, O (2017). Genetic Algorithm Essentials. Springer.
34. Krause, J., Ruxton, G. D., & Krause, S. (2010). Swarm intelligence in animals and humans. *Trends in Ecology and Evolution, 25*(1), 28–34.
35. Li B. H., Qu, H. Y., Lin, T. Y., Hou, B. C., Zhai, X., Shi, G.Q., et al. (2017). A swarm intelligence design based on a workshop of meta-synthetic engineering. *Frontiers of Information Technology and Electronic Engineering 18*(1), 149–152.
36. Li, S., & Zheng, Y. (2015). A memetic algorithm for the multi-depot vehicle routing problem with limited stocks. In *Handbook of research on artificial intelligence techniques and algorithms* (pp. 411–445). IGI Global.
37. Majumder, A., & Majumder, A. (2015). Application of standard deviation method integrated PSO approach in optimization of manufacturing process parameters. In *Handbook of research on artificial intelligence techniques and algorithms* (pp. 536–563). IGI Global.
38. Marler, R. T., & Arora, J. S. (2004). Survey of multi-objective optimization methods for engineering. *Structural and Multidisciplinary Optimization, 26*(6), 369–395.
39. Mavrovouniotis, M., Li, C., & Yang, S. (2017). A survey of swarm intelligence for dynamic optimization: Algorithms and applications. *Swarm and Evolutionary Computation, 33,* 1–17.
40. Mitchell, M. (1998). *An introduction to genetic algorithms*. MIT Press.
41. Mohan, U. (2008). Bio inspired computing. BSc. Seminar. Division of CS SOE. CUSAT.
42. Nino, J. (2007). *An introduction to programming and object-oriented design using java*. Wiley.
43. Pan, Y. (2016). Heading toward artificial intelligence 2.0. *Engineering, 2*(4), 409–413.
44. Parpinelli, R. S., & Lopes, H. S. (2011). New inspirations in swarm intelligence: A survey. *International Journal of Bio-Inspired Computation, 3*(1), 1–16.
45. Ross, T. J. (2009). Fuzzy logic with engineering applications. John Wiley & Sons.

46. Roy, P. K. (2016). A novel evolutionary optimization technique for solving optimal reactive power dispatch problems. Sustaining Power Resources through Energy Optimization and Engineering (pp. 244–275). IGI Global.
47. Shah-Hosseini, H. (2009). The intelligent water drops algorithm: A nature-inspired swarm-based optimization algorithm. *International Journal of Bio-Inspired Computation, 1* (1–2), 71–79.
48. Stroustrup, B. (1988). What is object-oriented programming? *IEEE Software, 5*(3), 10–20.
49. Sullivan, K. A., & Jacobson, S. H. (2001). A convergence analysis of generalized hill climbing algorithms. *IEEE Transactions on Automatic Control, 46*(8), 1288–1293.
50. Vasant, P. M. (Ed.). (2012). *Meta-heuristics optimization algorithms in engineering, business, economics, and finance*. IGI Global.
51. Vasant, P. (Ed.). (2013). *Handbook of research on novel soft computing intelligent algorithms: Theory and practical applications.* IGI Global.
52. Vasant, P. (Ed.). (2016). *Handbook of research on modern optimization algorithms and applications in engineering and economics*. IGI Global.
53. Vemuri, V. R., & Rogers, R. D. (1993). Artificial neural networks-forecasting time series. *IEEE Computer Society Press*.
54. Whitley, D. (1994). A genetic algorithm tutorial. *Statistics and Computing, 4*(2), 65–85.
55. Xing, B., & Gao, W. J. (2014) *Innovative computational intelligence: A rough guide to 134 clever algorithms* (Vol. 62, p. 451). Cham: Springer.
56. Yager, R. R., & Filev, D. P. (1994). *Essentials of fuzzy modeling and control*. USA: Wiley.
57. Yager, R. R., & Zadeh, L. A (Eds.). (2012). *An introduction to fuzzy logic applications in intelligent systems* (Vol. 165). Springer Science & Business Media.
58. Yang, X. S. (2009). Firefly algorithms for multimodal optimization. In *International Symposium on Stochastic Algorithms* (pp. 169–178). Heidelberg: Springer.
59. Yang, X. S. (2010). *Nature-inspired metaheuristic algorithms*. Luniver Press.
60. Yang, X. S., & Deb, S. (2009). Cuckoo search via Levy flights. In *2009 World Congress on Nature & Biologically Inspired Computing, NaBIC 2009*. (pp. 210–214). IEEE.
61. Yegnanarayana, B. (2009). *Artificial neural networks*. Ltd: PHI Learning Pvt.

Domain Model Definition for Domain-Specific Rule Generation Using Variability Model

Neel Mani, Markus Helfert, Claus Pahl, Shastri L Nimmagadda and Pandian Vasant

Abstract The business environment is rapidly undergoing changes, and they need a prompt adaptation to the enterprise business systems. The process models have abstract behaviors that can apply to diverse conditions. For allowing to reuse a single process model, the configuration and customisation features can support the design improvisation. However, most of the process models are rigid and hard coded. The current proposal for automatic code generation is not devised to cope with rapid integration of the changes in business coordination. Domain-specific Rules (DSRs) constitute to be the key element for domain specific enterprise application, allowing changes in configuration and managing the domain constraint with-in the domain. In this paper, the key contribution is conceptualisation of the do-main model, domain model language definition and specification of domain model syntax as a source visual modelling language to translate into domain specific code. It is an input or source for generating the target language which is do-main-specific rule language (DSRL). It can be applied to adapt to a process constraint configuration to fulfil the domain-specific needs.

N. Mani (✉) · M. Helfert
School of Computing, ADAPT Centre for Digital Content Technology,
Dublin City University, Dublin, Ireland
e-mail: neel.mani@computing.dcu.ie

M. Helfert
e-mail: markus.helfert@computing.dcu.ie

C. Pahl
Faculty of Computer Science, Free University of Bozen-Bolzano, Bolzano, Italy
e-mail: Claus.Pahl@unibz.it

S. L. Nimmagadda
School of Information Systems, Curtin Business School (CBS), Perth, WA, Australia
e-mail: shastri.nimmagadda@curtin.edu.au

P. Vasant
Department of Fundamental and Applied Sciences, Universiti Teknologi PETRONAS,
Perak, Darul Ridzuan, Malaysia
e-mail: pvasant@gmail.com

I. Zelinka et al. (eds.), *Innovative Computing, Optimization and Its Applications*, Studies in Computational Intelligence 741,
https://doi.org/10.1007/978-3-319-66984-7_3

1 Introduction

We are primarily concerned with the utilisation of a conceptual domain model for rule generation, specifically to define a domain-specific rule language (DSRL) [1, 2] syntax, its grammar for business process model and domain constraints management. We present a conceptual approach for outlining a DSRL for process constraints [3]. The domain-specific content model (DSCM) definition needs to consider two challenges. The first relates to the knowledge transfer from domain concept to conceptual model, where model inaccuracies and defects may have been translated because of misunderstandings, model errors, human errors or inherent semantic mismatches (e.g., between classes). The other problem relates to inconsistency, redundancy and incorrectness resulting from multiple views and abstractions. A domain-specific approach provides a dedicated solution for a defined set of problems. To address the problem, we follow a domain model language approach for developing a DSRL and expressing abstract syntax and its grammar in BNF [4] grammar. A domain-specific language (DSLs) [5–7] refers to an approach for solving insufficient models by capturing the domain knowledge in a domain-specific environment. In the case of semantic mismatches/defects, a systematic DSL development approach provides the domain expert or an analyst with a problem domain at a higher level of abstraction.

DSL is a promising solution for raising the level of abstraction that is easier to understand or directly represent and analyse, thus, attenuating the technical skills required to develop and implement domain concepts into complex system development. Furthermore, DSLs are either textual or graphical language targeted to specific problem domains by increasing the level of automation, e.g. through rule and code generation or directly model interpretation (transform or translate), as a bridge, filling the significant gap between modeling and implementation. An increase in effectiveness (to improve the quality) and efficiency of system process is aimed at rather than general-purpose languages that associated with software problems. Behavioral inconsistencies of properties can be checked by formal defect detection methods and dedicated tools. However, formal methods may face complexity, semantic correspondence, and traceability problems. Several actions are needed for implementation of any software system. These are from a high-level design to low level execution. The enterprises typically have a high level of legacy model with various designs in a domain or process model. Automatic code generation [8–11] is a well-known approach for getting the execution code of a system from a given abstract model. The Rule is an extended version of code since code requires compiling and building, but the rule is always configurable. Rule generation is an approach by which we transform the higher-level design model as input and the lower level of execution code as output. It manages the above mentioned constraint.

We structure the modelling and DSL principles in Sect. 1. In Sect. 2, we discuss the State of-the-Art and Related Work. We give an overview of the global intelligent content processing in a feature-oriented DSL perspective, which offers the domain model and language definition in Sect. 3. Then, we describe the ontology-based conceptual domain model in Sect. 4. The description of the domain model and language expressed in terms of abstract and concrete syntax are given in Sect. 5. As a part of DSRL, the general design and language are presented in Sect. 6. Section 7 provides details regarding implementation of a principal architecture of DSRL generation and how it translates from the domain model to the DSRL. We discuss analysis and evaluation of the rule in Sect. 8. Finally, we conclude our work with future scope. Throughout the investigation, we consider the concrete implementation as a software tool. However, a full integration of all model aspects is not aimed at, and the implementation discussion is meant to be symbolic. The objective is to outline the principles of a systematic approach towards a domain model used, as source model and the domain specific rule language, as a target for content processes.

2 Related Work

The web application development is described as a combination of processes, techniques and from which web engineering professionals make a suitable model. The web engineering is used for some automatic web application methodologies such as UWE [12], WebML [13] and Web-DSL [14] approaches. The design and development of web applications provide mainly conceptual models [15], focusing on content, navigation and presentation models as the most relevant researchers expressed in [16, 17]. Now, the model driven approach for dynamic web application, based on MVC and server is described by Distante et al. [18]. However, these methods do not consider the user requirement on the variability model. To simplify our description, we have considered the user requirement and according to the need of the user, the user can select the feature and customize the enterprise application at the dynamic environment.

A process modeling language provides syntax and semantics to precisely define and specify business process requirements and service composition. Several graph and rule-based languages have been emerged for business process modeling and development, which rely on formal backgrounds. They are Business Process Modeling Notation (BPMN) [19], Business Process Execution Language (BPEL)/WS-BPEL, UML Activity Diagram Extensions [20], Event-Driven Process Chains (EPC) [21], Yet Another Work ow Language (YAWL) [22], WebSphere FlowMark Definition Language (FDL) [23], XML Process Definition Language (XPDL) [24], Java BPM Process Definition Language (jPDL) [25], and Integration Definition for Function Modeling (IDEF3) [26]. These languages focus on a different level of abstraction ranging from business to technical levels and have their weaknesses and strengths for business process modeling and execution. Mili et al. [27] survey the major business process modeling languages and provide a brief comparison of the languages, as well as guidelines to select such a language. In [28], Recker et al. present an overview of different business-process modeling techniques. Among the existing languages, BPMN and BPEL are widely accepted as de facto standards for business process design and execution respectively.

Currently, there is no such type of methodology or process of development for creating a rule-based system in a web application (semantic-based). Diouf et al. [29, 30] propose a process which merges UML models and domain ontologies for business rule generation. The solution used for semantic web has ontologies and UML; to apply to the MDA approach for generating or extracting the rules from high level of models. Although, the proposed combination of UML and semantic based ontologies is for extracting the set of rules in target rule engine, they only generate the first level of the abstraction of the rules.

Our approach provides the systematic domain-specific rule generation using variability model. The case study uses intelligent content processing. Intelligent content is digital that provides a platform for users to create, curate and consume the content in dynamic manner to satisfy individual requirements. The content is stored, exchanged and processed by a dynamic service architecture and data are exchanged, annotated with metadata via web resources.

3 Business Process Models and Constraints

We use the intelligent content (IC) [31] processing as a case study in our application. The global intelligent content (GIC) refers to digital content that allows users to create, curate and consume content in a way that fulfills dynamic and individual requirements relating to information discovery, context, task design, and language. The content is processed, stored and exchanged by a web architecture and the data are revised, annotated with metadata through web resources. The content is delivered from creators to consumers. The content follows a certain path that consists of different stages such as extraction and segmentation, named entity recognition, machine translation, quality estimation and post-editing. Each stage, in the process, comes with its challenges and complexities.

The target of the rule language (DSRL) is an extendable process model notation for content processing. Rules are applied at processing stages in the process mode.

The process model that describes activities remains at the core. It consists of many activities and sub-activities of reference for the system and corresponds to the properties for describing the possible activities of the process. The set of activities constitutes a process referred to as the extension of the process and individual activities in the extension are referred as instances. The constraints may be applied at states of the process to determine its continuing behaviour depending on the current situation. The rules combine a condition (constraint) on a resulting action. The target of our rule language (DSRL) is a standard business process notation (as shown in Fig. 1).

The current example is a part of digital content (processing) process model as shown in Fig. 1, a sample process for the rule composition of business processes

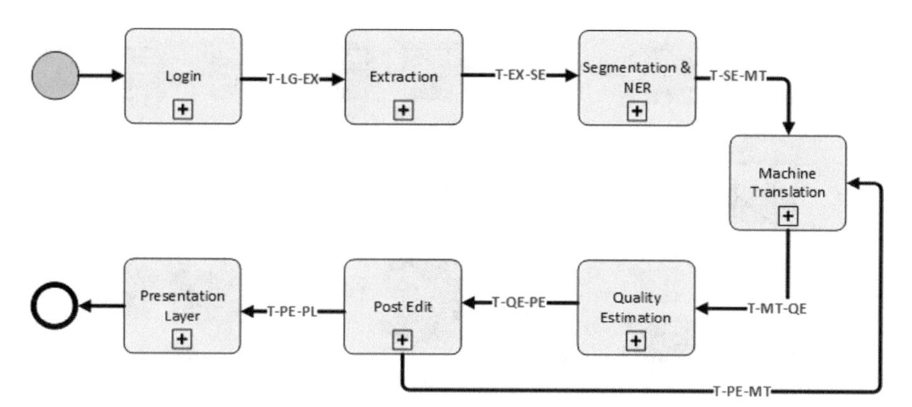

Fig. 1 Process model of global intelligent content

and domain constraints that conduct this process. The machine language activity translates the source text into the target language. The translated text quality decides whether further post-editing activity is required. Usually, these constraints are domain-specific, e.g., referring to domain objects, their properties and respective.

4 Ontology-Based Conceptual Domain Model

We outline the basics of conceptual domain modelling, the DSRL context and its application in the intelligent content context. The domain conceptual models (DCM) are in the analysis phase of application development, supporting improved understanding and interacting with specific domains. They support capturing the requirements of the problem domain and, in ontology engineering. A DCM is a basis for the formalized ontology. There are several tools, terminologies, techniques and methodologies used for conceptual modelling, but DCMs help better understanding, representing and communicating a problem situation in specific domains. We utilise the conceptual domain model to derive at a domain-specific rule language.

A conceptual model can define concepts concerning a domain model for a DSL, as shown in the class model in Fig. 2, which is its extended version described in [3]. A modeling language is UML-based language defined for a particular domain, defining relevant concepts as well as a relation (intra or inter-model) with a metamodel. The metamodel consists of the concrete syntax, abstract syntax and static semantics of the DSL. The abstract syntax defines modelling element such as classes, nodes, association, aggregation and generalisation, and relationships between the modelling elements [3].

A DSRL reuses domain model elements or define new modelling element depending on the domain concept and its relations with other elements. Modelling of elements is done with two fundamental types: concepts and relations. The concepts are

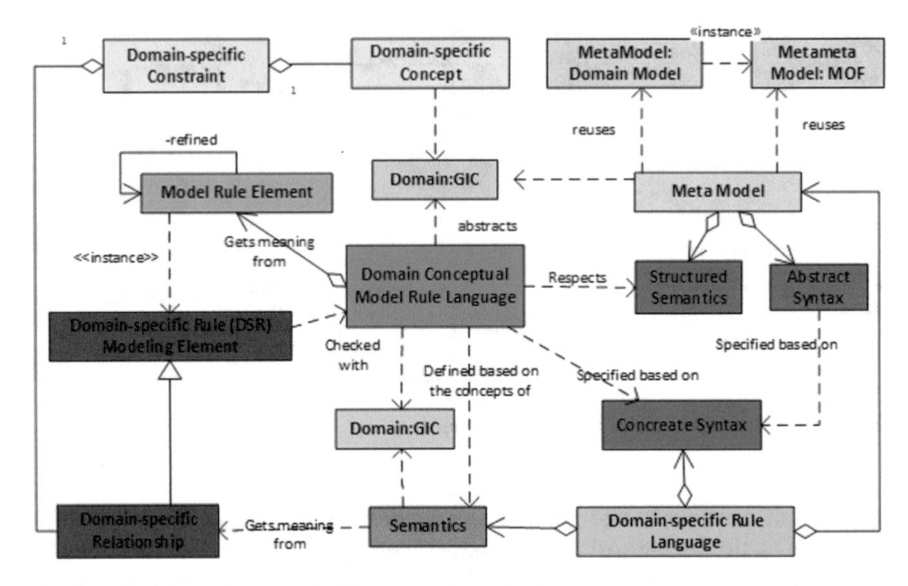

Fig. 2 A domain model based DSRL concept formalization

based on domain concepts such as entity, state, action, location, risk, menu and relations are used to connect elements. The relations are also of two types: generalisation and association. Aggregation is a special type of an association. The association has domain-specific relations like conditional flow, multiplicity and aggregation. Furthermore, relations may have properties such as symmetry, reflexivity, equivalence, transitivity and partial order.

5 Language Definition of a Domain Model

Domain model serves as the very basis of all types of Business Applications that run on a Domain, both individual as well as Enterprise applications. The objective is to define the language for Domain Model and recognises the internal data structures or schema used in it. It is easy to transform or translate the graphical Domain Model into Textual Rule language for a particular domain. In this scenario, the objective data structure refers to the storability of a domain model in a vulnerable environment, as in the case with a rule language. This is because the target of a language for mapping the translated domain model's knowledge into XML schema of DSRL that follows the rule paradigm.

1	*Domain*	*::=*	*<Domain model>*	*Domain definition*
2	*Concept*	*::=*	*<Concept>*	*Concept definition*
3	*Class*	*::=*	*<Attributes>,< Operation>, < Receptions>, <Template Parameters>, < Component>, <Constraints>, <Tagged Values>*	*Class Definition*
4	*<Relations>*	*::=*	*<Association>|<DirectedAssociation>|<Reflexive Association>|<Multiplicity>|<Aggregation>|<Composition>|<Inheritance/Generalization>|<Realization>*	*Class relationships*
5	*<Association>*	*::=*	'→' | '*' | '::'	*Structural relationship between objects (classes) of different type*
6	*<Type>*	*::=*	*<BuiltinType>|<UCase Ident >|<EnumType>*	*Domain model type concept type or extended type enumeation type list type*
7	*<PrimitiveTypes >*		*<String>,<Integer>,<Boolean>, ...,<Date>*	*Domain model primitive (built-in) types*

Fig. 3 Syntax definition of domain model language

5.1 Language Description

Metamodeling is used to accomplish specifications for the abstract syntax. We introduce the Domain Model language by analyzing its syntax definition (Fig. 3 shows in EBNF notation). The language with its basic notions and their relations are defined with structural constraints (for instance to express containment relations, or type correctness for associations), multiplicities, precise mathematical definition and implicit relationships (such as inheritance, refinement). The visual appearance of the domain specific language is accomplished by syntax specifications, which is done by assigning visual symbols to those language elements that are to be represented on diagrams.

5.2 Syntax

For describing the language in general, the rule language checks various kinds of activities. The primary requirement is to specify the concept of the syntax (i.e. abstract and or concrete syntax) and develop its grammar. The semantics is designed to define the meaning of the language. The activities are completed by concepting, designing and developing a systematic domain-specific rule language systems, defining the functions and its parameters, priorities or precedence of operators and its values, naming internal and external convention system. The syntaxes are expressed with certain rules, conforming to BNF or EBNF grammars that can be processed by rules or process engine to transform or generate the set of rules as an output. The generated rules follow the abstract syntax and gram-mar to describe the domain concepts and domain models because both the artefacts (abstract syntax and grammar) are reflected in the concrete syntax.

5.3 Abstract Syntax

The abstract syntax refers to a data structure that contains only the core values set in a rule language, with semantically relevant data contained therein. It excludes all the notation details like keywords, symbols, sizes, white space or positions, comments and color attributes of graphical notations. The abstract syntax may be considered as more structurally defined by the grammar and Meta model, representing the structure of the domain. The BNF may be regarded as the standard form for expressing the grammar of rule language, and some type describes how to recognize the physical set of rules. Analysis and downstream processing of rule language are the main usages of Abstract syntax. Users interact with a stream of characters, and a parser compiles the abstract syntax by using a grammar and mapping rules.

For example, we do process activities in our case domain (Global Digital Content): Extraction and Machine Translation (MT). The list of the process model, event and condition are following:

List of Process

```
<Process-ModelList>::=<gic:Extraction>|
<gic:MachineTranslation>
```

List of Event

```
<EventList> ::={
gic:Text-->SourceTextInput,
gic:Text-->SourceTextEnd,
gic:Text -->SourceTextSegmentation,
gic:Text-->SourceParsing,
gic:Text-->MTSourceStart,
gic:Text-->MTTargetEnd,
gic:Text-->TargetTextQARating,
gic:Text-->TargetTextPostEditing,
}
```

List of Conditions

```
<ConditionList>::=<gic:Extraction.Condition>|
            <gic:MachineTranslation.Condition>
<gic:Extraction.Condition>::=IF(<gic:Text.Length::=<L)|
    IF(<Source.Language::==Language_List>)
    IF(<Target.Language::==Language_List>)
```

```
     IF(<SingleLangugeDetection((gic:Text)::== True
                                                  |False>)
     IF(<MultiLanguageText(gic:Text)::== True|False>)
  <gic:MachineTranslation.Condition>::= IF (<gic:Translation
  (Source.Lang, TargetLang,gic:Text)   ::= True|False>)|
     IF (<gic:Translation.Memory ::= <TM)(Mem Underflow)|
     IF (<gic:Translation.Memory ::= >TM)(Mem Overflow)|
     IF(<gic:Translation(gic:TxtSource,Source.Lang)>
  gic:Translation(gic:TxtTarget,Target.Lang)>)
```

where L is length of text and TM is the specific memory size.

5.4 *Concrete Syntax*

Rule languages use textual concrete syntax, which implies that a stream of characters expresses the program syntax. The modelling languages traditionally have used graphical notations and primarily in modelling languages. Though textual domain-specific languages (and mostly failed graphic based general-purpose languages) have been in use for a long time only recently, the textual syntax has found a prominent use for domain-specific modelling. Textual, concrete syntax form have been traditionally used to store programs, and this character stream is transformed using scanners and parsers into an abstract syntax tree for further processing by the Programming languages. In the modelling languages, editors have found a major usage, as it directly manipulates the abstract syntax and uses projection to render the concrete syntax in the form of diagrams.

The concrete syntax of DSLs is expected to be textual by default. If good tool support is available, the textual support has been found to be adequate for comprehensive and complex software systems. The programmers write lesser code in DSL as compared to a GPL for expressing the same functionality—because the available abstractions are quite similar to the domain. An additional language module suitable for the domain is defined easily by the programmers.

6 Rule Language Definition

Now we go back to the full rule definition. The DSRL grammar [2] is defined as follows. We start with a generic skeleton and then map the globic domain model (gic).

```
  <DSRL Rules> ::= <EventsList>
                    <RulesList>
```

```
                  <ProcessModelList>
<EventLists> ::= <Event> | <Event> <EventLists>
<Event> ::= EVENT <EventName> IF <Expression> |
            EVENT <EventName> is INTERN or EXTERN
<RulesList> ::= <Rule> | <Rule> <RuleList>
<Rule> ::= ON<EventName>
           IF<Condition>DO<ActionList>
<ActionList> ::= <ActionName> |
                 <ActionName>,<ActionList>
<ProcessModelList> ::= <ProcessModel> |
                       <ProcessModel>,<ProcessModelList>
<ProcessModel> ::= ProcessModel <ProcessModelName>
               ::= <ProcessModelName>
                   [TRANSITION_(SEQUENCIAL(DISCARDDELAY))],
                   [TRANSITION_PARALLEL(DISCARD|DELAY)]
                   [INPUTS(<InputList>)]
                   [OUTPUTS(<OutputList>)]
```

TRANSITION_SEQUENCIAL and TRANSITION_PARALLEL denoted as transitions of a process model.

The description of DSRL contains lists of events, condition, an action of the rules and process model states. An event can be an internal or external (for rules generated as an action, it may be the INTERNAL or EXTERNAL term) or generated when the expression should have been satisfied by the condition or becomes true. An event name activates with ON syntax, which is a Boolean expression to determine the conditions that apply and the list of actions that should be per-formed when event and condition are matched or true (preceded by DO syntax). The process models contain the state name. A certain policy is decided in the process model when sequential, and parallel actions are performed or sent in that state. An action is an executable program or set of computation decussation. The action provides methods or function invocation, creating, modifying, updating, communicating or destroying an object. DISCARD allows discarding the instructions, and DELAY allows delaying the instructions, but one.

For example, the gic:Extraction is used in an event on Text Input by user as a source data.

```
EVENT IF TextInput_ON
EVENT gic:TextInput::BOOL IF TextInput_Get
EVENT gic:TextInput::BOOL IF TextInput_ON
ON presence
IF (gic:SourceLang:EN) DO
(
     ON presence
     IF (gic:TextLength <X) DO
gic:Translate(Text)
```

```
ELSE

 Notification to user (Text LENGTH LESS  THAN X)
)
ELSE
Notification to user(Source language is invalid)
```

7 Implementation of Principle Architecture of DSRL Generator

The principle architecture of a domain-specific rule (DSR) is the automated model to text generator on accessible domain models, extracting information from them, and translating it into output in a specific target syntax. This process follows the concept of Model-Driven Architecture which depends on the metamodel. The modeling language with its concepts, the source syntax, semantics and its rules are required by the domain-specific framework and target environment. We present the process architecture of a domain model translation and the target rule environment in Fig. 4.

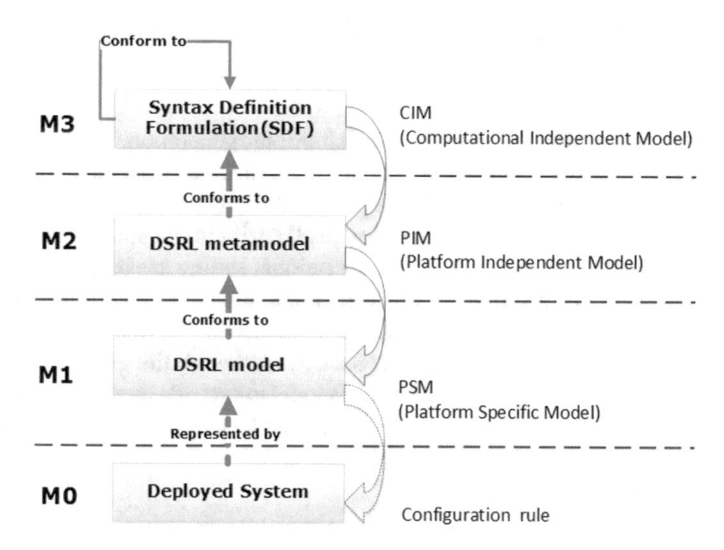

Fig. 4 MDA organisation view of models approach and artifacts of DSRL generator

7.1 Architecture

The architecture of the DSRL generator follows the MDA as four-level model organization, presented by Bzivin [32] as illustrated in Fig. 4. At the top level, the M3 is the Syntax Definition Formalism (SDF) metametamodel which is the grammar of the SDF. This level is also known as Computational Independent Model (CIM) or metametamodel as defined (and thus conforms to) itself [33]. A self-representation of the BNF notation takes some lines. This notation allows a defining infinity of well-formed grammars. A given grammar allows description of the infinity in syntactically correct DSR configuration.

At the M2 level, we describe the DSRL metamodel, i.e., the grammar of DSRL with ECA as defined in SDF and this level is called Platform Independent Model (PIM). The metamodel conforms to the metametamodel at level M3.

At the M1 level, we describe DSRL models for configuration applications. It is known as Platform Specific Model (PSM) consisting of entity and definitions. The model conforms to the metamodel at level M2. The bottom level is called M0, we define the configuration of BPM customization consisting of DSR and XML rules, which represent the models at the M1 level.

7.2 Mappings Domain Model and Domain-Specific Rule Language

A mapping is description of mapping rule definitions, generation, configuration and execution of order specification. Each mapping rule specifies what target model fragment is created for the given DSR. The mapping rule body contains one or more class of the domain model occurrences (Sect. 4) with all attribute and operational value set. Expressions for attribute, functional and operational setting are based on a specific source metamodel, Fig. 5 shows the corresponding domain model of gic:extraction sub type of digital content process used as a source metamodel to describe the DSRL conceptualization as illustrated in Fig. 2 (Sect. 3). Although the given source metamodel is completely translated into graphical model to text rule by using the grammar or language definition of source and target metamodel as given in Sects. 4 and 5, it is sufficient to show all basic mapping constructs.

7.3 Domain Model Translation into DSR

A rule generation is made automatic such that domain models are accessed in a way to extract information and translate it into output in a specific syntax based on feature model, as selected by the domain user. This process model is guided by the metamodel, the modeling language with its high level of concepts, syntactical,

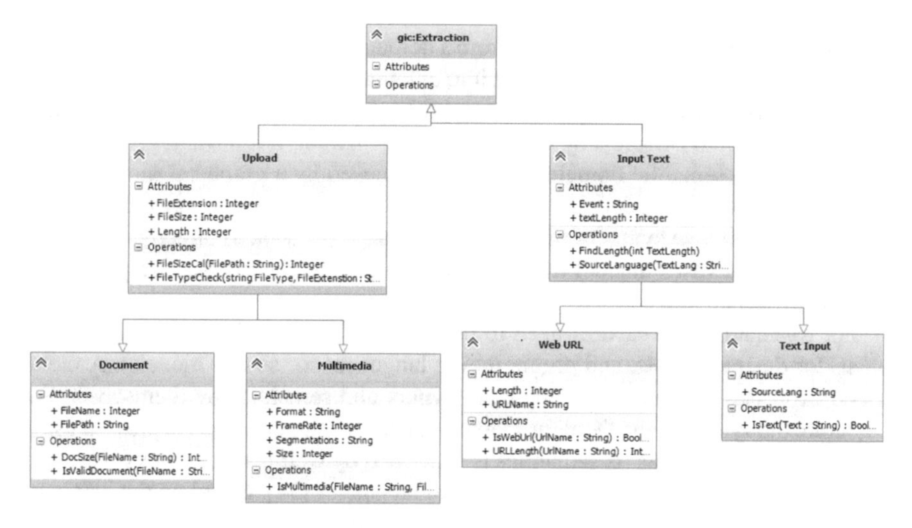

Fig. 5 Source metamodel of gic:Extraction as example DSRL used for mapping

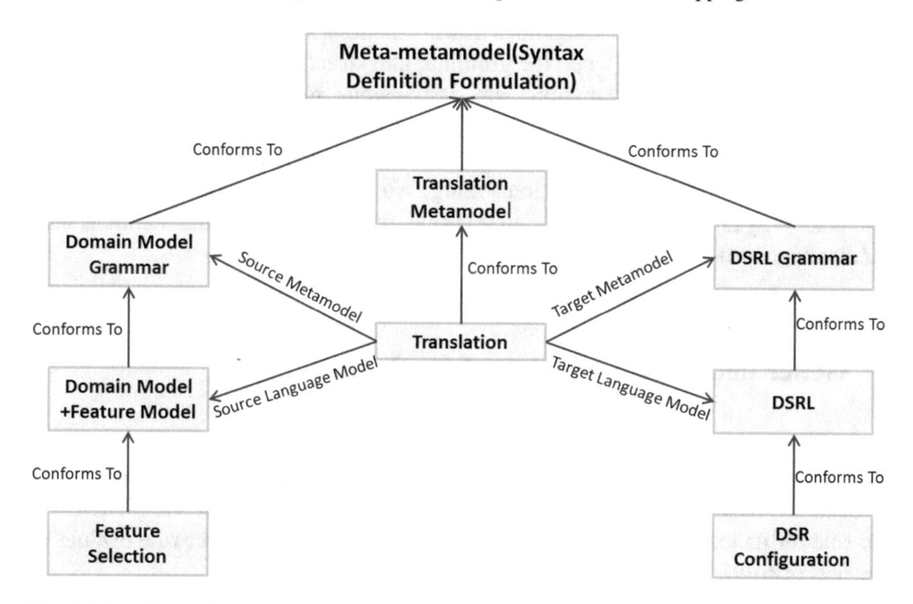

Fig. 6 Domain model to DSR translations

and semantics rules. The input required by the user (selection of the feature model) needs a domain to process the domain model, target environment as rule generation and configuration. We present the process of domain model translation and target rule environment in Fig. 6.

In an example of a rule generation from a domain model, we propose an approach: the domain model would be translated into a domain-specific rule language through a model transformation or translation, and then the DSRL meta-metamodel would be synthesized into DSR text by means of a rule generator. It is advantageous to have syntactical and semantic domain translation achieved by a graphical model to text model translation. It is a dedicated technology, because rule generators deal with the abstract and concrete syntaxes of the target language (in Sect. 3) directly. The entire process separates two distinct tasks (translation and synthesis) that are performed using appropriate tools.

For translation, the models need to be expressed in a modeling language (e.g., UML for design models, and programming languages for source models). A meta-mode expresses the modeling languages syntax and semantics by themselves. For example, the syntax of the Domain-metamodel has feature notations expressed using class diagrams, whereas its semantics is described by well-defined rules (expressed as OCL constraints) and a mixture of natural languages [34]. Based on the language in which the source and target models of a translations (grammar of model change) or transformation are expressed, a distinction is made between endogenous and exogenous transformations. The endogenous transformations are expressed between models in the same languages (when the grammar and structure are same).

The exogenous transformations are conversions made between models and expressed using different languages (the grammar and syntax are different), which is also known as translation. Essentially the same distinction was proposed in [35], but ported to a model transformation setting. We use the exogenous transformation that taxonomy and graphical model to text rule, whereas the term translation is used for an exogenous transformation.

8 Generated Rule Analysis and Evaluation

In rule generation evaluation, we validate the type of generated output concerning the correctness, completeness, output effectiveness and efficiency. Our primary goal is to have a proof of fully functional and operational correctness, and completeness of the rule for its feature requirement selected by the domain user. Our rule evaluation consists of following:

- Validation of rule generation concerning under- and over-generation.
  - Under generation—We define under generation as missing instance (for example events, actions etc.) at the time of generation or after generation.
  - Over generation—This is identified as an added information regarding syntax and semantics (functional and operational information).
- Evaluation of syntactical and semantical correctness of generated rule fulfills our goal. The above results imply that if one knows, for example, how to formulate partial correctness of a given deterministic algorithm in predicate mathematics,

the formulation of many other properties of the algorithm in predicate mathematics could have been straightforward. As a matter of fact, partial correctness has already been formulated in predicate mathematics and manual rule templates of many feature deterministic algorithms.

- Syntactical correctness means correct use of keys, functions, values, and grammatical rules.
- Semantical correctness is important for functional and operational point of view; here, we validate the correct order of generated rule and grammatical sequence.
- Grammatical correctness means the generated rules follow SDF grammar which is used in M3 level of MDA model as shown in Fig. 4.
- Comparing automated and hand-written rules.

- Evaluation of completeness of generated rule.
  - Completeness: Most rule languages are not designed as targets for rule generation, because of lack of functional and operational parameters in program fragments. Major challenging part in the rule sets has two constraints: dead-lock situation and live lock situation.
    Identification of rule deadlock.
    Identification of rule live lock.

It also needs to be identified if rules of the application require any additional set of rules to function as desired.

9 Discussion

In this paper, we have proposed a syntax definition for domain model language for rule generation and presented a DSR generation development for domain model through MDA approach using domain variability. We have presented a novel approach of handling knowledge transfer from domain concept (domain model) to conceptually configurable rule language, avoiding the inaccuracies and misunderstanding, model error, human error or semantic mismatches during translation of graphical abstract model to text rule. We have added adaptivity to the domain model. We provide a conceptual view of domain-specific rule generation and manage the domain model, and variability model using MDA. It helps in managing frequent changes of the business process along with variability schema of a set of structured variation mechanisms for the specification. The domain user can generate the DSRs and configure domain constraint in a dynamic environment. They can generate and configure DSRs without knowing any technical and programming skill. The novelty of our approach is a variability modelling usage as a systematic approach to transforming (generate) domain-specific rule from domain models.

We plan to extend this approach in combination with our existing work on business process model customization based on user requirement (feature model, domain model and process models) so that a complete development life cycle for the customization and configuration of the business process model is supported. We explore further research that focuses on how to define the DSRL concerning abstract and concrete syntactical description with grammar formation across different domains, converting conceptual models into generic domain-specific rule language which are applicable in other domains. So far, it is a model for text translation but has the potential to serve as a system that learns from existing rules and domain models, driven by the feature model approach with automatic constraints configuration that is resultant to an automated DSRL generation.

Acknowledgements This research is supported by Science Foundation Ireland (SFI) as a part of the ADAPT Centre for Digital Content Technology at Dublin City University (Grant No: 13/RC/2106) and EAI COMPSE 2016, Penang, Malaysia.

References

1. Mani, N., & Pahl, C. (2015). Controlled variability management for business process model constraints. *ICSEA 2015, The Tenth International Conference on Software Engineering Advances*. IARIA XPS Press.
2. Mani, N., Helfert, M., & Pahl, C. (2016). Business process model customisation using domain-driven controlled variability management and rule generation. *International Journal on Advances in Software*, *9*(3, 4), 179–190.
3. Tanrıöver, Ö. Ö, & Bilgen, S. (2011). A framework for reviewing domain specific conceptual models. Computer Standards. *Interfaces*, *33*(5), 448–464.
4. Knuth, D. E. (1964). Backus normal form vs. backus naur form. *Communications of the ACM*, *7*(12), 735–736.
5. Deursen, A. V., Klint, P., & Visser, J. (2000). Domain-specific languages: an annotated bibliography. *SIGPLAN Not.*, *35*(6), 26–36.
6. Fowler, M. (2010). Domain-specific languages. Pearson Education.
7. Mernik, M., Heering, J., & Sloane, A. M. (2005). When and how to develop domain-specific languages. *ACM Computing Surveys (CSUR)*, *37*(4), 316–344.
8. Hudak, P. (1997). *Domain-Specific Languages. Handbook of Programming Languages*, *3*, 39–60.
9. Ringert, J. O., et al. (2015). Code generator composition for model-driven engineering of robotics component connector systems. arXiv:1505.00904.
10. Edwards, G., Brun, Y., & Medvidovic, N. (2012). Automated analysis and code generation for domain-specific models. *2012 Joint Working IEEE/IFIP Conference on, Software Architecture (WICSA) and European Conference on Software Architecture (ECSA)*. IEEE.
11. Prout, A., et al. (2012). Code generation for a family of executable modelling notations. *Software Systems Modeling*, *11*(2), 251–272.
12. Koch, N., et al. (2008). UML-based web engineering. *Web Engineering: Modelling and Implementing Web Applications* (pp. 157–191). Springer.
13. Ceri, S., Fraternali, P., & Bongio, A. (2000). Web modeling language (WebML): a modeling language for designing web sites. *Computer Networks*, *33*(1), 137–157.
14. Groenewegen, D. M., et al. (2008). WebDSL: a domain-specific language for dynamic web applications. *Companion to the 23rd ACM SIGPLAN Conference on Object-Oriented Programming Systems Languages and Applications*. ACM.
15. Ceri, S., Fraternali, P., & Matera, M. (2002). Conceptual modeling of data-intensive Web applications. *IEEE Internet Computing*, *6*(4), 20–30.

16. Moreno, N., et al. (2008). Addressing new concerns in model-driven web engineering approaches. *International Conference on Web Information Systems Engineering*. Springer.
17. Linaje, M., Preciado, J. C., & Sánchez-Figueroa, F. (2007). Engineering rich internet application user interfaces over legacy web models. *IEEE Internet Computing*, *11*(6), 53–59.
18. Distante, D., et al. (2007). Model-driven development of web applications with UWA, MVC and JavaServer faces. *International Conference on Web Engineering*. Springer.
19. White, S. A. (2004). *Introduction to BPMN. IBM Cooperation, 2.*
20. Dumas, M., & Ter Hofstede, A. H. (2001). UML activity diagrams as a workflow specification language. *ł UML 2001—The Unified Modeling Language. Modeling Languages, Concepts, and Tools* (pp. 76–90). Springer.
21. Davis, R. (2001). Business process modelling with ARIS: a practical guide. Springer Science Business Media.
22. van der Aalst, W. M. P., & ter Hofstede, A. H. M. (2005). YAWL: yet another workflow language. *Information Systems*, *30*(4), 245–275.
23. IBM. (December 2010). WebSphere©MQ Workow FlowMareket©Definition Language (FDL).
24. Zeng, L., et al. (2004). Qos-aware middleware for web services composition. *IEEE Transactions on, Software Engineering*, *30*(5), 311–327.
25. Boss, J. (January 2008). jBPM Process Definition Language (jPDL).
26. Maker, R., et al. (1992). *IDEF3-Process Description Capture Method Report*. Information Integration for Concurrent Engineering (IICE), Armstrong Laboratory, Wright-Patterson AFB: OH.
27. Mili, H., et al. (2010). Business process modeling languages: Sorting through the alphabet soup. *ACM Computing Surveys (CSUR)*, *43*(1), 4.
28. Recker, J., et al. (2009). Business process modeling-a comparative analysis. *Journal of the Association for Information Systems*, *10*(4), 1.
29. Diouf, M., Maabout, S., & Musumbu, K. (2007). Merging model driven architecture and semantic web for business rules generation. *International Conference on Web Reasoning and Rule Systems*. Springer.
30. Musumbu, K., Diouf, M., & Maabout, S. (2010). Business rules generation methods by merging model driven architecture and web semantics. *2010 IEEE International Conference on Software Engineering and Service Sciences*. 2010. IEEE.
31. Pahl, C., Mani, N., & Wang, M. -X. (2013). A domain-specific model for data quality constraints in service process adaptations. *Advances in Service-Oriented and Cloud Computing* (pp. 303-317). Springer.
32. Bzivin, J. (2005). On the unification power of models. *Software Systems Modeling*, *4*(2), 171–188.
33. Visser, E. (1997). Syntax definition for language prototyping. Eelco Visser.
34. Group, O. M., Unified Modeling Language specification version 1.5. formal. 2003.
35. Visser, E. (2001). A survey of rewriting strategies in program transformation systems. *Electronic Notes in Theoretical Computer Science*, *57*, 109–143.

A Set-Partitioning-Based Model to Save Cost on the Import Processes

Jania Astrid Saucedo-Martínez, José Antonio Marmolejo-Saucedo and Pandian Vasant

Abstract Complying with standards as part of the process of importing products within the supply chain is important, as it is a requirement that must be met to continue with the flow of the chain. In addition, it ensures compatibility in operation, voltage, power, amperage in the home and business, etc. Issuance of certificates varies in cost as a result of the way the product family is grouped, tested and measured. The following research proposes a mathematical model, based on the Set Partitioning Problem, which makes groups by product families. Thus obtaining a significant reduction of stationery, tests and costs associated with this process for products to be imported and marketed in the country. Several test cases will be presented and a comparison will be made between the current empirical process and the proposed mathematical model.

1 Introduction

Due to the fact that in the company have been built departments that are increasingly influential in the decision-making, it is required that the supply chain management and its control allow different areas to work together, called "integration". In this way, each operational area (production, quality, transportation, sales, etc.) becomes a key element within the supply chain. However, achieving such integration is quite a challenge, making it work properly becomes a task that can determine the success

J.A. Saucedo-Martínez
Facultado de Ingeniería Mecánica y Eléctrica, Universidad Autónoma de Nuevo León,
Augusto Rodin 498, 03920 Ciudad de México, México
e-mail: jania.saucedo@gmail.com

J.A. Marmolejo-Saucedo (✉)
Facultad de Ingeniería, Universidad Panamericana,
Augusto Rodin 498, 03920 Ciudad de México, México
e-mail: jmarmolejo@up.edu.mx

P. Vasant
Faculty of Science and Information Technology, Universiti Teknologi Petronas,
Seri Iskandar, Malaysia
e-mail: pvasant@gmail.com

I. Zelinka et al. (eds.), *Innovative Computing, Optimization and Its Applications*, Studies in Computational Intelligence 741,
https://doi.org/10.1007/978-3-319-66984-7_4

or failure of today's companies. It is time to stop thinking only within the four walls of the company. That is, to think that more can be reduced within a company's operations it is better. In other words, not spending anything as well can be translated as better, or no production then no spending. This does not mean that the system usually used is not good, many companies use it and they succeed, but today we are facing global competition. How can we compete if products or services are more and more *commodities*? How easy can information be generated? How can information theft enable products of companies to have more similarity with existing products? So, working together with all departments of the company allows us to have an overview of the opportunities for improvement. In addition, the differentiation of the product or service, positions the companies above the others, with this way of seeing things the possibilities of improvement increase, since the limit to do so (providing customer service) is endless.

In this paper the *Set Partitioning Problem* (*SPaP*) is analyzed as a mathematical model to solve to solve the product families optimal allocation to help with the issuance of certificates and the processes it entails, tests and stationery. This is necessary for importing products and in compliance with the Mexican regulations. With this, the Government protects consumers by ensuring the correct functionality of the products, as well as its compatibility with the facilities at home, office, industry, company, etc. On the other hand, companies are also benefited, because with this document they underpin the quality and functionality of their products. When these products have a certification are prefer by the distributing companies and retailers. That is because the consumer trusts the tests performed on the certified production line.

So far, there is not documents where its explain how the certification processes and Mexican regulations can be optimized by a mathematical model. Then, using *SPaP* to analyze and optimize this process is an area of opportunity with great possibilities of implementation and development.

The *SPaP* has been studied by different authors such as Rezanova and Ryan [12], Bredström et al. [3], Vasant [16, 17]. The model adapts a current problem and uses cases given by a company of the same field. Then, these cases go through computational experiments, the results are compared with the current company process.

The proposed mathematical model will be transferable to other companies in the same process as long as they are in accordance with the corresponding parameters.

The article is divided in three sections: First, to explain the certification process a literature review is shown a literature review of the relationship among the certification process, normative and supply chain normativity. Subsequently, our proposed model is presented, which describes the current problem and the mathematical technique used, integer linear programming and *SPaP* bases. Finally, the proposed mathematical model, test cases and the conclusions obtained from the same one are described.

2 Background

2.1 Supply Chain

Creating a balance between customer services and service cost is quite a challenge. According to Ballou "Logistics and supply chain are a set of functional activities (transport, inventory control, etc.). These activities are repeated many times in the flow channel, in which the raw material is converted into finished products adding value to the final consumer" [2]. In this way, all the links must be strengthened in order to ensure that the supply chain will not collapse. Furthermore, the analysis according to the macro environment in the supply chain process makes the improving opportunities to increase. Hereunder, how this process is made step by step will be described in a general way (see Fig. 1).

Common process of supply chain:

1. The supplier of your supplier, usually it would be the supplier of raw materials or the first manufacturing process.
2. The supplier of your supplier sends goods to the supplier.
3. The supplier transforms the goods.
4. The supplier sends goods to the factory.
5. The product is transformed, or pieces are assembled.
6. The final product is sent from the factory to the distribution center or warehouses.
7. In the distribution center purchase orders that will be shipped are channeled or products are consolidated for their shipping.
8. Products are sent to the customer.
9. The customer receives the products.

2.2 Regulations

The Vice-Ministry for Competitiveness and Standardization implements policies with the purpose of strengthening competitiveness in the country. Thus, it gener-

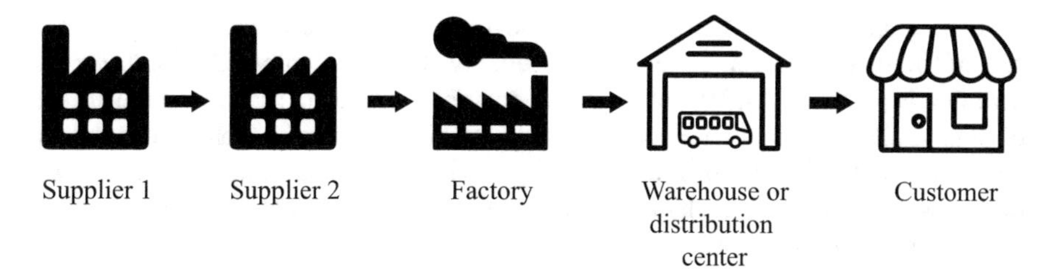

Fig. 1 Common process of the supply chain

ates efficient regulations and a simplified structure of rules that reduce costs in its compliance. Therefore, the objective of the Vice-Ministry is the strengthening of the national economic competitiveness and furthermore, the development of necessary actions that generate trust and income of foreign investment that the country needs. In addition, promotes the growth of economic activity through the development and implementation of clear, effective and simplified regulations.

Since begin of the last century, an effective control means was to characterize a service or a product in such a way that only those who did it originally could obtain an economic benefit. With such control, the market excludes any company who attempted to do something similar. For instance, the case of the gauge or track in railroad systems, where each company installing a railway system were forced to use the gauge that stipulated. Thus, that the acquisition of equipment such as trains could be done only through that manufacturer, which forced to change of train in each change of gauge.

However the need for connection between countries for commercial issues and convenience forced to standardize these differences. Then, in 1947 the International Organization for Standardization, better known as ISO was created. With this institute, the various member countries can negotiate to eliminate, create or validate standards for products or services that are exchanged between them. Although, in most cases standards issued by ISO are only recommendations that should be adopted by the member countries with the appropriate legislative treatment in their territories.

In Mexico, the standardization was influenced by the United States and its economy, which forced to issue laws and regulations to use certain features. For example, the case of 50 Hz voltages or alternating current, among others. The accumulation of such laws and regulations forcing manufacturers, producers and service providers to comply with a minimum of features in their products lasted 20 years. But in 1986 with the implementation in Mexico of the GATT [7] the Government commits to use the recommendations of ISO and other international organizations to create its own standards, which are specific to the Federal Law on Metrology and Standardization in its first version. Then, that obliges to use only one quantitative system of measurement. In specific, the so-called General System of Measurement Units, which is integrated with the International System of Units with those that not included in the International System that are accepted by law. In addition to a series of documents called rules, which standardize, in all the Mexican territory, certain characteristics of the products regarding these documents.

Two agencies of the Mexican Federal Government were created for that purpose: one of them is technical, called the National Metrology Center (CENAM, for its acronym in Spanish); and the other is administrative, called the General Directorate of Standardization (DGN, for its acronym in Spanish). Both of which are dependent on the so-called Ministry of Industry, today the Ministry of Economy, but for terms of representation of the Federal Government in international affairs both agencies depend on the Ministry of Foreign Affairs.

Thus, during the decades of 1980 and 1990 the General Directorate of Standardization emitted a series of rules based on recommendations from technical instances with both national and international scopes. They cover basically only products and services sold to the general public. For that reason, the use of rules begins to be adequate to prevent from going through the legislative branch. The law mentions different types of rule among which were found the Mexican Official Standards, Mexican rules, the emergency rules and rules of Ref. [8].

According to the Ministry of Economy [13] there are 10 National Standards Organizations. These are entities who have as main objective the development and issuance of Mexican standards related to the subjects they were registered with by the General Directorate for Standards. By the way, The National Standards Organization that currently serves these functions can be found on the official page of the Ministry of Economy and also the internet links to get more information about them.

2.3 Regulations and Supply Chain

The regulation process is linked to the supply chain. That is, in order to introduce a product into a particular country's market it is necessary to comply with the established guidelines set forth in the regulations of that country. Then, if these regulations are not complied, this can affects areas related to the product mobility. This information in the terms of logistics is required, where logistics according to the Council of Logistics Management (CSCMP, for its acronym in Spanish) [4] is:

> The part of the supply chain process that plans, implements and controls the efficient flow and storage of goods and services, as well as of the related information, from the point of origin to the point of consumption in order to meet the customers requirements.

In the previous definition, the CSCMP describes that logistics is also in charge of the stationery of products or services. Thus, to comply with guidelines or requirements the logistics department should carry out this role. However, they have to

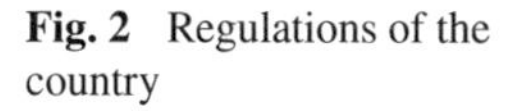

Fig. 2 Regulations of the country

work with institutions helping this process, such as are the Certification Bodies (see Fig. 2). In addition, they also have to work with testing calibration and verification laboratories, those that are endorsed by the Ministry of Economy. All this, in accordance with the assessment that carry out this process to comply with the regulations of the country.

2.4 Certification Process

It can be mentioned that the certification is proof to the customers that their demands have been met. In addition, Those requirements met in terms of certain national and international rules, which is carried out by an independent entity of the stakeholders. By other hand, there are different definitions for the product certification; they vary depending on the line of business arena or the business sector. To summarize, certification definition could be the following: it is a process that guaranty the product quality and its characteristics, according to what is established in a specific standard or other preset documents.

Furthermore, the compliance of the product certification, in order to demonstrate the quality is vital and usually becomes a way to enter the market. This means that is useful to communicate the quality of the certified products, to increase confidence in the final consumers, to be differentiated from the competitors and to ensure safety of the products.

According to the Ministry of Economy, in its official website, to comply with the country regulations it is necessary to perform, so to speak, an conformity assessment [14]. The conformity assessment is the determination of the compliance degree with the Official Mexican Standards or conformity with Mexican regulations, international standards or other specifications, requirements or characteristics. For which the following procedures have to be performed:

1. Sampling,
2. testing,
3. calibration,
4. certification and
5. verification.

Finally, it is considered that all products, processes, methods, facilities, services or activities must comply with the Official Mexican Standards. Thus, this should happen when the company is installed in the country. Then, when a product or service should comply with a specific Official Mexican Standard, other similar products or services must also comply with the established specifications of such rule (see the general certification process in Fig. 3). It is important mention that based on data obtained through results of an audit, the certification of the product is issued or.

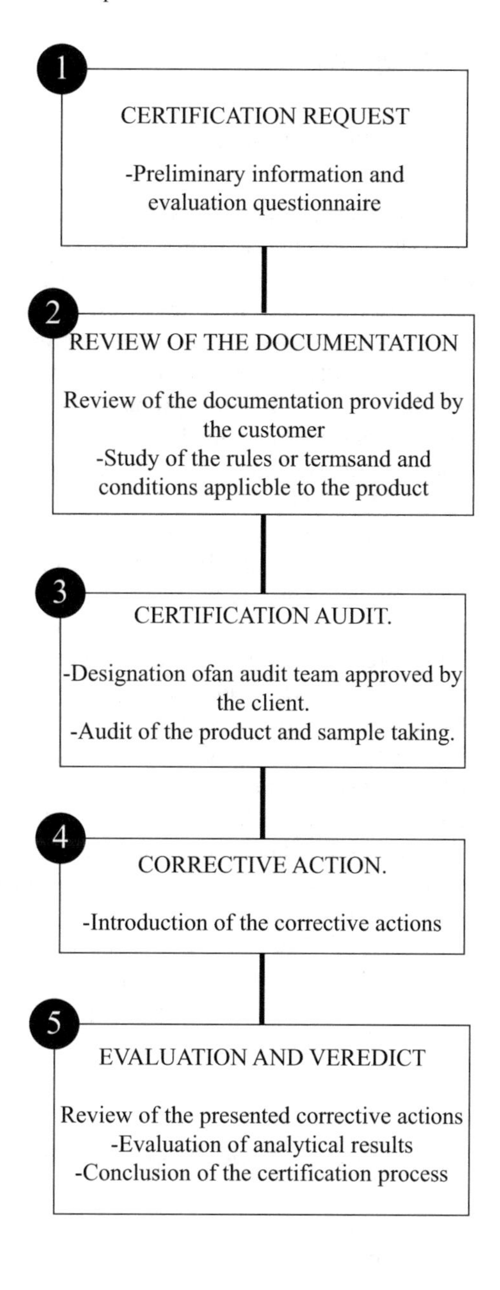

Fig. 3 Common certification process

3 Proposal

3.1 Current Problem

The compliance of regulations is mandatory, because is one of the requirements that the country demands to ensure the proper product operation. Moreover, it has a purpose to ensure these are within the safety standards or guidelines and parity of the region. Hence, failing to comply with the products regulations can provoke sanctions or simply the products shall be rejected in the country. Then, companies should be able to consult an expert who could help them to achieve their products certification. In order to fulfill these products standards, the expert's guide would be the best mean, because they know the subject and they are updated with changes made by relevant authorities, or whether necessary, they are able to make a research about comply regulations.

The certification process start when a list of products with their corresponding documentation and specific features to a company responsible of complying the rules' requirements. Then, the company receives this list and groups families of products per similar specific features to perform tests and necessary paperwork to certify products (as mentioned in the certification process). On the other hand, the same company, who is interested in the product importation, can perform such products allocation and carry out on their own the norms complying process. Nowadays, this is performed in a manual and empirical way. In order words, the sent product list is taken and based on experience; products are added in the corresponding family (products that have similar characteristics to each others). Therefore, the current process does not ensure that the grouping is the best, neither that it is the minor grouping. This could cause higher cost services, an increment in the number of tests, more paperwork, etc.

Finally, this leads to ask whether the way the product grouping is performed is optimal. Thence, in this work a mathematical model based on the *SPaP* will be developed, and it will allow us to group products in an optimal way into the minimum number of families.

3.2 Whole Applied Programming

The research operation includes the mathematical modeling, workable solutions, optimization and iterative calculations. It emphasizes that the correct definition of the problem is the most important phase (and the most difficult) to practice. It also stresses out that although the mathematical modeling is the cornerstone, when making the final decision it is necessary to take certain factors into consideration, such as the human behavior. In the research operation, there is no unique general technique to solve all the models that can arise in practice [15, 18].

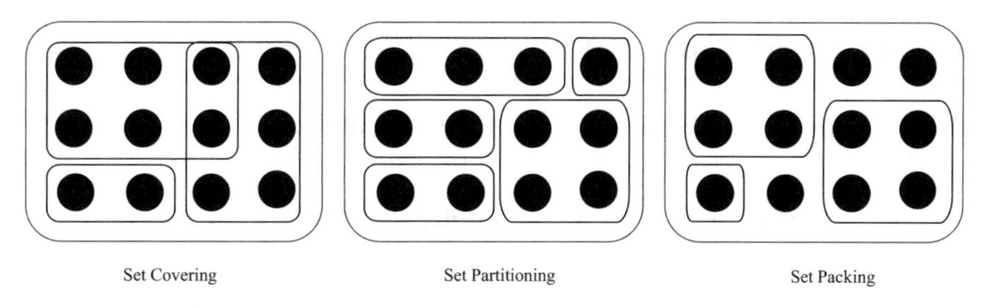

Fig. 4 Types of covering problems

The development of the linear programming has been classified as one of the most important mid-twentieth century scientific advances. Currently, it has saved thousands or millions of dollars to many companies or businesses, including small and medium-sized enterprises in different industrialized countries of the world. The most common type of application covers the general allocating problem consisting in assigning in the best way possible limited resources to activities that compete for them.

The linear programming uses a mathematical model to describe the problem, the adjective linear means that all math functions of the model must be linear functions. The linear programming involves planning activities for an optimal result. That is, the result that best reaches the determinate goal in accordance with the mathematical model among all feasible alternatives [10].

That is why the mathematical modeling using linear programming will allow us to find the optimal solution to group the products in families with similar characteristics (see Fig. 4). Using the *SPaP*, which consists in finding a set of solutions that allows us to reach the least amount of families of the subsets at the lowest cost. *SPaP* is a classic problem that comes from the Set Covering Problem (*SCP*), which along with the Set Packing Problem (*SPP*) belongs to the class **NP-Complete** [6, 9], where the input is given by several sets of elements or data that have an element in common.

In general, these problems consist in finding the minimum number of sets having a certain number of elements in all sets. In other words, consists in finding a set of solutions that allow to jointly covering a set of needs at the lowest cost. A set of needs corresponds to the rows, and the solution set is the selection of columns that cover in an optimal way the set of rows. These three problems have similar characteristics and they differ in their objective function and the way in which sets are grouped, this is reflected in their restrictions.

3.3 Set Partitioning Problem

A way of formally classifying the *SPaP* would be as follows:

$$min\left\{ \sum_{j=1}^{n} c_j x_j \middle| \sum_{j=1}^{n} A_j x_j = e, x_j \in \{0,1\}, \forall j = 1,\dots,n \right\}$$

where A is a matrix $m \times n$ of zeros and ones (binary), c is an arbitrary vector of n-dimensional integer, $e = (1,\dots,1)$ is a vector m, and $N = 1,\dots,n$.*SPaP* comes from the following interpretation: if A rows are associated with elements of the $M = 1,\dots,m$ set and each column a_j of A is associated with the subset M_j of those $i \in M$ so $a_{ij} = 1$, then *SPaP* is the problem to find the families with less quantity of each subset $M_j, j \in M$, which is a partition of M, each subset M_j of c_j [11]. Thus, the *Set Partitioning Problem* can be presented as follows:

$$\begin{aligned} min \quad & \textstyle\sum_{j=1}^{n} c_j x_j \\ s.a: \quad & \textstyle\sum_{j=1}^{n} A_j x_j = 1 \\ & x_j \in \{0,1\}, \quad \forall j = 1,\dots,n \end{aligned}$$

In *SPaP* there are multiple feasible solutions, but just one is optimal. *SPaP* has different uses. Some of them are: rail crew programming, truck deliveries, airline crew programming. Airline crew programming being one of the most used to be solved by this type of *SPaP* problem.

3.4 SPaP example

The planner of a small wedding wants to place the guests covering all possible tables with few amounts of attending families by making the rent cost of each table worthwhile. The following criteria is taken into consideration: families A, B,...,H can sit at tables $1,\dots,9$ (see Table 1).

$$\begin{aligned} min \quad & 100x_1 + 75x_2 + 80x_3 + 95x_4 + 50x_5 + 75x_6 + 55x_7 + 80x_8 + 100x_9 & \\ s.a: \quad & x_1 + x_3 + x_7 + x_9 = 1 & \text{Family A} \\ & x_4 + x_5 + x_8 = 1 & \text{Family B} \\ & x_2 + x_3 + x_6 + x_9 = 1 & \text{Family C} \\ & x_1 + x_4 + x_7 + x_8 = 1 & \text{Family D} \\ & x_1 + x_3 + x_6 + x_7 = 1 & \text{Family E} \\ & x_2 + x_4 + x_5 + x_8 = 1 & \text{Family F} \end{aligned}$$

Table 1 *SPaP* example

Family	Section 1				Section 2				
	1	2	3	4	5	6	7	8	9
A	x		x				x		x
B				x	x			x	
C		x	x			x			x
D	x			x			x	x	
E	x		x			x	x		
F		x		x	x			x	
G	x						x		x
H		x		x	x	x			
Price	100	75	80	95	50	75	55	80	100

$$x_2 + x_4 + x_5 + x_6 = 1 \qquad \text{Family G}$$
$$x_j \in \{0, 1\}, \qquad \forall j = 1, \ldots, 9$$

where $x_j = 1$ if table j is selected, 0, if not.

4 Proposed Mathematical Model

As it has been previously said, all products subject to be imported into Mexico and that are to be traded have to go through the process described above. With the *SPaP* model it is possible to adjust the criteria for allocating families according to the type of product and characteristics [5]. We use as sample a selection of lamp products, according to the certification body in order for a product to be able to belong to a family it is necessary to comply with the following characteristics: Same type of use (fixed, table, floor), power (from 60 to 1500 W, may vary depending on the amount of specifications of some lamps), voltage (127 or 220 V, is not applicable to all groups as the common voltage is 127 V) and type and material of the lamp holder (E26metallic, E26Ceramic, E26E26Phenolic, E26E26Metal, E26Plastic, E26Ceram, E26Phenol, E12Plast, E12Ceram, E12Phenol, etc.).

Sets

I Set of products $i = 1, 2, \ldots, m$.
J Set of characteristics related to the type of use $j = 1, 2, 3, \ldots, p$.
K Set of characteristics related to the type of material $k = 1, 2, 3, \ldots, q$.
L Set of characteristics related to the type of power $l = 1, 2, \ldots, r$.
O Set of characteristics related to the type of voltage $o = 1, 2, \ldots, s$.

Parameters

a_{ij} If the product i is allocated to the characteristic j.
b_{ik} If the product i is allocated to the characteristic k.
d_{il} If the product i is allocated to the characteristic l.
e_{io} If the product i is allocated to the characteristic o.

The decision variables are:

$$x_{ijklo} = \begin{cases} 1 & \text{if the product } i \text{ belongs to the family with the characteristics } j, k, l, o \\ 0 & \text{if not} \end{cases}$$

$$y_{jklo} = \begin{cases} 1 & \text{if exist the family products with the } j, k, l, o \text{ characteristics} \\ 0 & \text{if not} \end{cases}$$

$$min \quad \sum_{j=1}^{p} \sum_{k=1}^{q} \sum_{l=1}^{r} \sum_{o=1}^{s} y_{jklo} \tag{1}$$

$$s.a : \sum_{j=1}^{p} \sum_{k=1}^{q} \sum_{l=1}^{r} \sum_{o=1}^{s} a_{ij} b_{ik} d_{il} e_{io} x_{ijklo} = 1 \tag{2}$$

$$x_{ijklo} \leq y_{jklo} \tag{3}$$

$$x_{ijklo} \in \{0, 1\}, \qquad \forall i, j, k, l, o \tag{4}$$

$$y_{jklo} \in \{0, 1\}, \qquad \forall j, k, l, o \tag{5}$$

where (1) represents the objective function, which minimizes the amount of families with similar characteristics, the cost is not taken into account, as it is the same for any family to be certified because it is about lamps. The restriction 2 says that i products, are allocated to characteristics j, k, l, o. The restriction 3 says that product i is allocated to characteristics j, k, l, o. And finally, restrictions 4 and 5 are state variables pointing out that are binary.

The model is adapted to the characteristics of the certifying body, this model was adjusted according to the family grouping criteria for lamp type products, for more information, visit the ANCE website [1].

5 Computational Experimentation

5.1 Equipment

To test model 3 tests cases are proposed and were taken from the history of a company. To solve the cases GAMS was used with the CPLEX version 12 in Intel Xeon

E52697v2 2.7 GHz computer with 12 cores each, a ram memory of at least 64 GB, and a 1 TB hard drive, that is why the Dell PoweEdge T620 was selected.

5.2 *Test Cases*

All three cases that have been studied were provided by a local company whose line of business is to provide family allocation services, test lab, among other related services. Due to confidentiality issues names were not provided nor the names of the products. More cases will be provided to continue with the experiment, which will provide other types of products and different characteristics, so that the model will be adapted to these.

In Table 2, considerations that were taken into account for grouping families are shown, which are not the same for all cases because the products have different values or the material is not the same, N/A in the voltage feature means that there is only one type of voltage of 127 V. The first column represents the pilot case; the second, type of use for the lamp product; the third column, the type of material of the lamp

Table 2 Grouping criteria

Case	Family product (lamps) allocating criteria				
	Type of use	Type and material of the lamp holder	Voltage (V)	Power (W)	Amount of products to be grouped
1	Fixed, table, floor	E26Metalic, E26Ceramic, E26Phenolic, E26Plastic	60, 80, 100	127, 200	50
2	Fixed, table, floor	Ceramics, glass, fabric, metal, resin	60, 120, 180, 240, 300, 360, 420, 480, 540, 600, 660, 720, 780, 840, 900, 960, 1020	N/A	1573
3	Fixed, table, floor	E26Metalic, E26Ceramic, E26Phenolic, E12Plastic, E12Ceram, E12Phenol	60, 120, 180, 240, 300, 360, 420, 480, 540, 600, 660, 720, 780, 840, 900, 960, 1020, 1080, 1140, 1200, 1260, 1320, 1380, 1440, 1500, 1560, 1620	N/A	3267

holder; the fourth column, is the total power measured in watts of each product; and the fifth, is the grouping according to the product voltage, and finally the amount of products that were taken into consideration in each case.

5.3 *Analysis and Results*

In Table 3, the first column is the case in which products went through experimentation; the second column represents the quantity of the products; the third column represents the number of families that were created with the current process used by the company who provided the cases; the fourth column is the number of resulting families from the proposed model; the fifth column is the time it takes to assign each product to each family in accordance with the current process, and the last column is the time it takes for the model proposed to group families.

The results obtained with these first three cases, threw the same amount of families, where the products are grouped. The biggest difference lies in the total time in which the process took place. According to the process currently used, the time to group the products range from three days to 15 days, this refers to the time the company uses to perform this action, compared to the time result from the model by using the GAMS program that was only computational seconds. The time used for the certification process is requested during the course of arrival to the country, so it has to be as fast as possible, this arrival time is about 1 week. The solving time with the proposed model varies according to the grouping criteria, i.e. the diversity of families that can be created, plus the time related to the quantity of products that have to be assigned, to have an average it is necessary to perform more tests to compare the time of solution. On the other hand, the model ensures the minimum of families according to the criteria of the certifying body, which is compared to the quantity of certifications that a company would have to issue, because due to the fast introduction to the market they certify each product, causing a cost increase.

Table 3 Results of test cases

	Results of computational experimentation				
	Amount of products to be grouped	Total families with the current process	Total families with the proposed model	Total grouping time with the current process (days)	Total time with the proposed model (s)
1	50	4	4	3	0.03
2	1573	5	5	10	3.29
3	3267	17	17	15	5.69

5.4 *Conclusions*

The development of tools that allow us to optimize the available resources ensure the proper use and cost for the activity to be performed, implementing methodologies with mathematical basis guarantee and support the decision-making process.

The purpose of this research is to provide a linear programming tool for decision making in terms of product grouping, resulting in time and cost saving. The time is reflected within the logistic process, both for the company providing counselling, testing, grouping services, etc., because it has a faster response and with this it can start negotiations of their services, and for the company that wants to commercialize the products in Mexico because it reduces the amount of time of introducing their products to the country; this can be a great advantage for companies that sell their products either by season, fashion, technology, among others. And the cost-related saving is reduced in terms of certification process cost, which is an estimate of half million pesos per certificate plus the cost of services provided by the consultant company. One less certificate is a considerable saving, especially for companies whose products are below the profit margin, and savings are explicit in destruction tests performed on a product for companies whose products are in a high profit margin, that is, one more product that will enter the market.

6 Future Research Directions

A point that is important to emphasize is that in order to be issued a certificate it is necessary to select one product per family to be test (known as the head of the family). Usually the head of the family is the product that has the most unfavorable or the one more susceptible to fail the test. So we believe that is fundamental add this decision.

Acknowledgements We would like to express our acknowledgment for the First EAI International Conference on Computer Science and Engineering, November 11–12, 2016, Penang, Malaysia.

Appendix

When placed at the end of a chapter or contribution (as opposed to at the end of the book), the numbering of tables, figures, and equations in the appendix section continues on from that in the main text. Hence please *do not* use the `appendix` command when writing an appendix at the end of your chapter or contribution. If there is only one the appendix is designated "Appendix", or "Appendix 1", or "Appendix 2", etc. if there is more than one.

References

1. ANCE. (2015). Asociación de Normalización y Certificación, A.C.. In: website of the ANCE. http://www.ance.org.mx.
2. Ballou, R. H. (2004). *Cadena de Suministros: Logística Administración de la Cadena de Suministro*. México: Pearson Educación.
3. Bredström, D., Jörnsten, K., Rönnqvist, M., & Bouchard, M. (2013). Searching for optimal integer solutions to set partitioning problems using column generation. *International Transactions In Operational Research*, *21*(2), 177–197.
4. Bertsimas, D., & Tsitsiklis, J. N. (1997). *Introduction to linear optimization*. Dynamic Ideas.
5. CSCMP. (2014). De las normas del Consejo de Dirección Logística. In Council of Supply Chain Management Professionals. http://www.clm1.org.
6. Garey M. R., & Johnson D. S. (1979). *Computers and intractability: A guide to the theory of NP-completeness*. New York, USA.: W. H. Freeman & Co.
7. Geneva. (1986). México becomes a full member of the GATT. In website of the newspaper El País. http://elpais.com/diario/1986/07/26/economia/522712808_850215.html.
8. Huerta-Ochoa. C. (2011). Las Normas Oficiales Mexicanas en el Ordenamiento Jurídico Mexicano. In Biblioteca Jurídica Virtual UNAM. http://www.juridicas.unam.mx/publica/rev/boletin/cont/92/art/art4.htm.
9. Karp, R. M. (1972). Reducibility among combinatorial problems. *Complexity of Computer Computations*. In: The IBM Research Symposia Series (pp. 85–103). Ney York, USA: Springer.
10. Lieberman, F. S. (2010). *Introducción a la Investigación de Operaciones*. México: Mc Graw Hill Educación.
11. Pader, M. W., & Balas, E. (1976). Set partitioning: a survey. *SIAM*, *18*(4), 710–760.
12. Rezanova, N., & Ryan, D. M. (2010). The train driver problem a set partitioning based model and solution method. *Computers & Operations Research*, *37*(5), 845–856.
13. Secretaría de Economía. (2011). Secretaría de Economía. In: website of the Ministry of Economy. http://www.economia.gob.mx/comunidad-negocios/competitividad-normatividad/normalizacion/nacional/procesos-de-normalizacion/organismo-nacionales.
14. Secretaría de Economía. (2014). Evaluación de la conformidad. In: website of the Ministry of Economy. http://www.economia.gob.mx/comunidad-negocios/competitividad-normatividad/normalizacion/nacional/evaluacion-de-conformidad.
15. Taha, H. A. (2012). Investigación de operaciones. México: Pearson.
16. Vasant, P. M. (2013). *Meta-heuristics optimization algorithms in engineering, business, economics, and finance*. USA: IGI Global.
17. Vasant, P. M. (2014). *Handbook of research on novel soft computing intelligent algorithms: theory and practical applications*. USA: IGI Global.
18. Winston, W. L. (2005). Investigación de Operaciones: Aplicaciones y Algoritmos. México: Thomson 4ed.

Combining Genetic Algorithm with Variable Neighborhood Search for MAX-SAT

Noureddine Bouhmala and Kjell Ivar Øvergård

Abstract Variable Neighborhood Search (VNS) is a simple meta-heuristic that systematically changes the size and type of neighborhood during the search process in order to escape from local optima. In this paper, a variable-neighborhood-genetic-based-algorithm is proposed for the maximum satisfiability problem (MAX-SAT). Most of the work published earlier on VNS starts from the first neighborhood and moves on to higher neighborhoods without controlling and adapting the ordering of neighborhood structures. The order in which the neighborhood structures have been proposed in this work enables the genetic algorithm with a better mechanism for performing diversification and intensification. A set of benchmark problem instances is used to compare the effectiveness of the proposed algorithm against the standard genetic algorithm. This paper reports promising results when the proposed hybrid algorithm is compared with state-of-the art solvers.

1 Introduction

Combinatorial optimization is a lively field of applied mathematics, combining techniques from combinatorics, linear programming, and the theory of algorithms, to solve optimization problems over discrete structures. Utilizing classical methods of operations research often fail due to the exponentially growing computational effort. It is commonly accepted that these methods might be heavily penalized by the NP-hard nature of the problems and consequently will then be unable to solve large size instances of a problem. Therefore, in practice meta-heuristics are commonly used even if they are unable to guarantee an optimal solution. The driving force behind the high performance of meta-heuristics is their ability to find an appropriate balance between intensively exploiting areas with high quality solutions

N. Bouhmala (✉) · K.I. Øvergård
Department of Maritime Technology and Innovation, University College Southeast, Notodden, Norway
e-mail: noureddine.bouhmala@usn.no

I. Zelinka et al. (eds.), *Innovative Computing, Optimization and Its Applications*, Studies in Computational Intelligence 741,
https://doi.org/10.1007/978-3-319-66984-7_5

(the neighborhood of elite solutions) and moving to unexplored areas when necessary. The evolution of meta-heuristics has taken an explosive upturn. The recent trends in computational optimization move away from the traditional methods to contemporary nature-inspired meta-heuristic algorithms [33, 38, 39], though traditional methods can still be an important part of the solution techniques for small size problems. Several meta-heuristics incorporating different design choices (the existence or the absence of adaptive memory, the kind of neighborhood used, and the number of current solutions carried from one iteration to the next) have be been proposed.

Genetic Algorithms (GAs) are population-based approaches and have become hugely popular for several reasons. First, they are simple and easy to implement, and yet they can solve very diverse problems. Second, they offer a good advantage, especially when dealing with a multi-modal search space where a GA algorithm keeps track of several optima in parallel and maintains diversity in the population of solutions, with the aim of performing a better exploration of the search space for finding the global optimum. Third, the ergodicity of evolution operator makes GAs very effective at performing global search [33, 38].

Despite the significant progress made in the last few decades, however, premature convergence is an inherent characteristic of such classical genetic algorithms that makes them incapable of searching numerous solutions of the problem domain. In order to overcome this flaw which still remains a challenging task, it is a significant task to find some effective approaches for the maintenance of diversity for a longer period; in this way, the search space would be more covered for a longer period thereby increasing the probability of reaching a better solution quality. It is widely accepted within the GA community that the high diversity of a population greatly contributes to GA performance [25]. This lack of diversity often leads to stagnation, as genetic algorithms gets trapped in local optima, lacking the genetic diversity needed to escape. Favoring exploitation leads the premature convergence of GA, but has good capability to tune solutions when they are near the optimal solution. On the other hand, favoring exploration can make GA ineffective because strong exploration has the impact of not improving the solution quality. The work in [36] concluded that the optimization of the efficiency and effectiveness of genetic algorithms depends on maintaining a balance between the exploitation of existing solutions by using crossover in order to improve them, and the exploration of the solution space by using mutation so as to increase the probability of finding the optimal solution. The strategy is based on managing this balance by parameters control settings in which the parameter values (crossover rate, mutation rate) are subject to change during a GA run.

Hansen and Mladenovic have proposed a meta-heuristic called variable neighborhood search (VNS for short) [28]. The key idea of this method is to use various neighborhoods during the search. Considering exploration ability of GAs and exploitation capability of variable neighborhood search (VNS), the purpose of this paper is to introduce a hybrid approach combining the strengths of both GA and VNS for solving combinatorial optimization problems and illustrating its concept using MAX-SAT. The rest of the paper is organized as follows. Section 2 defines the maximum satisfiability problem. Section 3 describes the basic variable neighborhood search algorithm. Section 4 explains the hybrid approach. Experimental results for MAX-SAT are presented in Sect. 5. Finally, Sect. 6 draws conclusions and addresses future research directions.

2 Maximum Satisfiability Problem (MAX-SAT)

The MAX-SAT problem which is known to be NP-complete [13] is defined as follows. Given a positive constant k, a propositional formula $\Phi = \bigwedge_{j=1}^{m} C_j$ with M clauses and N Boolean variables. Each Boolean variable, $x_i, i \in \{1, \ldots, n\}$, takes one of the two values, ***True*** or ***False***. Each clause C_j, in turn, is a disjunction of Boolean variables and has the form:

$$C_j = \left(\bigvee_{l \in I_j} x_l \right) \vee \left(\bigvee_{l \in \bar{I}_j} \bar{x}_l \right),$$

where $I_j, \bar{I}_j \subseteq \{1, \ldots, n\}, I \cap \bar{I}_j = \emptyset$, and $\bar{x}_i$ denotes the negation of x_i.

As a simple example let $\Phi(x)$ be the following formula containing 4 variables and 3 clauses:

$$\Phi(x) = (x1 \vee \neg x4) \wedge (\neg x1 \vee x3) \ldots \wedge (\neg x1 \vee x4 \vee x2).$$

The task is to determine whether or not there exist a truth assignment for Φ that satisfies the maximum number k of clauses. One of the earliest local search for solving MAX-SAT is GSAT [31]. The GSAT algorithm operates by changing a complete assignment of variables into one in which the maximum possible number of clauses are satisfied by changing the value of a single variable. Another widely used variant of GSAT is the WalkSAT based on a two stage selection mechanism originally introduced in [30]. Several state-of-the-art local search algorithms are enhanced versions of GSAT and WalkSAT algorithms. Examples include GSAT/Tabu [26], WalkSAT/Tabu [27], Novelty+ and R-Novelty+ heuristics [18], G2WSAT [23]. The main difference between meta-heuristics relies in the way neighborhood structures are defined and explored. While the aforementioned meta-heuristics, work only with a single neighborhood structure, other meta-heuristics choose to operate on a set of different neighborhood structures giving rise to variable neighborhood search algorithms [15, 28]. They aim at finding a tactical interplay between diversification and intensification to overcome local optimality using a combination of a local search with systematic changes of neighborhood. To avoid manual parameter tuning, methods have been designed to automatically adapt parameter settings during the search [17, 24] and results have shown their effectiveness for a wide range of problems. As the quality of the solution improves when larger neighborhood is use, the work proposed in [40] uses a restricted 2 and 3-flip neighborhoods and better performance has been achieved compared to the 1-flip neighborhood for structured problems. Clause weighting based algorithms [11, 14] have been proposed to solve SAT and MAX-SAT problems. The key idea is associate the clauses of the given CNF formula with weights. Although these clause weighting algorithms differ in the manner clause weights should be updated (probabilistic or deterministic) they all choose to increase the weights of all the unsatisfied clauses as soon as a local minimum is encountered. The genetic local search algorithm (GASAT) [20] is considered to be the best known genetic algorithm for MAX-SAT problems. GASAT is a hybrid algorithm that combines a specific crossover and a tabu search procedure. Experiments have shown that GASAT provides very competitive results compared with state-of-art MAX-SAT algorithm. Strategies based on an automatic procedure for integrating selected components from various existing solvers have been devised in order to build new efficient algorithms that draw the strengths of multiple algorithms [21, 37]. The work conducted in [41] proposed an adaptive memory based local search algorithm that exploits various strategies in order to guide the search to achieve a suitable trade-off between intensification and diversification. The computational results show that it competes favorably with some state-of-the-art MAX-SAT solvers. Multilevel algorithms [5, 7, 8] have recently been proposed for solving MAX-SAT. The idea involves recursive coarsening to create a hierarchy of smaller problems which are hopefully much easier to solve. An initial solution is computed at the coarsest level and then iteratively refined at each level. Projection operators are applied to transfer the solution from one level to another Finally, Learning Automata has been introduced as a mechanism for enhancing stochastic local algorithms for the satisfiability problem [3, 9, 10] with promising results.

3 Basic Variable Neighborhood Search (VNS)

Variable Neighborhood Search (VNS) [15, 16, 28] is a meta-heuristic jointly proposed by Mladenovi and Hansen for solving combinatorial and global optimization problems. VNS aims at finding a tactical interplay between diversification and intensification [2] to overcome local optimality using a combination of a local search with systematic changes of neighborhood. Unlike many standard meta-heuristics where only a single neighborhood is employed, VNS systematically changes different neighborhoods within a local search. The idea is that a local optimum reached within one neighborhood structure is not necessarily the same local optimum of another neighborhood structure, thus the search can systematically explore different search areas which are defined by different neighborhood structures. The basic scheme of VNS is as follows: Let N_k represents a finite set of pre-selected neighborhood structures ($k = 1, \ldots, k_{max}$) and with $N_k(s)$ the set of solutions in the kth neighborhood of the solution s. Let S_{start} denotes a random initial solution. VNS starts using the so-called shaking step. This step involves generating a random solution S_{rand} from the neighborhood N_1 ($S_{rand} \in N_1(S_{start})$). Let S_{local} denotes the reached local optimum when a local search is used with S_{rand} as input. If S_{local} is better compared to S_{rand}, the solution is updated and a new round of local search with a random solution from $N_1(S_{local})$ is performed. If the test fails, k is incremented and VNS moves to the next neighborhood structure. The effectiveness of VNS is strongly affected by the ordering in which a given type of neighborhood is considered [19]. Both the choice and the order of neighborhood structures are critical for the performance of the algorithm. Most of the work published earlier on VNS starts from the first neighborhood and moves on to higher neighborhoods without controlling and adapting the ordering of neighborhood structures.

4 A Variable Neighborhood Search Structure Based Genetic Algorithm (VNS-GA)

Genetic Algorithms [22] are stochastic methods for global search and optimization and belong to the group of evolutionary algorithms. They simultaneously examines and manipulates a set of possible solution. A gene is part of a chromosome, which is the smallest unit of genetic information. Every gene is able to assume different values called allele. All genes of an organism form a genome which affects the appearance of an organism called phenotype. The chromosomes are encoded using a chosen representation and each can be thought of as a point in the search space of candidate solutions. Each individual is assigned a score (fitness) value that allows assessing its quality. The members of the initial population may be randomly generated or by using sophisticated mechanisms by means of which an initial population of high quality chromosomes is produced. The reproduction operator selects (randomly or based on the individual's fitness) chromosomes from the population to be parents and enters them in a mating pool. Parent individuals are drawn from the mating pool and combined so that information is exchanged and passed to off-springs depending on the probability of the cross-over operator. The new population is then subjected to mutation and enters into an intermediate population. The mutation operator acts as an element of diversity into the population and is generally applied with a low probability to avoid disrupting cross-over results. Finally, a selection scheme is used to update the population giving rise to a new generation. The individuals from the set of solutions which is called population will evolve from generation to generation by repeated applications of an evaluation procedure that is based on genetic operators. Over many generations, the population becomes increasingly uniform until it ultimately converges to optimal or near-optimal solutions. The proposed VNS-GA is described in Algorithm 1.

- **Neighborhood Construction**
  The variable neighborhood search algorithm proposed in thsi work was first introduced in [4]. Let L denotes the set of variables of the problem to be solved. The first phase of the algorithm consists in constructing a set of neighborhood satisfying the following property: $N_1(x) \subset N_2(x) \subset \ldots N_{k_{max}}(x)$. The starting (default) neighborhood with $k = 1$ consists of a move based on the flip of a single variable. A flip means assigning the opposite state to a variable (i.e. change True $\rightarrow$ False or

Table 1 Default neighborhood N_1

x1	x2	x3	x4	x5	x6	x7	x8	x9	x10	x11	x12

Table 2 First neighborhood N_2

x1, x7	x2, x5	x3, x8	x4, x11	x10, x12	x6, x9

Table 3 Second neighborhood N_3

x1, x7, x4, x11	x2, x5, x6, x9	x3, x8, x10, x12

False → True). The first neighborhood N_2 is constructed from L by merging variables. The merging procedure is computed using a randomized algorithm. The variables are visited in a random order. If a variable l_i has not been matched yet, then a randomly unmatched variable l_j is selected, and a cluster CL_k consisting of the two variables l_i and l_j is created. The set N_2 consists of moves based on flipping predefined clusters each having 2^1 variables. The new formed clusters are used to define a new and larger neighborhood N_3 and recursively iterate the process until the desired number of neighborhood (k_{max}) is reached (lines 3, 4, 5 of Algorithm 1). This process, is graphically illustrated in Tables 1, 2 and 3 using an example with 12 variables. The construction of the neighborhoods uses two phases in order to generate the neighborhood N_3. N_1 corresponds to default neighborhood. The random coarsening procedure is used to merge randomly the variables in pairs leading to the first neighborhood N_2 having 6 clusters. This process is repeated leading to the second neighborhood N_3 having 3 clusters. This neighborhood has already been combined with a simple local search and proved to improve its performance [6].

- **Initial Population**
  A representation is a mapping from the state space of possible solutions to a state of encoded solutions within a particular data structure. The chromosomes which are assignments of values to the variables are encoded as strings of bits, the length of which is the number of variables. The values ***True*** and ***False*** are represented by 1 and 0 respectively. In this representation, a chromosome X corresponds to a truth assignment and the search space is the set $S = \{0, 1\}^n$. A random initial population is generated from the largest neighborhood ($N_{k_{max}}$) (line 7 of Algorithm 1). The random initial population consists in assigning to each chromosome of the population a random value (True or False) to each cluster and all the variables within that cluster will get the same value.
- **Fitness Function**
  The notion of fitness is fundamental to the application of genetic algorithms. It is a numerical value that expresses the performance of an individual (solution) so that different individuals can be compared. The fitness of a chromosome (individual)

Algorithm 1: VNS-GA

input : Problem in CNF Format
output: Number of unsatisfied clauses

1 /*Select the set of neighborhood structures N_k (k=1,2,...,k_{max}) /* ;
2 k := 0 ;
3 **while** *(Not reached the desired set of neighborhood)* **do**
4 N_{k+1}:= Define (N_k) ;
5 $k := k + 1$;
6 /*Generate a random initial population from the k^{th}_{max} neighborhood */ ;
7 $S^k_{current} \leftarrow RandomSolution$ () ;
8 Evaluate the fitness of each individual in the population $S^k_{current}$;
9 **while** *($k \geq 0$)* **do**
10 **while** *(Not Stop)* **do**
11 Select individuals from $S^k_{current}$ according to a scheme to reproduce ;
12 Breed if necessary each selected pairs from $S^k_{current}$ through crossover;
13 Apply mutation operator if necessary to each offspring in $S^k_{current}$;
14 Evaluate the fitness of the intermediate population ;
15 $S^k_{new} \leftarrow S^k_{current}$;
16 $S^{k-1}_{current} \leftarrow ChangeNeighborhood(S^k_{new})$;
17 $k \leftarrow k - 1$;
18 return (Best-Individual (S^{k+1}_{new}));

in the population is equal to the number of clauses that are unsatisfied by the truth assignment represented by the chromosome.

- **Starting Neighborhood**
The most crucial aims at selecting the different neighborhoods according to some strategy for the effectiveness of the search process. The strategy adopted in this work is to let VNS-GA start the search process from the largest neighborhood $N_{k_{max}}$ and continues to move towards smaller neighborhood structures (lines 9–15 of Algorithm 1). The motivation behind this strategy is that the order in which the neighborhood structures have been selected offers a better mechanism for performing diversification and intensification. The largest neighborhood N_{max} allows GA to view any cluster of variables as a single entity leading the search to become guided in far away regions of the solution space and restricted to only those configurations in the solution space in which the variables grouped within a cluster are assigned the same value. As the switch from one neighborhood to another implies a decrease in the size of the neighborhood, the search is intensified around solutions from previous neighborhoods in order to reach better ones.
- **Parent Selection**
New solutions are created by combining pairs of individuals in the population and then applying a crossover operator to each chosen pair. Combining pairs of individuals can be viewed as a matching process. In the version of GA used in this work, the individuals are visited in random order. An unmatched individual i_k is matched randomly with an unmatched individual i_l (line 11 of Algorithm 1).

Table 4 Two parents before the cross-over operator

x1:0	x2:0	x3:1	↓ **x4 : 0**	↓ **x5 : 1**	↓ **x6 : 0**	↓ **x7 : 1**	x8:1	x9:0	x10
x1:0	x2:1	x3:0	↑ **x4 : 1**	↑ **x5 : 0**	↑ **x6 : 1**	↑ **x7 : 0**	x8:0	x9:1	x10

Table 5 Results of the cross-over operator

x1:0	x2:0	x3:1	↓ **x4 : 1**	↓ **x5 : 0**	↓ **x6 : 1**	↓ **x7 : 0**	x8:1	x9:0	x10
x1:0	x2:1	x3:0	↑ **x4 : 0**	↑ **x5 : 1**	↑ **x6 : 0**	↑ **x7 : 1**	x8:0	x9:1	x10

- **Cross-Over Operator**
  The task of the crossover operator (line 12 of Algorithm 1) is to reach regions of the search space with higher average quality. The two-point crossover operator (Tables 4 and 5) is applied to each matched pair of individuals. The two-point crossover selects two randomly points within a chromosome and then interchanges the two parent chromosomes between these points to generate two new offspring. Recombination can be defined as a process in which a set of configurations (solutions referred as parents) undergoes a transformation to create a set of configurations (referred as offspring). The creation of these descendants involves the location and combinations of features extracted from the parents. The reason behind choosing the two point crossover are the results presented in [35] where the difference between the different crossovers are not significant when the problem to be solved is hard. The work conducted in [32] shows that the two-point crossover is more effective when the problem at hand is difficult to solve. In addition, the author propose an adaptive mechanism in order to have evolutionary algorithms choose which forms of crossover to use and how often to use them, as it solves a problem.
- **Mutation**
  The purpose of mutation is to generate modified individuals by introducing new features in the population. By mutation, the alleles of the produced child have a chance to be modified, which enables further exploration of the search space. The mutation operator takes a single parameter p_m, which specifies the probability of performing a possible mutation. Let $C = c_1, c_2, \dots, c_m$ be a chromosome represented by a binary chain where each of whose gene c_i is either 0 or 1. In our mutation operator, each gene c_i is mutated through flipping this gene's allele from 0 to 1 or vice versa if the probability test is passed. In case of a large neighborhood ($k > 0$), the mutation is applied to a cluster of variables. The mutation probability ensures that, theoretically, every region of the search space is explored. If on the other hand, mutation is applied to all genes, the evolutionary process will degenerate into a random search with no benefits of the information gathered in preceding generations. The mutation operator prevents the searching process form being trapped into local optimum while adding to the diversity of the population

and thereby increasing the likelihood that the algorithm will generate individuals with better fitness values.

- **Survivor Selection**
  The selection acts on individuals in the current population. Based on each individual quality (fitness), it determines the next population. In the roulette method, the selection is stochastic and biased toward the best individuals. The first step is to calculate the cumulative fitness of the whole population through the sum of the fitness of all individuals. After that, the probability of selection is calculated for each individual as being $P_{Selection_i} = f_i / \sum_1^N f_i$, where f_i is the fitness of individual i (Line 15 of Algorithm 1).
- **Change of Neighborhood**
  Once GA has reached the convergence criterion with respect to a neighborhood N_i, It switches to another neighborhood and the assignment reached on the neighborhood N_i must be projected on its parent neighborhood N_{i-1}. The projection algorithm (line 10 of Algorithm 1) is simple; if a cluster $c_i \in N_m$ is assigned the value of true then the merged pair of clusters that it represents, c_j, $c_k \in N_{m-1}$ are also assigned the true value. Finally, the algorithm WalkSAT is applied at the default neighborhood (line 12 of Algorithm 1).

5 Experimental Results

5.1 *Test Suite and Parameter Settings*

The performance of VNS-GA is evaluated against GA using a set of random problems (MAX-2SAT, MAX-3SAT, MAX-4SAT, MAX-5SAT). This set is taken from the Ninth MAX-SAT 2014 evaluation (http://maxsat.ia.udl.cat/benchmarks/) organized as an affiliated event of the 17th International Conference on Theory and Applications of Satisfiability Testing (SAT 2014). Due to the randomization nature of both algorithms, each problem instance was run 50 times with a cut-off parameter (max-time) set to 15 min. The tests were carried out on a DELL machine with 800 MHz CPU and 2 GB of memory. The code was written in C++ and compiled with the GNU C compiler version 4.6. The following parameters have been fixed experimentally and are listed below:

- k_{max}: The cardinality of the neighborhood is set such that the number of the formed clusters is 10% of the size of the problem instance (i.e., a problem with 100 variables will lead to k_{max} equals to 4 (N_1, N_2, N_3, N_4)).
- GA is assumed to have reached convergence and switch to a smaller neighborhood if the fitness of the fittest chromosome remains unchanged during five consecutive generations.

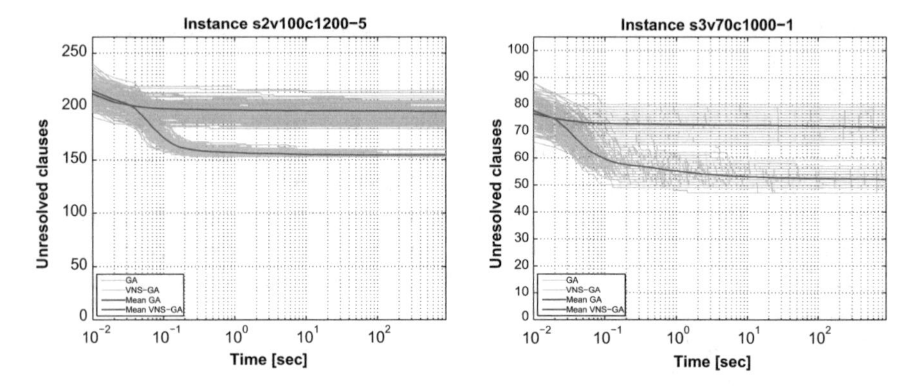

Fig. 1 (Left) 2SAT-V100-C1200-5, (right) 3SAT-V70-C1000-1—time development for 50 runs in 15 min

5.2 *Observed Behavior*

Giving few highlights and comments on the plots described in Figs. 1 and 2 which show the general trend observed when GA is compared with VNS-GA. During the early stage of the search, the two mean of the two curves overlay each other closely and are almost indistinguishable up to a certain point before the two curves start showing no cross-over and VNS-GA soon shows better performance than GA throughout the run. The variable neighborhood scheme appears to enhance the asymptotic convergence of GA. A comparison between the individual runs of GA and VNS-GA, reveals that the overlapping is only restricted during the early stages of the search, and becomes significantly less apparent as the time increases. VNS-GA offers a cost effective solution strategy considering the amount of time required. Its strong component relies on coupling the cross-over operator process across different neighborhoods. This paradigm offers two main advantages which enables GA to become much more powerful in the variable neighborhood context. The two-points cross-over operator is viewed as a pure unstructured operator to generate new solutions. By allowing the gene of each individual representing a cluster of variables at different neighborhoods, the cross-over operator becomes guided and restricted to only those configurations in the solution space in which the variables grouped within a cluster are assigned the same value. The switch from one neighborhood to another allows the possibility for both cross-over and mutation operators of exploring different regions in the search space in a structured manner while intensifying the search by exploiting the solutions from previous neighborhoods in order to reach better solutions.

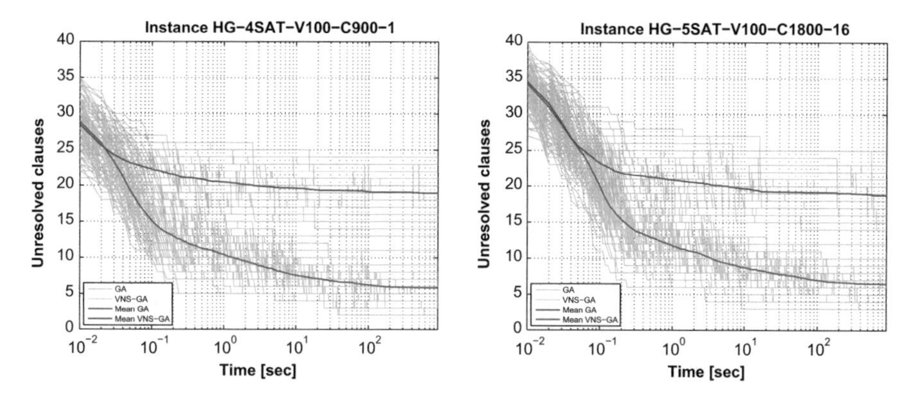

Fig. 2 (Left) HG-4SAT-V100-C900-1, (right) HG-5SAT-V100-C1800-16—time development for 50 runs in 15 min

5.3 *Statistical Analysis*

The results showing the statistical comparison between the two algorithms are presented in Tables 6, 7 and 8. The mean (x) and the standard deviation (σ) of unsolved clauses for VNS-GA and GA algorithms are reported. The range of solutions from each algorithm is also presented in order to show the overlap between solution spaces for any given instance. Statistical inferential analysis was done with an independent samples t-test which compares the difference in means between the two groups. Comparison using the non-parametric Mann-Whitney U-test gave identical results. The non-parametric effect size measure $\hat{A}_{12}$ [1, 34] was used to evaluate the relative dominance of one algorithm over the other. The $\hat{A}_{12}$ effect size measure is calculated using the rank sum which is a common component in any non-parametric analysis such as the Mann-Whitney U-test [1]. Calculating $\hat{A}_{12}$ is done according to the following formula:

$$\hat{A}_{12} = (R_1/m - (m+1)/2)/n. \tag{1}$$

where R1 is the rank sum of algorithm VNS-GA, m is the number of observations in the first data sample, and n is the number of observations in the second data sample. Calculating $\hat{A}_{12}$ results in a number between 0 and 1 which represent the probability that VNS-GA will yield a better solution than GA. If the two algorithms are equivalent then $\hat{A}_{12} = 0.5$, while a complete dominance of algorithm VNS-GA over GA would entail $\hat{A}_{12} = 1$. $\hat{A}_{12}$ is more easily interpreted than the more common parametric Cohen's d [12] which represents the mean difference between two groups in standard deviations for several reasons. First, Cohen's d assumes that the observed samples are normally distributed [1]. Second, when dealing with solutions to optimization problems, a researcher or practitioner would only be interested in the single best solution given a sample of different solutions from one or more algorithms.

Table 6 2SAT: VNS-GA versus GA. *Notes* x = mean, σ = standard deviation, Min = minimum observed value, Max = maximal observed value

Instance	GA		VNS-GA	
	x(σ)	Min-max	x(σ)	Min-max
s2v100c1200-1	210.6 (6.2)	192–229	171.5 (3.4)	169–180
s2v100c1200-2	203.1 (6.3)	186–217	163.9 (1.3)	163–169
s2v100c1200-3	212.8 (5.5)	199–226	173.7 (2.7)	172–182
s2v100c1200-4	207.2 (6.2)	191–223	171 (2.2)	169–178
s2v100c1200-5	207.2 (6.2)	191–223	171 (2.2)	169–178
s2v100c1200-6	195.5 (7.3)	179–215	154.5 (1.8)	153–160
s2v100c1200-7	204 (5.9)	188–221	167.8 (0.9)	167–172
s2v100c1200-8	207.9 (5.9)	192–222	172.5 (2.3)	171–181
s2v100c1200-9	206.5 (5.6)	195–221	173.8 (1.6)	171–179
s2v100c1200-10	206.7 (6.7)	184–224	163.5 (3.2)	162–176
s2v120c1200-1	203.2 (5.4)	185–217	163 (2.5)	161–171
s2v120c1200-2	200.1 (5.7)	184–212	161.2 (1.9)	159–167
s2v120c1200-3	204.6 (4.8)	192–217	162.8 (4.7)	160–176
s2v120c1200-4	201.2 (6.3)	185–219	158.2 (1.5)	157–164
s2v120c1200-5	188.1 (6.7)	172–211	144.2 (1.1)	143–148
s2v120c1200-6	207 (5.6)	194–225	170.2 (2.2)	167–176
s2v120c1200-7	201.2 (5.3)	188–215	164.2 (2.7)	162–170
s2v120c1200-8	203.6 (5.7)	192–224	167.4 (2.1)	165–173
s2v120c1200-9	195 (6.2)	178–209	148.4 (0.7)	148–151
s2v120c1200-10	196.6 (4.8)	186–207	154.3 (1.4)	154–165
s2v140c1300-1	163.8 (1.8)	162–169	163.6 (1.7)	162–168
s2v140c1300-2	171.9 (1.5)	171–179	172.6 (2.2)	171–181
s2v140c1400-1	183.7 (2.2)	182–192	184.1 (2.3)	182–197
s2v140c1400-2	179 (2.9)	178–190	179 (3.1)	178–192
s2v140c1500-1	205 (0.7)	205–208	205.6 (1)	205–211
s2v140c1500-2	201 (1.7)	199–207	200.9 (1.7)	199–208

Hence, using an effect size measure that indicates the probability that one algorithm would lead to a better solution than another (given the same amount of time) would be more informative and more easily interpretable for an optimization practitioner. The 95% confidence intervals of $\hat{A}_{12}$ shown in Tables 9, 10 and 11 (where applicable) are calculated using a bootstrapping procedure [29] which is used to estimate the 95% confidence interval of $\hat{A}_{12}$. The procedure uses a computer intensive step-by-step process that consists of the following three steps:

1. Random re-sampling with replacement from the original observations to create new data sets.
2. Calculation of the rank sum of VNS-GA for each new data set.

Table 7 3SAT: VNS-GA versus GA. *Notes* x = mean, σ = standard deviation, Min = minimum observed value, Max = maximal observed value

Instance	GA		VNS-GA	
	x(σ)	Min-max	x(σ)	Min-max
s3v70c1000-1	71.6 (3.5)	65–80	52 (2.6)	47–59
s3v70c1000-2	69.1 (4.3)	57–80	48.2 (4.3)	43–60
s3v70c1000-3	71.1 (4)	62–80	51 (3.6)	45–60
s3v70c1000-4	72.2 (3.8)	63–81	52.9 (3)	47–61
s3v70c1000-5	67.8 (4.1)	56–77	45 (2.9)	42–52
s3v70c1000-6	71.9 (3.6)	63–81	54.4 (2.3)	51–61
s3v70c1000-7	70.9 (2.9)	62–78	53.7 (2.1)	49–58
s3v70c1000-8	69.5 (3.8)	60–80	51.8 (2.9)	48–60
s3v70c1000-9	70.9 (3.9)	60–83	52.7 (2.6)	49–61
s3v70c1000-10	69.7 (4.4)	61–86	48.8 (3)	45–61
s3v80c1000-1	47.8 (2.9)	44–55	47.6 (3.2)	44–57
s3v80c1000-2	46.4 (2.1)	43–52	47.5 (2.7)	43–55
s3v80c1000-3	44.7 (4)	39–57	45.3 (4)	39–54
s3v80c1000-4	49.3 (2.6)	45–57	50 (2.9)	45–60
s3v80c1000-5	45.4 (3.1)	41–53	45.4 (3.4)	41–55
s3v80c1000-6	44.8 (3.5)	40–57	44.4 (3.3)	40–52
s3v80c1000-7	43.2 (3.1)	40–52	43.1 (2.9)	40–56
s3v80c1000-8	45.4 (2.8)	41–58	45.9 (2.7)	41–53
s3v80c1000-9	41.1 (3)	38–55	41.5 (2.7)	38–49
s3v80c1000-10	42.8 (2.8)	39–51	43.4 (3.3)	39–55

3. Using the rank sum to calculate $\hat{A}_{12}$ with the Eq. 1. The three steps are then repeated 1000 times and the resulting statistic $\hat{A}_{12}$ is saved to create a sampling distribution of the statistic $\hat{A}_{12}$.

VNS-GA algorithm dominates GA for all MAX2SAT instances with less than 1300 clauses. However, for one instance (s2v140c1300-2) GA algorithm shows better performance, although the best solution (171 unsolved clauses) is identical for both algorithms. For the five remaining instances there is no statistically significant difference between the two algorithms. VNS-GA dominates GA algorithm on all MAX3SAT instances with 70 variables (s3v70c1000-1 to s3v70c1000-10). GA algorithm is significantly better on one instance (s3v80c1000-2) with no statistical difference between the remaining algorithms with 80 variables (s3v80c1000-1 to s3v80-1000-10). VNS-GA dominates GA algorithm on all MAX4SAT instances. VNS-GA dominates GA on all MAX5SAT instances. There is no overlap between the output samples from the two algorithms- with one exception on instance HG-5SAT-V100-C1800-14 where a single run of GA was better compared to that of VNS-GA. When VNS-GA is found to be better than GA, the difference in quality is huge with most instances showing no overlap between the solutions given by the algorithms. When

Table 8 3SAT: VNS-GA versus GA. *Notes* x = mean, σs = standard deviation, Min = minimum observed value, Max = maximal observed value

Instance	GA		VNS-GA	
	x(σ)	Min-max	x(σ)	Min-max
HG-4SAT-V100-C900-1	19 (2.2)	13–24	5.8 (1.9)	2–11
HG-4SAT-V100-C900-2	18.7 (2.5)	13–28	5.6 (1.8)	2–11
HG-4SAT-V100-C900-3	18.6 (2.3)	13–26	5.4 (1.8)	2–11
HG-4SAT-V100-C900-4	19 (2.2)	14–25	6.2 (1.6)	3–12
HG-4SAT-V100-C900-5	18.9 (2.3)	13–25	5.3 (1.9)	1–10
HG-4SAT-V100-C900-10	19.1 (2.1)	13–24	5.5 (1.5)	2–10
HG-4SAT-V100-C900-11	19.2 (2.1)	14–26	6 (1.6)	2–12
HG-4SAT-V100-C900-12	18.5 (2.2)	15–25	5.8 (1.9)	2–10
HG-4SAT-V100-C900-13	19 (2.3)	14–25	5.9 (1.5)	2–10
HG-4SAT-V100-C900-14	19 (2.4)	14–24	5.9 (1.6)	3–11
HG-4SAT-V100-C900-15	19.3 (2.2)	14–25	5.8 (1.5)	3–10
HG-4SAT-V100-C900-16	18.8 (2)	14–23	5.7 (1.7)	2–10
HG-4SAT-V100-C900-17	19 (2.4)	13–24	6 (1.7)	3–11
HG-4SAT-V100-C900-18	18.6 (2.3)	12–26	5.5 (1.7)	1–11
HG-4SAT-V100-C900-19	19.2 (2.6)	10–25	5.8 (1.6)	2–10
HG-4SAT-V100-C900-100	18.4 (2)	13–23	5.6 (1.7)	2–10
HG-5SAT-V100-C1800-2	18.9 (2.1)	15–24	7 (1.8)	3–11
HG-5SAT-V100-C1800-3	19.6 (2.2)	15–26	7.1 (1.8)	3–12
HG-5SAT-V100-C1800-4	18.9 (2.3)	14–25	6.9 (1.7)	2–11
HG-5SAT-V100-C1800-5	18.8 (2.3)	14–24	6.5 (1.9)	2–11
HG-5SAT-V100-C1800-10	19.2 (2.1)	14–25	6.9 (1.8)	3–13
HG-5SAT-V100-C1800-11	19 (2.2)	14–25	7.1 (1.7)	3–11
HG-5SAT-V100-C1800-12	19.3 (2.1)	15–24	6.7 (1.8)	3–13
HG-5SAT-V100-C1800-13	19 (2.4)	15–26	6.5 (1.8)	3–11
HG-5SAT-V100-C1800-14	19.2 (2.1)	15–25	6.9 (2.1)	3–16
HG-5SAT-V100-C1800-15	19.5 (2.3)	14–25	6.7 (1.9)	2–12
HG-5SAT-V100-C1800-16	18.8 (2.1)	14–25	6.5 (1.7)	3–12

GA is significantly better the difference with respect to best solutions is marginal or non-existent. VNS-GA has a much lower variation of results across the test suite and we take this to mean that the variable neighborhood scheme stabilizes GA in some way. Table 12 compares VNS-GA with two highly efficient incomplete solvers used at the 2014 MAX-SAT competition. For each solver, we report the number of unsatisfied clauses, while the number between parenthesis denotes the time in seconds. CCLS2akms and ISAC+2014-ms gives similar solution quality while the main differences resides on the time required to produce such quality. VNS-GA is capable of delivering the same solution quality as these two solvers in 18 cases out of 27. When compared to CCLS2akms, VNS-GA converges faster in 8 cases. The time

Table 9 Statistical comparison of the solutions given by VNS-GA and GA. *Notes* Δ = mean difference, CI = confidence interval, p = p-value, PS = probability of superiority (i.e. that VNS-GA will have less unsolved clauses than GA for each instance), $*** = p < 0.001$. $** = p < 0.01$, $* \ p < 0.05$, c = not possible to calculate CI of PS due to no overlap between VNS-GA and GA. The 95% CI for PS is calculated in Excel using a bootstrapping procedure (random selection with replacement) performed 1000 times

Prob	Δ [95%$CIof\Delta$]	p	PS	[95% CI of PS]
s2v100c1200-1	39.2 [37.3, 41]	***	1	c
s2v100c1200-2	39.3 [37.6, 41]	***	1	c
s2v100c1200-3	39 [37.4, 40.6]	***	1	c
s2v100c1200-4	36.2 [34.5, 37.9]	***	1	c
s2v100c1200-5	41 [39, 42.9]	***	1	c
s2v100c1200-6	36.1 [34.6, 37.7]	***	1	c
s2v100c1200-7	35.4 [33.8, 37.1]	***	1	c
s2v100c1200-8	32.8 [31.3, 34.3]	***	1	c
s2v100c1200-10	43.2 [41.2, 45.1]	***	1	c
s2v120c1200-1	40.2 [38.6, 41.7]	***	1	c
s2v120c1200-2	38.9 [37.4, 40.5]	***	1	c
s2v120c1200-3	41.8 [40.1, 43.6]	***	1	c
s2v120c1200-4	43 [41.3, 44.7]	***	1	c
s2v120c1200-5	43.8 [42.1, 45.6]	***	1	c
s2v120c1200-6	36.9 [35.3, 38.4]	***	1	c
s2v120c1200-7	36.9 [35.4, 38.5]	***	1	c
s2v120c1200-8	36.2 [34.6, 37.8]	***	1	c
s2v120c1200-9	46.5 [44.9, 48.2]	***	1	c
s2v120c1200-10	42.3 [41, 43.6]	***	1	c
s2v140c1300-1	0.2 [−0.5, 0.8]	0.526	0.528	[0.456, 0.607]
s2v140c1300-2	−0.7 [−1.4, 0]	0.014*	0.422	[0.351, 0.494]
s2v140c1400-1	−0.4 [−1.3, 0.4]	0.176	0.431	[0.356, 0.509]
s2v140c1400-2	0 [−1.1, 1.1]	0.944	0.523	[0.465, 0.579]
s2v140c1500-1	0 [−0.3, 0.3]	0.844	0.534	[0.465, 0.601]
s2v140c1500-2	0 [−0.6, 0.7]	0.901	0.496	[0.418, 0.577]

of VNS-GA in these cases lies between 10 and 77% of the time of CCLS2akms. For the the remaining cases, the time of CCLS2akms is between 31 and 86% of the time of VNS-GA. The comparison between VNS-GA and ISAC+2014-ms reveals that VNS-GA converges faster in 12 cases. The time of VNS-GA lies between 19 and 89% of the time of ISAC+2014-ms. The time of ISAC+2014-ms is between 10 and 82% in the remaining 6 cases. VNS-GA was beaten by these two solvers in 9 cases. The difference in quality expressed as the number of unsatisfied clauses never exceeds 1.

Table 10 Statistical comparison of the solutions given by VNS-GA and GA. *Notes* Δ = mean difference, CI = confidence interval, p = p-value, PS = probability of superiority (i.e. that VNS-GA will have less unsolved clauses than GA for each instance), $*** = p < 0.001$. $** = p < 0.01$, $* \; p < 0.05$, c = not possible to calculate CI of PS due to no overlap between VNS-GA and GA. The 95% CI for PS is calculated in Excel using a bootstrapping procedure (random selection with replacement) performed 1000 times

Prob	$\Delta[95\% CI of \Delta]$	*p*	*PS*	[95% CI of PS]
s3v70c1000-1	19.6 [18.4, 20.7]	***	1	c
s3v70c1000-2	20.9 [19.3, 22.5]	***	0.999	[0.998, 1]
s3v70c1000-3	20.2 [18.8, 21.5]	***	1	c
s3v70c1000-4	19.3 [18, 20.6]	***	1	c
s3v70c1000-5	22.8 [21.4, 24.1]	***	1	c
s3v70c1000-6	17.6 [16.4, 18.7]	***	1	c
s3v70c1000-7	17.2 [16.2, 18.1]	***	1	c
s3v70c1000-8	17.6 [16.4, 18.9]	***	1	[0.998, 1]
s3v70c1000-9	18.2 [17, 19.4]	***	1	[0.998, 1]
s3v70c1000-10	20.9 [19.5, 22.3]	***	1	[0.998, 1]
s3v80c1000-1	0.2 [−0.9, 1.3]	0.708	0.528	[0.451, 0.611]
s3v80c1000-2	−1.1 [−2, −0.2]	0.002**	0.383	[0.307, 0.458]
s3v80c1000-3	−0.6 [−2.1, 0.9]	0.279	0.452	[0.372, 0.530]
s3v80c1000-4	−0.8 [−1.8, 0.3]	0.055	0.417	[0.344, 0.494]
s3v80c1000-5	0.1 [−1.1, 1.3]	0.879	0.514	[0.438, 0.590]
s3v80c1000-6	0.4 [−0.9, 1.6]	0.42	0.527	[0.447, 0.606]
s3v80c1000-7	0.1 [−1, 1.2]	0.852	0.496	[0.418, 0.570]
s3v80c1000-8	−0.4 [−1.4, 0.6]	0.274	0.44	[0.367, 0.522]
s3v80c1000-9	−0.4 [−1.4, 0.7]	0.242	0.444	[0.373, 0.522]
s3v80c1000-10	−0.6 [−1.7, 0.6]	0.191	0.46	[0.381, 0.538]

6 Conclusions

In this work, a hybrid approach combining VNS with GA has been described. VNS follows a simple principle that is based on systematic changes of neighborhood within the search. The set of neighborhood proposed in this paper can easily be incorporated into any meta-heuristic when dealing with various combinatorial optimization problems. Starting the search from the largest neighborhood and moving systematically towards the smallest neighborhood is a better strategy for performing diversification and intensification. The approach has been tested on MAX-SAT using random instances. The results indicate that the variable neighborhood search strategy can enhance the convergence behavior of GA. It appears clearly from the results that the performance of VNS-GA is better compared to that of GA. The larger the problem, the larger the size of the neighborhood is needed, and consequently the more efficient is GA at different neighborhoods. Finally, VNS-GA was capable of

Table 11 Statistical comparison of the solutions given by VNS-GA and GA. *Notes* Δ = mean difference, CI = confidence interval, p = p-value, PS = probability of superiority (i.e. that VNS-GA will have less unsolved clauses than GA for each instance), $*** = p < 0.001$. $** = p < 0.01$, $* \ p < 0.05$, c = not possible to calculate CI of PS due to no overlap between VNS-GA and GA. The 95% CI for PS is calculated in Excel using a bootstrapping procedure (random selection with replacement) performed 1000 times

Prob	$\Delta[95\% CI of \Delta]$	*p*	*PS*	[95% CI of PS]
HG-4SAT-V100-C900-1	13.2 [12.4, 13.9]	***	1	c
HG-4SAT-V100-C900-2	13.1 [12.3, 13.9]	***	1	c
HG-4SAT-V100-C900-3	13.1 [12.4, 13.9]	***	1	c
HG-4SAT-V100-C900-4	12.8 [12.1, 13.5]	***	1	c
HG-4SAT-V100-C900-5	13.6 [12.8, 14.4]	***	1	c
HG-4SAT-V100-C900-10	13.6 [12.9, 14.3]	***	1	c
HG-4SAT-V100-C900-11	13.2 [12.5, 13.9]	***	1	c
HG-4SAT-V100-C900-12	12.7 [12, 13.5]	***	1	c
HG-4SAT-V100-C900-13	13.1 [12.3, 13.8]	***	1	c
HG-4SAT-V100-C900-14	13.1 [12.4, 13.9]	***	1	c
HG-4SAT-V100-C900-15	13.5 [12.8, 14.2]	***	1	c
HG-4SAT-V100-C900-16	13.2 [12.5, 13.8]	***	1	c
HG-4SAT-V100-C900-17	12.9 [12.2, 13.7]	***	1	c
HG-4SAT-V100-C900-18	13.1 [12.3, 13.8]	***	1	c
HG-4SAT-V100-C900-19	13.5 [12.7, 14.2]	***	1	[0.999, 1]
HG-4SAT-V100-C900-100	12.8 [12.1, 13.5]	***	1	c
HG-5SAT-V100-C1800-2	12 [11.2, 12.7]	***	1	c
HG-5SAT-V100-C1800-3	12.5 [11.7, 13.2]	***	1	c
HG-5SAT-V100-C1800-4	12 [11.3, 12.8]	***	1	c
HG-5SAT-V100-C1800-5	12.3 [11.5, 13.1]	***	1	c
HG-5SAT-V100-C1800-10	12.3 [11.6, 13]	***	1	c
HG-5SAT-V100-C1800-11	11.9 [11.2, 12.6]	***	1	c
HG-5SAT-V100-C1800-12	12.6 [11.9, 13.3]	***	1	c
HG-5SAT-V100-C1800-13	12.5 [11.8, 13.3]	***	1	c
HG-5SAT-V100-C1800-14	12.3 [11.5, 13.1]	***	0.999	[0.997, 1]
HG-5SAT-V100-C1800-15	12.8 [12, 13.6]	***	1	c
HG-5SAT-V100-C1800-16	12.3 [11.6, 13]	***	1	c

producing competitive results than the top ranked solvers for MAX-SAT. The random procedure used to build the different neighborhoods does not exploit the information structure of the problem. The author believe that VNS-GA might benefit from further research into merging strategies used to construct the neighborhoods. A better strategy would be to construct the different neighborhoods based on merging variables by exploiting the number of clauses in the case of MAX-SAT.

Table 12 Comparing VNS-GA with state-of-art incomplete solvers

Problem	CCLS2akms	ISAC+2014-ms	VNS-GA
s2v120c1200-1	161 (7.43)	161 (14.74)	161 (11.48)
s2v120c1200-2	159 (8.54)	159 (27.95)	159 (7.77)
s2v120c1200-3	160 (3.88)	160 (7.88)	160 (8.66)
s2v120c1200-4	157 (3.98)	157 (7.01)	158 (6.71)
s2v120c1200-5	143 (2.05)	143 (5.32)	144 (34.53)
s2v120c1200-6.	167 (8.66)	167 (16.18)	169 (3.79)
s2v120c1200-7	162 (9.24)	162 (12.23)	162 (65.42)
s2v120c1200-8	165 (27.84)	165 (34.01)	165 (7.16)
s2v120c1200-9	148 (2.23)	148 (4.80)	148 (7.67)
s2v120c1200-10	154 (2.58)	154 (8.40)	154 (10.08)
s2v120c1300-1	180 (18.74)	180 (22.93)	180 (13.57)
s2v120c1300-2	172 (6.53)	172 (14.79)	172 (16.94)
s2v120c1300-3	173 (8.42)	173 (20.16)	173 (12.69)
s2v120c1300-5	168 (3.32)	168 (11.90)	169 (37.63)
s2v120c1300-7	169 (2.56)	169 (12.31)	169 (2.75)
s2v120c1300-9	186 (39.54)	186 (59.20)	186 (12.31)
s2v140c1200-1	144 (12.29)	144 (16.98)	144 (13.11)
s2v140c1200-2	155 (153.63)	155 (243.53)	156 (31.04)
s2v140c1300-3	168 (51.45)	168 (70.52)	169 (13.89)
s2v140c1400-1	182 (56.99)	182 (113.89)	182 (13.23)
s2v140c1400-2	178 (32.54)	178 (47.53)	178 (53.85)
s2v140c1400-3	193(142.84)	193 (385.21)	193 (69.41)
s2v140c1400-4	184 (35.14)	184 (73.44)	184 (50.51)
s2v140c1500-1	205 (200.67)	205 (170.26)	205 (138.31)
s2v140c1500-2	199 (203.13)	199 (317.05)	200 (178.10)
s2v140c1500-3	212 (588.53)	212 (1098)	213 (300.10)
s2v140c1500-4	197 (71.49)	197 (89.68)	198 (91.10)

Acknowledgements We would like to address a particular warm thank to the members of the organizing committee and scientific committee for making the First EAI International Conference on Computer Science and Engineering, NOVEMBER 11–12, 2016, PENANG, MALAYSIA a great success.

References

1. Arcuri, A., & Briand, L. (2011). A Hitchhiker's guide to statistical tests for assessing randomized algorithms in software engineering. Technical report, Simula research laboratory, number 13/2011.

2. Blum, C., & Roli, A. (2003). Meta-heuristics in combinatorial optimization: Overview and conceptual comparison. *ACM Computing Surveys*, *35*(3), 268–308.
3. Bouhmala, N., & Oseland, M. (2017). Antelnd bradland. WalkSAT based-learning automata for MAX-SAT. In *Advances in Intelligent Systems and Computing: Recent Advances in Soft Computing (MENDEL)* (Vol. 576). Springer, Czech Republic.
4. Bouhmala, N. (2016). A simple and efficient variable neighborhood structure for the satisfiability problem. In *Proceedings of 6th International Conference on Meta-heuristics and Nature* (pp. 126–133), Marrakech.
5. Bouhmala, N. (2015). A multilevel learning automata for MAX-SAT. *International Journal of Machine Learning & Cybernetics*. Heidelberg: Springer. https://doi.org/10.1007/s13042-015-0355-4.
6. Bouhmala, N., Hjelmervik, K., & Øvergård, K. (2015). A generalized variable neighborhood search for combinatorial optimization problems. *Electronic Notes in Discrete Mathematics*, *47*, 45–52.
7. Bouhmala, N., & Cai, X. (2009). A multilevel approach for the satisfiability problem. *ISAST Transactions on Computers and Intelligent Systems*, *2*(1), 29–37.
8. Bouhmala. N. (2012). A multilevel memetic algorithm for large SAT-encoded problems. *Evolutionary Computation*, *20*(4), 641–664.
9. Bouhmala, N., & Granmo, O. C. (2010). Stochastic learning for SAT-encoded graph coloring problems. *International Journal of Applied Meta-heuristic Computing*, *1*(3), 1–19.
10. Bouhmala, N., & Granmo, O. C. (2010). Combining finite learning automata with GSAT for the satisfiability problem. *Engineering Applications of Artificial Intelligence*, *23*(5), 715–726.
11. Cha, B., & Iwama, K. (1995). Performance tests of local search algorithms using new types of random CNF formula. In *Proceedings of IJCAI95* (pp. 304–309). Morgan Kaufmann Publishers.
12. Cohen, J. (1988). *Statistical power analysis for the behavioral sciences* (2nd ed.). Lawrence Erlbaum.
13. Cook, S. A. (1971). The complexity of theorem-proving procedures. In *Proceedings of the Third ACM Symposium on Theory of Computing* (pp. 151–158).
14. Frank, J. (1997). Learning short-term clause weights for GSAT. In *Proceedings of IJCAI97* (pp. 384–389). Morgan Kaufmann Publishers.
15. Hansen, P., Jaumard, B., Mladenovic, N., & Parreira, A. D. (2000). Variable neighborhood search for maximum weighted satisfiability problem. Technical Report G-2000-62, Les Cahiers du GERAD, Group for Research in Decision Analysis.
16. Hansen, P., & Mladenovic, N. (1999). An introduction to variable neighborhood search. In S. Voss, S. Martello, I. H. Osman, & C. Roucairol (Eds.), *Meta-heuristics: Advances and trends in local search paradigms for optimization* (pp. 433–458). Boston: Kluwer.
17. Hoos, H. (2002). An adaptive noise mechanism for WalkSAT. In *Proceedings of AAAI-2002* (pp. 655–660).
18. Hoos, H. (1999). On the run-time behavior of stochastic local search algorithms for SAT. In *Proceedings of AAAI-99* (pp. 661–666).
19. Hu, B., & Raidl, R. (2006). Variable neighborhood descent with self-adaptive neighborhood-ordering. In C. Cotta, A. J. Fernandez, & J. E. Gallardo (Eds.), *Proceedings of the 7th EU/MEeting on Adaptive, Self-Adaptive, and Multi-Level Meta-heuristics*, Malaga, Spain.
20. Jin-Kao, H., Lardeux, F., & Saubion, F. (2003). Evolutionary computing for the satisfia-bility problem. In *Applications of Evolutionary Computing*, LNCS (Vol. 2611, pp. 258–267). England: University of Essex.
21. KhudaBukhsh, A. R., Xu, L., Hoos, H., & Leyton-Brown, K. (2009). SATenstein: automatically building local search SAT solvers from components. In *Proceedings of the 25th International Joint Conference on Artificial Intelligence (IJCAI-09)*.
22. Lardeux, F., Saubion, F., & Hao, J. K. (2006). GASAT: A genetic local search algorithm for the satisfibility problem. *Evolutionary Computation*, *14*(2), 223–253.
23. Li, C. M., & Huang, W. Q. (2005). Diversification and determinism in local search for satisfiability. In *Proceedings of the Eighth International Conference on Theory and Applications of Satisfiability Testing (SAT-05)*, Lecture Notes in Computer Science (Vol. 3569, pp. 158–172).

24. Li, C. M., Wei, W., & Zhang, H. (2007). Combining adaptive noise and look-ahead in local search for SAT. In *Lecture notes in computer science* (Vol. 4501, pp. 121–131).
25. Lozano, M., Herrera, F., & Cano, R. (2008). Replacement strategies to preserve useful diversity in steady-state genetic algorithms. *Information Sciences, 178*(23), 4421–4433.
26. Mazure, B., *Saïs*, L., & *Grégoire*, E. (1997). Tabu search for SAT. In *Proceedings of the Fourteenth National Conference on Artificial Intelligence (AAAI-97)* (pp. 281–285).
27. McAllester, D., Selman, B., & Kautz, H. (1997). Evidence for invariants in local search. In *Proceedings of the Fourteenth National Conference on Artificial Intelligence (AAAI-97)* (pp. 321–326).
28. Mladenović, N., & Hansen, P. (1997). Variable neighborhood search. *Computer and Operations Research, 24*, 1097–1100.
29. Mooney, C. Z., & Duval, R. D. (1993). *Bootstrapping—A nonparametric approach to statistical inference*. Sage University Press.
30. Selman, B., Kautz, H. A., & Cohen, B. (1994). Noise strategies for improving local search. In *Proceedings of AAAI'94* (pp. 337–343). MIT Press.
31. Selman, B., Levesque, H., & Mitchell, D. (1992). A new method for solving hard satisfiability problems. In *Proceedings of AAA92* (pp. 440–446). MIT Press.
32. Spears, W. (1995). Adapting crossover in evolutionary algorithms. In *Proceedings of the Fourth Annual Conference on Evolutionary Programming* (pp. 367–384). MIT Press.
33. Talbi, E. G. (2009). *Meta-heuristics: From design to implementation*. Wiley.
34. Vargha, A., & Delaney, H. D. (2000). A critique and improvement of the CL common language effect size statistics of McGraw and Wong. *Journal of Educational and Behavioral Statistics, 25*(2), 101–132.
35. Vrajitoru, D. (1999). Genetic programming operators applied to genetic algorithms. In *Proceedings of the Genetic and Evolutionary Computation Conference*, Orlando (FL) (pp. 686–693). Morgan Kaufmann Publishers.
36. Wong, Y., Lee, Y., Leung, K., & Ho, C. (2003). A novel approach in parameter adaptation and diversity maintenance for genetic algorithms. *Soft Computing, 7*, 506–515.
37. Xu, L., Hutter, F., Hoos, H., & Leyton-Brown, K. (2008). SATzilla: Portfolio-based algorithm selection for SAT. *Journal of Artificial Intelligence Research (JAIR), 32*, 565–606.
38. Yang, X. S., & Gandomi, A. H. (2012). Bat algorithm: A novel approach for global engineering optimization. *Engineering Computations, 29*(5), 464–483.
39. Yang, X. S., & Deb, S. (2010). Eagle strategy using *Lévy* work and firefly algorithms for stochastic optimization. In *Nature Inspired Cooperative Strategies for Optimization (NICSO2010)* (pp. 101–111). Springer.
40. Yagiura, M., & Ibaraki, T. (2001). Efficient 2 and 3-flip neighborhood search algorithms for the MAX SAT: Experimental evaluation. *Journal of Heuristics, 7*, 423–442.
41. Zhipeng, L., & Jin-Kao, H. (2012). Adaptive memory-based local search for MAX-SAT. *Applied Soft Computing*.

Enzyme Classification on DUD-E Database Using Logistic Regression Ensemble (Lorens)

Heri Kuswanto, Jainap N. Melasasi and Hayato Ohwada

Abstract Discovery of drugs has been a complex process, time-consuming and expensive until an alternative of making drug has been found i.e. using in silico method to discover potential inhibitor. During the process of drug design, compound classification is carried out through docking score steps. The aim of this research is to predict the docking score results using proper methods for classification i.e. a computationally based method and a standard statistical method. This research examined three target enzymes listed in DUD-E database i.e. aofb, cah2 and hs90a. Each enzyme consists of different compounds that will be classified as good inhibitor (ligand) and bad inhibitor (decoy). In this research, the docking score step is conducted by binary logistic regression and logistic regression ensemble (Lorens). Binary logistic regression yields on 90.4% of accuracy for aofb, 91.7% for cah2 and 94% for hs90a enzyme. Meanwhile, logistic regression ensemble (Lorens) results on the accuracy levels of 88.95, 92.1 and 100% for aofb, cah2 and hs90a consecutively. This paper showed that logistic regression ensemble method outperforms standard logistic regression to be used for the inhibitor classification.

Keywords Lorens · Ensemble · Enzyme · Classification

H. Kuswanto (✉) · J.N. Melasasi
Department of Statistics, Institut Teknologi Sepuluh Nopember,
Kampus ITS Sukolilo, 60111 Surabaya, Indonesia
e-mail: heri_k@statistika.its.ac.id

J.N. Melasasi
e-mail: jainapniken@gmail.com

H. Ohwada
Department of Industrial Administration, Faculty of Science
and Technology, Tokyo Universty of Science, Chiba, Japan
e-mail: ohwada@rs.tus.ac.jp

I. Zelinka et al. (eds.), *Innovative Computing, Optimization and Its Applications*, Studies in Computational Intelligence 741,
https://doi.org/10.1007/978-3-319-66984-7_6

1 Introduction

Discovery of drugs is a complex process, time consuming and expensive. In general, the period to develop newly prospective drug may take about 5 years. Moreover, clinical test phase to become commercial drugs may take 7 years and costs more than 700 million US dollar [1]. In silico method has been introduced in the drug design process and it has been proven to be much more effective. Filtering in silico related to drug discovery is necessary to be conducted in order to find potential inhibitor. In the process of producing drugs, enzyme compound classification is performed using docking software as the main tool to simulate the bond from the mixture (newly inhibitor candidate) with targeted enzyme based on the molecule structure [2].

Docking score process in this research is applied to 3 out of 102 types of enzyme from DUD-E Database as a standard database to simulate docking. This database consists of three types of data i.e. target enzyme, ligand and decoys. This research classifies three types of target enzymes based on the degree of the underlying compounds namely aofb, cah2 and hs90a. The data has characteristic of a high dimensional data shown by a large number of compound. A study on classifying enzyme based on the calculation of docking score with Support Vector Machine (SVM) has been done by [3]. The SVM introduced by Cortes and Vapnik [4] in 1995 yields on 99% classification accuracy, but it has a weakness i.e. unable to give a balance between sensitivity and specivity compared to other classification methods [5]. Machine learning approaches have been widely applied to support works in medical field (see [6] for a comprehensive overview).

This research aims to perform docking score steps with proper statistical method i.e. binary logistic regression. The binary logistic regression is a standard approach which works well to be applied for a standard case in term of the data size. Lim et al. [5] stated that classification using logistic regression for high dimensional data requires feature selection when the number of predictors is too large. The weakness of logistic regression can be overcome by logistic regression ensemble method (Lorens) developed by [7, 14] which includes of Classification by Ensembles from Random Partition (CERP) method. A research which discusses the application of logistic regression ensemble (Lorens) has been carried out by Lim et al. [5] to study of gene expression for child health science. Lorens has been applied also by Kuswanto et al. [8] to classify the customer defection, and they found that Lorens outperformed standard logistic regression for the case of a very large number of observations. The multinomial case of the logistic regression ensemble has also been developed by Lee et al. [9].

This research discusses the result of enzyme compound classification for DUD-E database using two approaches i.e. binary logistic regression and logistic regression ensemble (Lorens). The Lorens performance is evaluated under hold out as well as cross validation approaches.

2 Literature Review

2.1 Binary Logistic Regression

Binary Logistic Regression is a method of data analysis to find the relation between the binary response variable (y) with one or more predictors [10]. Logistic regression model can be written as follows,

$$\pi(x) = \frac{e^{(\beta_0 + \beta_1 x_1 + \ldots + \beta_p x_p)}}{1 + e^{(\beta_0 + \beta_1 x_1 + \ldots + \beta_p x_p)}} \tag{1}$$

where p is the number of predictor variable. In order to relax the assumption about the regression parameters, the above logistic regression model can be written as a logit transformation of $\pi(x)$ to obtain this equation.

$$g_x = ln\left(\frac{\pi(x)}{1 - \pi(x)}\right) = \beta_0 + \beta_1 x_1 + \ldots + \beta_p x_p \tag{2}$$

Parameters in logistic regression can be estimated using maximum likelihood method and assume that the data follows a certain distribution. In logistic regression, each observation follows Bernoulli process, therefore the likelihood function could be determined. Likelihood function is easier to be maximized using log $l(\beta)$ as follow.

$$L(\beta) = \log l(\beta) = \sum_{j=0}^{p}\left(\sum_{i=1}^{n} y_i x_{ij}\right)\beta_j - \sum_{i=1}^{n} \log\left(1 + e^{\sum_{j=0}^{p} \beta_j x_{ij}}\right) \tag{3}$$

The significance of the obtained β coefficient is tested by partial test using statistical test as follow.

$$W^2 = \frac{\beta_i^2}{SE(\beta_i)^2} \tag{4}$$

where the statistic follows chi-square distribution. Meanwhile, simultaneous test is done by G statistic which is Likelihood Ratio test as follows.

$$G = -2\ln\frac{\left(\frac{n_1}{n}\right)^{n_1}\left(\frac{n_0}{n}\right)^{n_0}}{\sum_{i=1}^{n} \hat{\pi}^{y_i}(1 - \hat{\pi})^{(1 - y_i)}} \tag{5}$$

The Odd ratio is defined as the tendency of response variable to have certain value given if x = 1 compared to x = 0. It said that there is no relation between the response variable and predictor variable if odd ratio (ψ) = 1. If odds ratio (ψ) < 1, therefore between predictor variable has negative influence on the response.

Otherwise, if odds ratio (ψ) > 1 therefore between predictor variable and response variable have positive relation [11].

2.2 Logistic Regression Classification by Ensembles from Random Partitions (LR-CERP)

LR-CERP is the basis of Lorens' application. The first step is to choose θ indicating the distance between predictor variable. To minimize the correlation among groups of a classifier, θ is randomly partitioned into k sub spaces $(\theta_1, \theta, \ldots, \theta_k)$ with the same size. The CERP combines multiple logistic regression results to increase the accuracy from prediction using majority voting from group of a classifier or the average of prediction values. To increase further the CERP performance, therefore the majority voting is investigated between group of ensembles. Figure 1 depicts the CERP illustration where the predictors are allocated into k sub-spaces with the same number.

LR-CERP is the basis of Lorens' application. The first step is to choose θ indicating the distance between predictor variable. To minimize the correlation

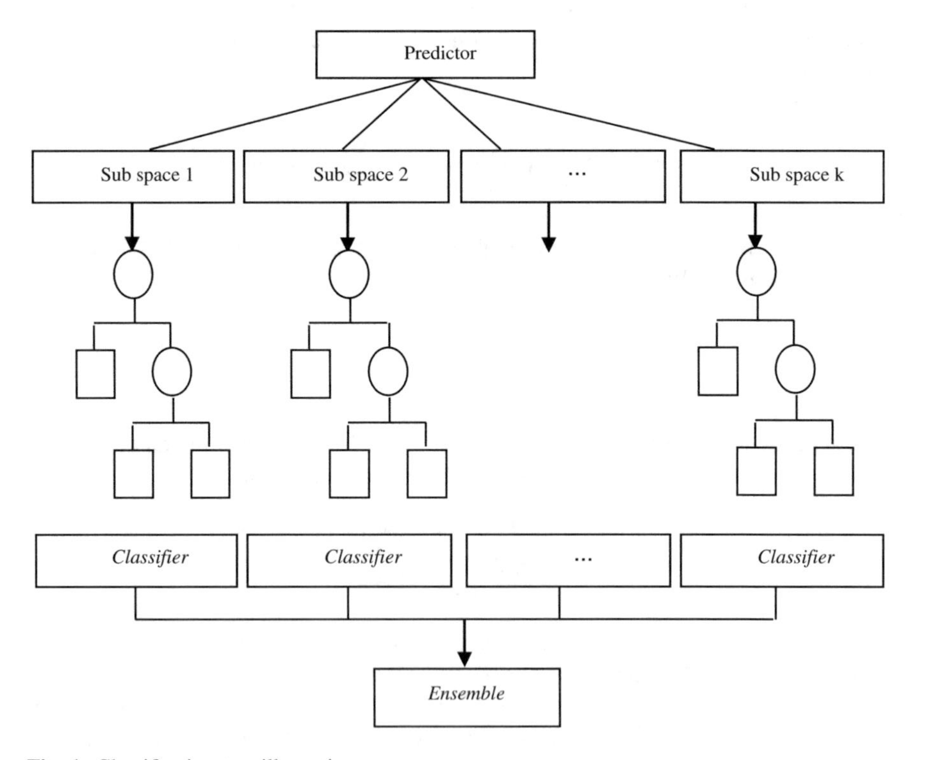

Fig. 1 Classification tree illustration

among groups of a classifier, θ is randomly partitioned into k sub spaces $(\theta_1, \theta, \ldots, \theta_k)$ with the same size. The CERP combines multiple logistic regression results to increase the accuracy from prediction using majority voting from group of a classifier or the average of prediction values. To increase further the CERP performance, therefore the majority voting is investigated between group of ensembles. Figure 1 depicts the CERP illustration where the predictors are allocated into k sub-spaces with the same number.

The CERP performance depends on how many predictors in a partition determined by the number of optimal partition. The optimum partition is determined with the following equation.

$$K = \frac{6 \times p}{n} \tag{6}$$

where p is number of predictor and n is number of observation. If n is larger than p, the optimal partition could be obtained by splitting data into $\binom{n}{i}$ where ii is a random integer less than n. Optimal partition is normally chosen from $k = \binom{n}{i}$ accuracy values which yield on the highest accuracy [7].

2.3 *Logistic Regression Ensemble (Lorens)*

Using CERP (*Classification by Ensemble from Random Partition*) algorithm, Lorens is developed using logistic regression model as the classifier. Lorens combines prediction value from multiple logistic regression models to increase the classification accuracy by averaging the ensemble. Prediction value is the combination of classifier (logistic regression model) which is averaged and further be classified into 0 and 1 using optimum threshold. Standard logistic regression uses threshold of 0.5 for binary response. Furthermore, the classification accuracy will be biased if the proportion class 1 and 0 is unbalanced. In order to balance the sensitivity and specificity, Lorens finds optimum threshold using the following formula.

$$Threshold = \frac{r + 0.5}{2} \tag{7}$$

where r is the positive response proportion in the observation.

Table 1 Two categorical binary response table classification

Actual	Prediction		Total
	1	2	
1	n_{11}	n_{12}	N_1
2	n_{21}	n_{22}	N_2
Total	N_1	N_2	N

Fawcet [12] proposed a method to calculate the classification accuracy namely apparent error rate (APER). APER value is the representation of misclassified sample proportion. This research uses two binary response categories to calculate the misclassification error from the following table (Table 1).

The APER value is calculated as follows,

$$APER(\%) = \frac{n_{12} + n_{21}}{N} \times 100\% \tag{8}$$

$$\text{Classification accuracy} = 1 - APER \tag{9}$$

Sensitivity value is obtained from $n_{11}/(n_{11}/n_{21})$ and *specificity* is calculated by $n_{22}/(n_{12} + n_{22})$. Meanwhile, the accuracy is calculated from $(n_{11} + n_{22})/(n_{11} + n_{12} + n_{21} + n_{22})$, where

n_{11}: Total of observation from class 1 predicted correctly as class 1
n_{21}: Total of observation from class 2 mispredicted as class 1
n_{12}: Total of observation from class 1 mispredicted as class 2
n_{22}: Total of observation from class 2 predicted correctly as class 2
N_1: Total of observation from class 1
N_2: Total of observation from class 2
N: Total of observation.

2.4 Cross Validation

The standard method to predict false rate is stratified 10-fold cross-validation. In 10-fold cross-validation, the data is partitioned into 10 parts with the same proportion, therefore nine of ten of the data are used as training data and one of ten of the data is used as testing. This procedure is repeated 10 times. Finally, 10 false predictions are averaged to obtain overall false prediction. Machine learning and data mining society argued that this evaluation procedure is the best approach and hence, 10 folds cross validation become a standard practical method. A study also showed that stratification increases the prediction accuracy [13].

3 Materials and Method

3.1 Data Source

The data analyzed in this research is a secondary data about enzyme structure with docking simulation classified into ligand and decoy. The data is obtained from DUD-E database which consists of 102 types of enzyme from each enzyme structure, however, this research only uses three types of enzyme.

3.2 Variable

This research uses 3 types of enzyme i.e. aofb, cah2 and hs90a. Predictor variable consists of constituent enzyme compound, while response variable consists of enzyme classification i.e. good inhibitor (ligand) and bad inhibitor (decoy) with the detail given in Table 2 as follows:

Table 2 Observation variable

Enzyme	Var.	Variable name
aofb	y(0)	*Decoy* (Bad inhibitor)
	y(1)	*Ligand* (Good inhibitor)
	x1	*A Log P*
	x2	*A Log P MR*
	x3	*A Log P98*
	⋮	⋮
	x69	*Molecular_3D_PolarSASA*
	x70	*Molecular_3D_SAVol*
cah2	y(0)	*Decoy* (Bad inhibitor)
	y(1)	*Ligand* (Good inhibitor)
	x1	*A Log P*
	x2	*A Log P MR*
	x3	*A Log P98*
		⋮
	x69	*Molecular_3D_PolarSASA*
	x70	*Molecular_3D_SAVol*
hs90a	y(0)	*Decoy* (Bad inhibitor)
	y(1)	*Ligand* (Good inhibitor)
	x1	*A Log P*
	x2	*A Log P MR*
	x3	*A Log P98*
		⋮
	x69	*Molecular_3D_Polar SASA*
	x70	*Molecular_3D_SAVol*

3.3 Analysis Steps

The step of data analysis in this research is described as follows:

1. Do the data partition into 90% training data and 10% testing data
2. Perform binary logistic regression analysis
3. Perform logistic regression ensemble analysis (Lorens) using hold out method
4. Perform logistic regression ensemble analysis using 10-fold cross validation evaluation
5. Compare classification accuracy from each combination of data training and testing
6. Compare the classification accuracy rate between binary logistic regression ensemble (Lorens)
7. Performance evaluation.

4 Results and Discussion

This section describes the results of analyzing three enzymes on DUD-E database to be classified by logistic regression model and logistic regression ensemble (Lorens). The analysis is conducted by splitting the data into training and testing dataset with two different compositions i.e. 90%:10% and 85%:15%.

4.1 Characteristic of Enzymes on DUD-E Database

The three enzymes that we analyzed are aofb (Monoamine Oxidase), cah2 (Carbonic Anhidrase) and hs90a (Heat shock protein). The aofb consists of 672 observations with 70 compounds, cah2 has 3340 observations with 71 compounds while hs90a consists of 500 observations with 69 compounds. Each observation is classified into two responses i.e. good inhibitor (ligand) and bad inhibitor (decoy). The percentage of positive response on aofb enzyme is 25% and the decoy is 75%. The same proportion holds also for the other two enzymes. Those three enzymes have five kinds of basic compounds i.e. A log P, specific structure, surface weight, energy and other compounds.

4.2 Binary Logistic Analysis for Enzymes on DUD-E Database

The logistic regression analysis is carried out by combining training and testing data as 90%:10%. A standard analytical approach such as simultaneous test followed by

Table 3 Fitness of simultaneous model

Enzyme	-2 log likelihood (G)	df
Aofb	673.2	12
cah2	3357.5	8
hs90a	507.2	4

partial test have been conducted. Forward step is applied to obtain the best model. Table 3 below summarizes the fitness of the simultaneous model for each enzyme.

In order to test the goodness of fit of the model, the statistic G is compared with Chi-square value. The statistic G for aofb is 673,2 which is greater than $\chi^2_{(12:0,05)} =$ 21,026 meaning that at least one variable is significant in the model. The Chi-square values for cah2 and hs90a are $\chi^2_{(8:0,05)} = 15{,}507$ and $\chi^2_{(4:0,05)} = 9{,}488$ consecutively, which are also greater than statistic G in Table 3 for the corresponding enzyme.

Table 4 Partial test of the logistic regression

Enzyme	Predictors	B	Wald	df	p-value	Exp(B)
aofb	Constant	−2,462	109,023	1	0,000	
	A Log P98 unknown	0,586	14,134	1	0,000	1,796
	Is chiral	0,513	7,329	1	0,007	1,671
	Average bond length	−1,041	21,925	1	0,000	0,353
	HBD Count	0,887	6,966	1	0,008	2,428
	Num explicit atoms	5,584	25,269	1	0,000	266,03
	Num aromatic bonds	3,604	5,333	1	0,021	36,745
	Num_Aromatic rings	−4,179	7,736	1	0,005	0,015
	Num chains	−7,727	47,876	1	0,000	0,000
	Num H acceptors	−1,454	22,905	1	0,000	0,234
	Num H donors	−1,535	12,214	1	0,000	0,215
	Num H donors Lipinski	−0,817	7,680	1	0,006	0,442
	Rad of gyration	1,649	20,894	1	0,000	5,204
cah2	Constant	−2,634	0,003	1	0,956	
	Molecular mass	2,346	168,725	1	0,000	10,445
	HBA count	1,369	147,752	1	0,000	3,930
	N Plus O Count	−2,032	158,830	1	0,000	0,131
	Num Rings6	1,316	179,631	1	0,000	3,729
	Num_Stereo Bonds	−1,116	101,976	1	0,000	0,328
	Num Aaom classes	−3,194	294,515	1	0,000	0,041
	Num_H_acceptors	−2,913	0,000	1	0,996	0,054
	Molecular 3D Polar SASA	2,684	352,392	1	0,000	14,639
hs90a	Constant	−12,469	0,000	1	0,992	
	Is chiral	−10,615	0,000	1	0,992	0,000
	Num_Rings 9 Plus	7,968	0,000	1	0,997	0,003
	Num_H acceptors	1,903	30,430	1	0,000	6,706
	Rad of gyration	−1,981	36,004	1	0,000	0,138

In order to investigate which variables are significant, a partial test is conducted by evaluating the results performed in Table 4.

The table listed the best model obtained by forward method. For aofb, there are 12 predictors in the model, obtained after 12 iterations. These predictors are *A Log P98 Unknown, IsChiral, LogD, HBD Count, Num Explicit Atoms, Num AromaticBonds, Num AromaticRings, Num Chains, Num H Acceptors, Num H Donors, Num H Donors Lipinski dan Rad Of Gyration* and all of them are significant as the *p-value* $< \alpha$. The logistic regression model for cah2 indicates that there are 8 predictors i.e. *Molecular Mas, HBA Count, N Plus O Count, Num Rings6, Num_Stereo Bonds, Num Atom Classes, Num_H_Acceptors* and *Molecular 3D Polar SASA*. Among these 8 predictors, H_acceptors is not significant in the model as the *p-value* $> \alpha$. Meanwhile, only four predictors in the model for hs90a, where two of them are significant.

The Exp(B) indicates odd ratio, which can be interpreted as the tendency of a specific predictor to be a ligand or decoy. Odds ratio greater than one observed for *A Log P98 Unknown, IsChiral, HBD Count, Num ExplicitAtoms, Num Aromatic and Rad Of Gyration* indicates that these predictors have positive influence to the model, which also means that it tends to form a good ligand. In a practical work, as these predictors are compounds to form the enzyme, therefore they have to be treated simultaneously.

The logistic regression model for each enzymes involving selected predictors can be written as follow:

$$g(x) = \ln\left(\frac{\pi(x)}{1-\pi(x)}\right) = -2,462 + 0,58x_4 + 0,513x_7 - 1,041x_8 + 0,887x_{15} + 5,584x_{21}$$
$$+ 3,604x_{29} - 4,179x_{32} - 7,727x_{41} - 1,454x_{51} - 1,535x_{52} - 0,817x_{54} + 1,649x_{67}$$
$$g(x) = \ln\left(\frac{\pi(x)}{1-\pi(x)}\right) = -263 + 2,34x_{11} + 1,37x_{14} - 2,03x_{16}$$
$$+ 1,31x_{37} - 1,12x_{44} - 3,19x_{46} - 2,91x_{51} + 2,68x_{70}$$
$$g(x) = \left(\frac{\pi(x)}{1-\pi(x)}\right) = -12,47 - 10,61x_7 + 7,37x_{39} + 1,9x_{50} - 1,98x_{66}$$

The models above are used to classify the inhibitors.

The goodness of fit of the model can be validated through the values listed in Table 5.

From the table, the p-values of aofb and cah2 are less than 0.05 indicating that the models do not fit well, while the logistic regression model for hs90a fits well as

Table 5 Goodness of fit test

Enzyme	Chi-square	df	p-value	Nagelkerke R square
Aofb	26,087	8	0,001	70,5
cah2	45,022	8	0,000	73,9
hs90a	1,748	8	0,988	83

Table 6 Enzyme classification using logistic regression

Enzyme		Obs.	Prediction		Total	1-APER
			0	1		
Training data	*aofb*	0	436	21	457	90,4
		1	37	111	148	
	cah	0	2169	96	2265	91,7
		1	154	587	741	
	hs90a	0	319	18	337	94
		1	9	104	113	
Testing data	*aofb*	0	46	5	51	91,04
		1	1	15	16	
	cah	0	230	19	249	91,3
		1	10	75	85	
	hs90a	0	38	0	38	100
		1	0	12	12	

the P-value is less than 0.05. The Nagelkerke R Square can be interpreted as the same way to interpret coefficient of determination in multiple linear regression. The value 83 for hs90a indicates that 83% of the variance of hs90a can be explained by the regression model involving four predictors. The results for aofb and cah2 can be seen as a misleading case of logistic regression because the models are supposed to be the best one, but they do not fit well. This is one of the weaknesses of logistic regression applied to high dimensional data. Regardless of this fact, the classification accuracy in Table 6 is calculated from the models.

We see that the accuracy of the logistic regression model for all enzymes is greater than 90%, which is very good. The accuracy for hs90a on testing data reached 100%.

4.3 Analysis Logistic Regression Ensembles (Lorens) to Classify DUD-E Database

The logistic regression ensembles (Lorens) analysis is an ensemble method developed to acquire high classification result on high dimensional data. The Lorens analysis in this study is conducted by the splitting dataset into training and testing data with the composition of 90%:10%. One of the differences between standard logistic regression with Lorens is about the threshold, in which the Lorens assigns optimum threshold depending on the proportion of class response. Table 7 provides the optimum threshold used for classification with Lorens.

The threshold for those three different cases are slightly different, but they are significantly different with the threshold used in logistic regression (0.5). The classification accuracy obtained from the majority voting in Lorens are given in

Table 7 Threshold optimum

Enzyme	Ligand	Decoy	Threshold
aofb	148	457	0,3723
cah2	741	2265	0,3732
hs90a	113	337	0,3755

Table 8 Random Partition

Partition	Variable	Ensemble									
		1	2	3	4	5	6	7	8	9	10
2	*A Log P*	2	1	1	1	2	1	1	1	1	2
	A Log P MR	1	2	1	1	1	1	1	2	1	2
	A Log P98	2	1	1	1	1	1	2	1	2	2
	⋮	⋮	⋮	⋮	⋮	⋮	⋮	⋮	⋮	⋮	⋮
	Molecular_3D_SAVol	2	2	2	1	2	2	2	2	1	2
3	*A Log P*	2	3	1	2	1	1	1	3	3	2
	A Log P MR	2	3	3	3	3	3	1	3	1	3
	A Log P98	1	3	1	3	1	3	3	2	2	1
	⋮	⋮	⋮	⋮	⋮	⋮	⋮	⋮	⋮	⋮	⋮
	Molecular_3D_SAVol	3	3	1	2	3	2	1	2	3	3
4	*A Log P*	3	4	4	3	2	4	4	1	2	1
	A Log P MR	1	3	3	3	4	1	4	4	3	3
	A Log P98	3	2	4	4	4	3	1	2	2	1
	⋮	⋮	⋮	⋮	⋮	⋮	⋮	⋮	⋮	⋮	⋮
	Molecular_3D_SAVol	1	1	2	2	2	1	4	2	2	3
10	*A Log P*	3	9	9	2	8	2	6	3	9	1
	A Log P MR	9	3	10	2	4	3	3	4	9	4
	A Log P98	1	7	5	6	7	3	10	10	3	5
	⋮	⋮	⋮	⋮	⋮	⋮	⋮	⋮	⋮	⋮	⋮
	Molecular_3D_SAVol	1	4	6	6	4	8	6	9	7	5

Table 8. The classification by Lorens did a partition to the variables into 9 sub-spaces, 4 sub-spaces and 5 sub-spaces for aofb, cah2 and hs90a consecutively. Note also that Lorens uses all variables or predictors in the procedure, which is different with standard logistic regression.

4.4 Random Partition and Model Building

The partition of random variables is conducted by minimizing the correlation on an ensemble where the random partition is chosen with same distribution so that it is assumed that there is no bias within each partition. The steps of random partition, model building and calculating the classification accuracy are the same for those three enzymes. The variables in each sub-space are assigned randomly with the same number of variables in each sub-space. The steps of random partition and model building with 10 ensembles for aofb enzyme is described in Table 8.

From the table, we see that if we determine the number of partition equals to 2, thus 70 predictors in aofb enzyme will be allocated into 2 sub-spaces, i.e. 1st partition and 2nd partition. In each ensemble, there will be 2 models and hence, in total, we have 20 models for the whole 10 ensembles. In the first ensemble, the A Log P variable is allocated to the second partition, the A Log P MR is allocated to the first partition, the A Log P98 is allocated to the second partition, and so forth. Similar procedure is applied to the partition number of 4, 5, 6, 7, 8, 9 and 10. After each variable has been allocated according to its partition, logistic regression is built from each partition in an ensemble. For instance, given that we set 2 partitions, the logistic models in 10 ensembles can be seen in Table 9.

Table 9 shows the models that are formed from the allocation of predictors into 2 sub-spaces (partition 1 and partition 2). In partition 1 and partition 2, the models involve of 35 predictors. In total we have 20 models with 2 partitions, 30 models for 3 partitions, 40 models for 4 partitions, 50 models for 5 partitions and so on. These logistic models are used to classify the data.

After the models have been built in each partition, the classification is done for 90% of training data and 10% testing data. The classification is conducted by substitute the training and testing data into the models on the partition in an ensemble. The substitution results on training and testing data in each model are averaged to obtain a probability. These probability values are compared with the threshold listed in Table 7. If the probability exceeds the threshold, then the data is classified into good inhibitor (ligand) and otherwise. The same procedure is applied to other ensembles. The final classification is obtained by majority voting for each observation. If the majority of 10 ensemble classify the observation into good inhibitor (ligand), then the decision is to classify the observation into ligand. Meanwhile, if the majority of ensembles classify the observation into bad inhibitor (decoy), then the observation is a decoy.

The following table listed the classification accuracy for aofb. The classification results for other enzymes are omitted for the sake of space (Table 10).

From the table, the optimum partition is 9 by considering classification accuracy for testing data. The optimum partition is chosen by looking at the significant increment resulted from increasing 1 partition. Summary of the optimum accuracy are performed in Table 11.

Comparing the values in Tables 6 and 11 leads to the conclusion that Lorens outperforms logistic regression in most cases, with an exception for aofb enzyme. This conclusion holds for the testing dataset as well.

Table 9 Logistic regression model

Ensemble	Logistic regression model partition 1	Logistic regression model partition 2
1	$g(x) = ln\left(\frac{\pi(x)}{1-\pi(x)}\right) = -2,8 + 2,47x_1 + 1,2x_3 + 0,47x_4 + \cdots - 0,799x_{69}$	$g(x) = ln\left(\frac{\pi(x)}{1-\pi(x)}\right) = -3,9 + 9,57x_2 - 1,16x_5 + 302,4x_6 + \cdots + 2,26x_{70}$
2	$g(x) = ln\left(\frac{\pi(x)}{1-\pi(x)}\right) = -3,7 - 1,06x_1 + 2,45x_3 + 1,05x_6 + \cdots - 1,82x_{68}$	$g(x) = ln\left(\frac{\pi(x)}{1-\pi(x)}\right) = -3,15 - 3,56x_2 + 0,75x_4 + 0,04x_5 + \cdots + 3,82x_{70}$
3	$g(x) = ln\left(\frac{\pi(x)}{1-\pi(x)}\right) = -3,7 + 0,9x_4 + 346,6x_6 + 0,28x_9 + \cdots + 1,2x_{69}$	$g(x) = ln\left(\frac{\pi(x)}{1-\pi(x)}\right) = -2,9 - 9,81x_1 + 2,32x_2 + 1,5x_3 + \cdots + 3,35x_{70}$
⋮		
10	$g(x) = ln\left(\frac{\pi(x)}{1-\pi(x)}\right) = -2,95 - 0,89x_1 + 2,27x_2 + 1,97x_3 + \cdots + 14,7x_{68}$	$g(x) = ln\left(\frac{\pi(x)}{1-\pi(x)}\right) = -2,95 + 0,83x_4 + 0,83x_5 + \cdots + 0,64x_{69}$

Table 10 Classification accuracy for aofb

Partition	Training			Testing		
	Sensitivity	Specificity	Accuracy	Sensitivity	Specificity	Accuracy
2	79,74	95,07	91,07	84,21	91,67	89,55
3	76,13	93,33	88,9	83,33	89,80	88,06
4	76,35	92,34	88,4	77,78	87,76	85,07
5	74,34	92,27	87,76	78,95	89,58	86,57
6	71,62	90,81	86,11	77,78	87,76	85,07
7	71,23	90,41	85,78	88,24	90,00	89,55
8	70,92	89,65	85,28	83,33	89,80	88,06
9*	70,71	89,46	85,12	88,24	90,00	89,55
10	72,79	89,55	85,78	82,35	88,00	86,57

**optimum partition*

Table 11 Classification accuracy using Lorens

Data	Enzyme	Sensitivity	Specificity	Accuracy
Training data	*aofb*	70,71	89,46	85,12
	cah2	86,98	96,74	94,24
	hs90a	88,10	99,38	96,22
Testing data	*aofb*	88,24	90,00	89,55
	cah2	89,77	93,90	92,81
	hs90a	100	100	100

4.5 Classification by Cross Validation

The classification performance of Lorens can be evaluated also by Cross Validation which is expected to improve the accuracy. Cross Validation is also a way to conduct sensitivity analysis. In this research, the Cross Validation is applied to several different partition numbers i.e. 2, 3, 4, 5, 6, 7, 8, 9, and 10. The number of ensemble is set to be 10 ensembles, where with 10 fold Cross Validation means that each fold will consist of 10 models. With 2 partitions will result on 200 models, 3 partitions will yield on 300 models, and 10 partitions will have 1000 models. The 10 folds Cross Validation process will divide the observations into 10 groups with same number, and furthermore the first group will be used as training and the rests

Table 12 Optimum threshold by cross validation

Fold	Optimum threshold	Fold	Optimum threshold
1	0,375	6	0,376
2	0,375	7	0,371
3	0,376	8	0,378
4	0,376	9	0,375
5	0,375	10	0,373

Table 13 Classification performance by cross validation for aofb

Partition	Sensitivity	Specificity	Accuracy
2	73,89	92,89	87,8
3	73,71	92,15	87,35
4*	72,83	91,58	86,76
5	71,10	90,98	85,86
6	71,18	90,64	85,71
7	70,30	89,74	84,97
8	69,64	89,74	84,82
9	70,00	89,06	84,52
10	70,89	89,11	84,82

**chosen optimum partition*

Table 14 Classification evaluation using cross validation

Enzyme	Sensitivity	Specificity	Accuracy
aofb	72,83	91,58	86,76
cah2	84,34	94,62	92,07
hs90a	89,86	99,72	97,00

are testing dataset. The classification process is carried out with the same procedure as standard Lorens. In each fold, we obtain an optimum threshold as listed in Table 12 (case of aofb enzyme partitioned into 2 subspaces).

With threshold of 0.375 means that observation with probability greater than 0.375 will be classified into good inhibitor (ligand) and observation with probability below 0.375 will be classified into decoy. The majority voting in each ensemble is still be applied in this case. After the models for each fold are formed, they are substituted into training data to find the classification accuracy in each fold. Furthermore, the classification accuracy in each fold is fused to obtain classification performance for each enzyme. The classification performance by Cross Validation for aofb with different partition numbers are performed in Table 13.

Based on the table, the optimum partition is 4 subspaces chosen by considering the significant decrease of the sensitivity, specificity and accuracy measures. Summary of the optimum (Table 14).

We see that those three criterias showed a good performance of Lorens. We noticed also that the performance of using hold out and cross validation are similar, indicating that Lorens has consistent good performance in classifying the compounds into inhibitor.

5 Conclusion

Based on the results presented in the previous section, we conclude that Lorens outperforms standard logistic regression model. Examining those three enzymes which have characteristic of high dimensional data showed that the logistic regression has a weakness i.e. inconsistency in the model building between the best model and goodness of fit test. Meanwhile, classification using Lorens is very flexible in the sense that it uses all predictors without doing a variable selection. Moreover, no testing hypothesis is necessary to be carried out within Lorens' application. Another issue is about the threshold choice, where Lorens offer a rational way to assign the threshold. Using threshold 0.5 for the case of imbalance response rate (e.g. 75%:25% such as in this study) violates the basic assumption of standard logistic regression. Finding the dominant predictors through Lorens is a subject of future research.

Acknowledgements The authors gratefully acknowledge the financial support from The Ministry of Research, Technology and Higher Education Indonesia through Research Grant for International Collaboration and Scientific Publication. Moreover, the authors would like to thank also to the First EAI International Conference on Computer Science and Engineering, NOVEMBER 11–12, 2016, PENANG, MALAYSIA as well as the anonymous refrees.

References

1. DiMasi, J. A., Hansen, R. W., & Grabowski, H. G. (2003). The price of innovation: new estimates of drug development costs. *Journal of Health Economics, 22*(2), 151–185.
2. Jenwitheesuk, E. H. (2008). Novel paradigms for drug discovery computational multitarget screening. *Trends in Pharmacological Sciences, 29,* 62–71.
3. Okada, M., Ohwada, H., & Aoki, S. (2013). Docking score calculation using machine learning with an enhanced inhibitor database. *Bioinformatics and Computational Biology*, 1.
4. Cortes, C., & Vapnik, V. (1995). Support vector networks. *Machine Learning, 20,* 273.
5. Lim, N., Ahn, H., Moon, H., & Chen, J. J. (2010). Classification high dimensional data with ensemble of logistic regression models. *Journal of Biopharmaceutical Statistics*, 20, 160–17.
6. Pombo, N., Garcia, N., Bousson, K., & Felizardo, V. (2015). Machine learning approaches to automated medical decision support systems. In *Handbook of Research on Artificial Intelligence Techniques and Algorithms. Chapter, 6,* 183–203.
7. Lim, N. (2007). Classification by ensembles from random partitions using logistic models. In *Applied Mathematics and Statistics*. Stony Brook University.
8. Kuswanto, H., Asfihani, A., Sarumaha, Y., & Ohwada, H. (2015). Logistic regression ensemble for predicting customer defection with very large sample size. *Procedia Computer Science, 72,* 86–93.
9. Lee, K., Ahn, H., Moon, H., Kodell, R. L., & Chen, J. J. (2013). Multinomial logistic regression ensembles. *Journal of Biopharmaceutical Statistics, 23*(3), 681–694.
10. Hosmer, Watson D., & Lemeshow, S. (1995). *Applied logistic regression*. New York: Wiley.
11. Agresti, A. (1990). *Categorical Data Analysis*. New York: Wiley.
12. Fawcett, T. (2006). An introduction to ROC analysis. *Pattern Recognition Letters, 27*(8), 861–874.
13. Witten, I. H., Frank, E., & Hall, M. A. (2011). *Data mining: Practical machine learning tools and techniques* (3rd ed.). Burlington: Morgan Kaufmann.
14. Ahn, H., Moon, H., Fazzari, M. J., Lim, N., Chen, J. J., & Kodell, R. L. (2006). Classification by ensemble from random partitions of high-dimensional data. *Computational Statistic and Data Analysis*, 4–6.

Consolidation of Host-Based Mobility Management Protocols with Wireless Mesh Network

Wei Siang Hoh, Bi-Lynn Ong, R. Badlishah Ahmad and Hasnah Ahmad

Abstract The number of mobile devices increases exponentially and it becomes the trends and needs of human. Presently, the network infrastructures have the coverage issues in specific areas such as underground facilities. The cost of upgrade causes high budget and it is less profitable based on business and market point of view. Somehow, in coverage areas, the mobile devices still operate by the traditional Mobile Internet Protocol version 6 (MIPv6) for mobility management in inter network scenario. MIPv6 operation mechanisms frequently trigger the signaling overhead problem. Thus, these increase the end-to-end delay and lower the network throughput performance. Having known this issues, we consolidate MIPv6, HMIPv6, FMIPv6 and FHMIPv6 with Wireless Mesh Network (WMN) into one environment. The reason of constructing WMN is because WMN caters rural areas. We identify, analyze, and compare the performance of Host-Based mobility management protocols integrate with WMN in terms of latency, throughput and packet loss ratio. Finally, it is proven that the design and development of FHMIPv6 with WMN performs better as compared to the others Mobile Internet Protocols over the Internet using NS-2 Network Simulation software. Having implemented the FHMIPv6 with WMN, the MAP mechanism allow mobile node does not need to inform the highest hierarchical node upon the handover process. For fast handover mechanism, when the mobile node senses lower signal strength, mobile node advertises to the neighbour network for the need to attach to the new higher signal strength access point. Mobile node informs the new access point of the need to change to the new access point before the process of handover. Thus, this two

W.S. Hoh (✉) · B.-L. Ong · R.B. Ahmad · H. Ahmad
School of Computer and Communication Engineering, Universiti Malaysia Perlis (UniMAP), 02600 Arau, Perlis, Malaysia
e-mail: weisiangkelvin1990@gmail.com

B.-L. Ong
e-mail: drlynn@unimap.edu.my

R.B. Ahmad
e-mail: badli@unimap.edu.my

I. Zelinka et al. (eds.), *Innovative Computing, Optimization and Its Applications*, Studies in Computational Intelligence 741,
https://doi.org/10.1007/978-3-319-66984-7_7

mechanism can reduce the handover latency and increase the network throughput. In future, in intra network scenario also can implement FHMIPv6 to improve the network performance.

Keywords Host-Based mobility management protocols · MIPv6 · HMIPv6 FMIPv6 · FHMIPv6 · Wireless mesh network (WMN)

1 Introduction

The Internet consolidated itself as a very powerful platform that has forever changed the way human communicates and their behavior. The mobile communications technology had made it possible for much greater reach of the Internet and increase the number of Internet users through the mobile devices in wireless environment. The dependency of human toward the Internet has raise the usage of Internet which causes congestion and intermittent connection issues that rise rapidly. Present telecommunication infrastructures have some coverage issues as it cannot cover specific areas in its coverage area such as alpine area, underground facilities and forest region. Furthermore, people from rural areas have been receiving limited coverage from entire major telecommunication service providers. Cost of upgradation of this phase can cause a fortune and it is less profitable based on business and market point of view.

Finding the ways to solve the congestion and intermittent connection problems, Internet Engineering Task Force (IETF) introduces the mobility management protocols. The mobility management is the essential part for mobile devices that are automatically connect to Internet while simultaneously can roam freely without disturbing the communication. Mobility management provides routing support and permits Internet Protocol (IP) nodes using either IPv4 or IPv6 to seamlessly roam among IP subnetworks and media types. Host-Based mobility management protocols are in IPv6 and it is also one of the main parts of mobility management protocols. It includes Mobile Internet Protocol version 6 (MIPv6) [1] and its enhancement such as Fast Handover Mobile Internet Protocol version 6 (FMIPv6) [2], Hierarchical Mobile Internet Protocol version 6 (HMIPv6) [3] and Fast Handover for Hierarchical Mobile Internet Protocol version 6 (FHMIPv6) [4].

The Wireless Mesh Network (WMN) is selected as wireless environment that implemented all the Host-Based mobility management protocols. The WMN can be connected to wireless networks such as Worldwide Interoperability for Microwave Access (WiMAX), generic Wireless Fidelity (Wi-Fi); cellular and sensor networks. Third (3G) and Fourth (4G) Generation networks include all Internet Protocol (IP) which are wired and wireless networks interworks together as heterogeneous networks [5]. However, the challenge is to connect to Host-Based mobility management. Host-Based Mobility Management protocols rely on the good performance of an infrastructure-based network [6]. However, a typical WMN topology tends to be an unplanned graphs and routes of it dynamically changes [7]. Mobility management provides an undisrupted support of real-time and non-real-time services to mobile network users. Additionally, mobility management also facilitates the maintenance of connections for users on the move when they change their points of attachment from one access point (AP) to another. Host-Based mobility allows a Mobile Node (MN) to change its point of attachment to the network, without interrupting IP packet delivery to or from the node [8]. The current location of all the MNs in the network is maintained by Access Network Procedures [9]. On the other hand, the WMN primary advantages lie in its inherent fault tolerance against network failure, broadband capability and simplicity of setting up network. The WMN is reliable and offers redundancy [10]. When one node of WMN is failure, the rest of the nodes can still able to communicate with each other, directly or through one or more intermediate nodes.

In this research paper, all the Host-Based Mobility Management Protocols are investigated firmly on Wireless Mesh Network topology environment. MIPv6, HMIPv6, FMIPv6 and FHMIPv6 are developed and analyzed in Wireless Mesh Network (WMN) environment which are considering the performance parameters: Packet Delivery Ratio (PDR), delay/latency and throughput. The Wireless Mesh Network (WMN) topology are developed by using network simulation software and the result obtained are analyzed to agree the best mobility management protocols to handle the inter domain mobility with Wireless Mesh Network (WMN) topology.

2 Related Works

Lee et al. [11] had investigated a simulation research on analytical comparison of IPv6 mobility management protocols handover scheme. The researchers compared the Host-Based mobility management protocols and Network-Based mobility management protocols to identify the optimized routing protocol for mobile network. The Host-Based mobility management protocols include Mobile IPv6 and its extensions such as Fast Mobile IPv6 and Hierarchical Mobile IPv6 while Network-Based mobility management protocols include Proxy Mobile IPv6 (PMIPv6) and Fast Proxy Mobile IPv6 (FPMIPv6). These mobility management protocols have been standardized. The existing IPv6 mobility management protocols are developed by the IETF and have been analyzed and compared in terms of handover latency, handover blocking probability, and packet loss. The conducted analysis results can be used to identify each mobility management protocol's characteristic and performance indicators. The results obtained are used to facilitate decision making in development a new mobility management protocol.

Vasu et al. [12], had investigated a survey and comparative analysis for MIPv6 protocols. The researchers had performed various mobility management protocols in terms of handover latency and the number of hops is needed to evaluate these protocols. The IPv6 mobility management protocols such as MIPv6, FMIPv6 (Reactive), FMIPv6 (Predictive), HMIPv6, PMIPv6, FPMIPv6 (Reactive), and FPMIPv6 (Predictive) are analyzed and compared in terms of average hop delay, wireless link delay, wired part delay, binding update and registration delay.

PMIPv6 and FPMIPv6 have been compared with the Host-Based mobility management protocols to make a decision that suits the future networks. The conclusion that the authors made among these protocols: reactive mode protocols performs better in terms of delay compare to predictive based protocols. The performance are measured in terms of delay during AP to MAG/AR, and binding update/registration components. Whereas predictive based protocols performs better performance in term wireless link delay for faster radio access technologies and performs rather slower performance for slower radio access technologies.

Sun et al. [13], had investigated the mobility management techniques for next generation wireless networks. The researcher had performed macro and micro mobility protocols in terms of handover performance. The macro and micro mobility protocols such as Mobile Internet Protocol version 6 (MIPv6), Fast Handover Mobile Internet Protocol version 6 (FMIPv6), Hierarchical Mobile Internet Protocol version 6 (HMIPv6) and Fast Handover for Hierarchical Internet Protocol version 6 (FHMIPv6) and Proxy Mobile Internet Protocol version 6 (PMIPv6). These protocols are analyzed and compared in term of handover latency. The conclusion that the authors made that the best handover latency is achieved by FHMIPv6. The result is signalling load reduction, improvement in latency and less packet losses apart from aiding the handover process.

Sko et al. [14] had investigated a simulation research based on analytical comparison of Mobile IPv6 handover schemes. The authors have done a comparison for four most common handover schemes in term of the cost of packet delivery of Mobile IPv6 that is MIPv6, FMIPv6, HMIPv6 and FHMIPv6. The access network for using in this research is based on IEEE 802.11b and the transport core network is Ethernet—IEEE 802.3 100BaseT. The researchers used analytical methods for the comparison. The researchers used network simulation software to run the simulation. The researchers have taken in count these two performance matrices during the comparison that is, the handover cost and handover latency. The authors concluded that the hierarchical Internet Protocol consists of HMIPv6 and FHMIPv6 performed well comparing to other Internet Protocol in terms of packet delivery cost. The problem of the Intra network issue is not addressed detail in this research. As conclusion, the WMN can enhance the Mobile Internet Protocols to perform better in handover latency either in Intra condition or Inter condition compares to basic wireless network like 802.11b.

Makaya et al. [15] had investigated an analytical framework for performance evaluation of IPv6-based mobility management protocols. The researchers have developed an analytical framework for performance analysis of IPv6 mobility management protocols. MIPv6, FMIPv6, HMIPv6, and a combination of FMIPv6 and HMIPv6 have been compared and evaluated in terms of signaling cost, binding refresh cost, packet delivery cost, required buffer space, and handover latency. The researchers presented the effect of subnet residence time, packet arrival rate, and wireless link delay to the different IPv6 mobility management protocols. The packet delivery ratio, throughput and the delay are considered in this research.

Fu et al. [16] had done an investigation on handover latency in SIGMA, FMIPv6, HMIPv6, and FHMIPv6 protocols. The researchers had designed a new scheme called Seamless IP diversity based Generalized Mobility Architecture (SIGMA) which can provide low latency and low packet loss mobility. The researcher compared the handover latency of SIGMA with FMIPv6, HMIPv6, and FHMIPv6. The researches have taken in count various parameters such as layer 2 handover/setup latency, layer 2 beacon period, mobile host moving speed, and IP address resolution latency. The software Network Simulator version 2 (NS-2) is used to run the simulation. The researcher concluded handover latency of SIGMA is lower than that of MIPv6 enhancements under various simulated scenarios. SIGMA could also seamlessly handle relatively high speed movement. Having studied previous research, this research is not considering the handover latency of SIGMA but it is considering the latency for network-based and host-based mobility managements. The reason is because cellular network is not one of the considerations of this research.

Zhang et al. [17] had done research on hierarchical mobile IPv6 with fast handover. In the study, the authors have compared four types of mobile routing protocol to identify the best routing protocol for mobile network, that are MIPv6, FMIPv6, HMIPv6 and FHMIPv6. NS-2 has been used to conduct the simulations. Performance metrics that have been taken in count are handover delay and jitter. This research review has been useful to our research as the authors mentioned about performance metrics jitter and delay which is similar to our initial research plan to consider those performance metrics. Delay is crucial to any performance investigation as it is the most demanding performance metric in the field of networking as speed is everything in networking. At the end of research, it's been concluded that FHMIPv6 performed extremely better compared to other MIP. The reason of low performance of the other MIP protocols like MIPv6, FMIPv6 and HMIPv6 are not presented in the research. Thus, in this research, investigation is carried out to find the reason of low performance.

Murtadha et al. [18] has proposed a fully distributed mobility management scheme for future heterogeneous wireless network. The researchers develop Distributed Mobility Management (DMM) scheme based on the PMIPv6 and compare with the Centralized Mobility Management (CMM) scheme PMIPv6. The proposed approach removes any central anchor node and disable the signaling between MN and the access networks. The performance metrics include the signaling cost, handover latency and packet loss. The DMM scheme reduced the signaling cost, handover latency and packet loss which compare to the CMM scheme PMIPv6. The main advantages of the DMM scheme are effectively reduced the processing requirement and the power consumption of the MN.

3 Terminology

3.1 Host-Based Mobility Management Protocols

Host-Based mobility management protocols include Mobile Internet Protocol version 6 (MIPv6), and its enhancement such as Fast Handover Mobile Internet

Protocol version 6 (FMIPv6), Hierarchical Mobile Internet Protocol version 6 (HMIPv6) and Fast Handover for Hierarchical Mobile Internet Protocol version 6 (FHMIPv6). Host-Based mobility management protocols are deployable in wireless mobile communication infrastructures, communication service providers and standards development organizations [19]. These mobility management protocols have identified that such conventional solutions for mobility service are not suitable; in particular, for telecommunication service. The reason is because the mobile node (MN) is required to perform mobility functionalities at its network protocol stack inside, and thus, modifications or upgrades of the MN are needed. It obviously increases the operation expenses and complexity for the MN. Hence, the extension of Mobile IPv6 (MIPv6) had been introduced to overcome the handover latency problem. The extension of MIPv6 includes HMIPv6, FMIPv6 and FHMIPv6.

Mobile Internet Protocol Version 6 (MIPv6)

Internet Engineering Task Force (IETF) brought into use of Mobile Internet Protocol version 6 (MIPv6) to allow mobile nodes (MN) to be reachable and maintained on-going connection while changing location within topology without changing the allocated IP address [20]. The operation of MIPv6 is as illustrated in Fig. 1.

The operation begins as MN detects movement to a Foreign Agent (FA) and auto-configures itself with a New Care of Address (NCoA) using either stateful or stateless method [21]. MN sends Binding Update (BU) to its Home Agent (HA) to notify the new address and HA returns back Binding Acknowledgment (BAck). Then, all packets are tunneled to MN's NCoA with the help of HA as HA encapsulates packets and sends to MN's NCoA and MN decapsulates the packets received from HA. However, the tunnelling also causes drawback due to the long path between mobile node (MN) and correspondent node (CN). This leads the air interface-traffic overhead high and it's also causes high tunnelling overhead at MN. Hence, an additional mode for MIPv6 is Route Optimization (RO). It allows the datagrams to be delivered using shortest path. This process requires MN to register its current Binding to CN. This allows CN to triangulate datagrams to be delivered

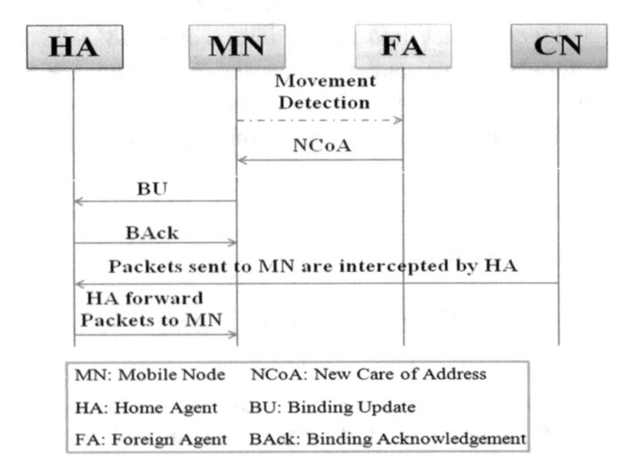

Fig. 1 MIPv6 flow diagram

to MN without concerning HA. This measure deceases the signalling overhead and handover latency between MN and CN and it also reduces congestion at MN's HA and Home Link.

Hierarchical Mobile Internet Protocol Version 6 (HMIPv6)

The Internet Engineering Task Force (IETF) has introduced the HMIPv6 based on its predecessor MIPv6 and has implemented new technologies to it to ensure the increment in the performance of mobile networking. One of the latest features in HMIPv6 is Mobility Anchor Point (MAP). The introduction of MAP in HMIPv6 has improved the handover latency and reduced the amount of signaling between the Mobile Node (MN), its Correspondent Nodes (CNs), and its Home Agent (HA) [22]. The Fig. 2 illustrates the process flow of HMIPv6.

Operation of HMIPv6 involves in three phases that are MAP discovery, MAP registration and packet forwarding. The first step is the MAP discovery procedure which is to obtain a successful connection. Normally, MAP is a router which is located in a network that is visited by the MN. When the visit point router starts to advertise, the discovery begins. There are two discovery options in HMIPv6 which are Static Configuration and Dynamic MAP Discovery.

The HMIPv6 operation starts when the MN obtains its MAP IP address through the discovery, then it calculates the distance of MAP from the current AR. This is to verify the connection strength between AR and MAP. The second step is MAP registration that is to register the MN to MAP. Visit point router is assigned as Regional Care-of-Address (RCoA) which is obtained by the MN from the visited network. RCoA is formed using prefix advertised by visit point router. MN assigns as On Link Care-of-Address (LCoA). LCoA is configured on a MN's interface based on the prefix advertised by its default router (AR). After that, MN creates a binding between RCoA and LCoA at MAP. Then, MN sends Local Binding Update

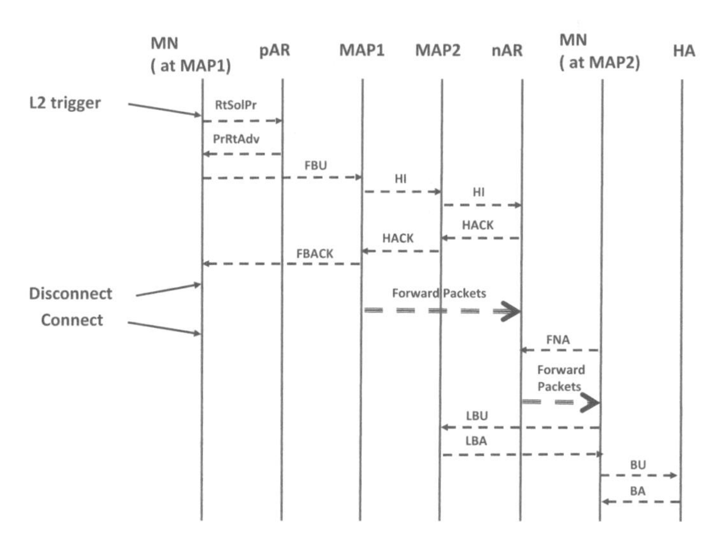

Fig. 2 HMIPv6 flow diagram

(LBU) to newly discovered MAP. Next, HA performs Duplicate Address Detection (DAD) and updates the binding cache. Then, MAP sends Binding Acknowledgement (BAck) to MN. After this process, MN sends binding update to its HA and active CN's with RCoA as its source address and HA. CN's address is set as destination address. The final step is to forward packets which are performed after the MAP discovery and MAP registration. A Bi-directional tunnel between MAP and MN is established to allow the packet forwarding. All packets sent by the MN are tunnelled through MAP and also all packets destined to the MN's RCoA are intercepted by MAP and tunnelled to MN's LCoA.

As simplified to figure out the HMIPv6 mechanism, when an MN moves in MAP domain, MAP acts as HA of the MN locally and the MN registers only to its location information to MAP. Therefore, HMIPv6 is smaller location update cost than MIPv6 which updates an MN's location information to the HA and CNs. Thus, HMIPv6 is more efficient compared to the previous MIPv6 in terms of the handover and broadcasting. The technology also improves the inter network connection and smoothen the users' intermittent connections.

Fast Handover Mobile Internet Protocol Version 6 (FMIPv6)

FMIPv6 is another initiative by the Internet Engineering Task Force (IETF) to improve the mobile network for the mobile users. The FMIPv6 is also designed based on the previous version of MIPv6. The FMIPv6 handover schemes introduce three signaling messages which involved in the anticipation phase that are Router Solicitation for Proxy Advertisement (RtSolPr), Proxy Router Advertisement (PrRtAdv) and Fast Binding Update (FBU) [23]. The Fig. 3 shown the FMIPv6 Predictive Fast Handover flow diagram. The Fig. 4 shown the FMIPv6 Reactive Fast Handover flow diagram.

For FMIPv6 predictive fast handover scheme, the MN obtains the new CoA before actual movement to new subnet through newly defined messages: Router Solicitation for Proxy Advertisement (RtSolPr) and Proxy Router Advertisement (PrRtAdv) [24].

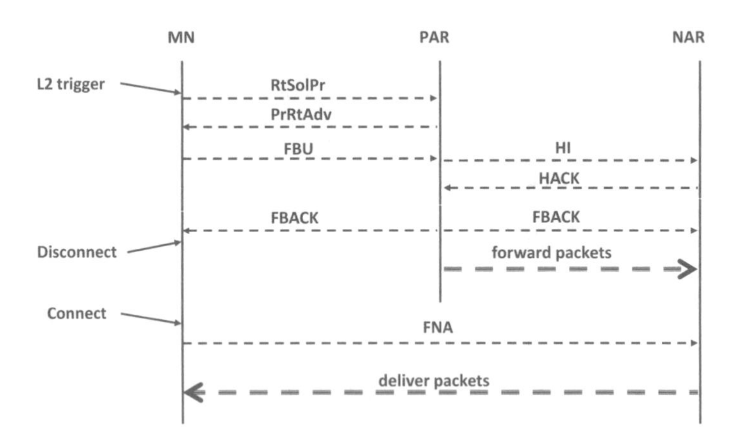

Fig. 3 FMIPv6 predictive fast handover scheme flow diagram

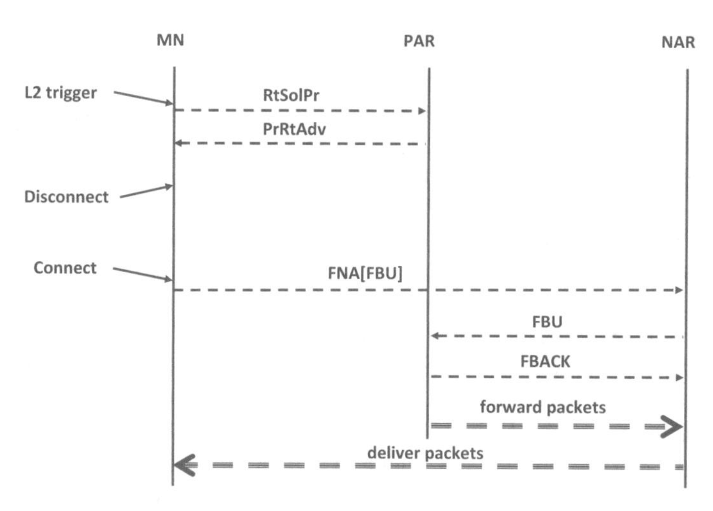

Fig. 4 FMIPv6 reactive fast handover scheme flow diagram

Next, when MN sends Fast Binding Update (FBU) to the pAR and requesting pAR to send a Handover Initiate (HI) messages to the nAR, this is to obtain the New Care of Address (NCoA). So that, all the packets arriving to the Previous Care of Address (PCoA) can be tunnelled to the NCoA. The nAR performs Duplicate Address Detection (DAD) and returns Handover Acknowledgement (HACK) to the pAR with the tunnel establishment. Next, pAR sends Fast Binding Acknowledgement (FBack) to MN and nAR. In the last step, the MN sends the Fast Neighbour Advertisement (FNA) to the nAR. This informs that the MN is in the nAR subnet and the nAR returns the FNA Acknowledgement (FNA-Ack) to the MN.

When the predictive fast handover scheme is failed or not possible, FMIPv6 can operate in reactive fast handover scheme. The reactive fast handover scheme almost same as the predictive fast handover scheme. The main difference is in predictive fast handover scheme, it allows the MN to send FBU even before it is attached to the nAR. But in reactive fast handover, it only allows MN to send FBU after it is attached to the nAR.

As summary, when MN moves to the new subnet and connect with the new link, it can receive the forwarded packets from pAR. The buffers exist in pAR and new Access Router (nAR) for protecting packet loss. Therefore, it reduces the service disruption duration and handover latency. The MN must need to update the HA and CNs. As conclusion, the introduction of the FMIPv6 has minimize the packet loss and latency due to handover process, thus these have improved the network connection and smoothen the users' intermittent connections.

Fast Handover for Hierarchical Mobile Internet Protocol Version 6 (FHMIPv6)

FHMIPv6 is the combination of two mechanisms that are Fast Handover Mobile Internet Protocol version 6 (FMIPv6) and Hierarchical Mobile Internet Protocol version 6 (HMIPv6). Fast Handover for Hierarchical Mobile IPv6 (FHMIPv6) reduces the signaling overhead and Binding Update (BU) delay during handover by using HMIPv6 procedures [25]. Furthermore, movement detection latency and new CoA configuration delay during handover are reduced by utilizing FMIPv6 processes.

The FHMPv6 contains the Router Solicitation for Proxy Advertisement (RtSolPr), Proxy Router Advertisement (PrRtAdv) and Fast Binding Update (FBU) technology from the FMIPv6 mechanism and also the Mobility Anchor Point (MAP) technology from the HMIPv6 mechanism which are combined into one single technology namely FHMIPv6. When the MN associates with a new MAP domain, HMIPv6 procedures are performed with the HA and the Mobility Anchor Point (MAP). If the MN moves from a previous Access Router (pAR) to a new Access Router (nAR) within the domain, it follows the local BU process of HMIPv6. Packets are sent to the MN by the CN during handover which are tunneled by the MAP enroute for the nAR. Figure 5 shown the FHMIPv6 flow diagram.

Based on the FHMIPv6 operation flow figure, the MN sends RtSolPr message containing the information of nAR to MAP. Continuingly, the MAP sends out PrRtAdv message to the MN, which contains information of New Link Care of Address (NLCoA) for MN to be used in nAR region. Next, the MN sends out the Fast Binding Update (FBU) to the MAP, which encloses Previous Link Care of Address (PLCoA) and IP address of the nAR. Once the MAP receives FBU from the MN, MAP sends out Handover Initiate (HI) to nAR. In response to the HI message, nAR sets up a host route for the MN's PLCoA and responds with a Handover Acknowledge (HACK) message. It means that a bi-directional tunnel between MAP and nAR is established. After that, MAP sends out Fast Binding Acknowledgement (FBACK) toward MN over pAR and nAR. Then, MAP begins to forward data packets destined to MN to the nAR by using the established tunnel.

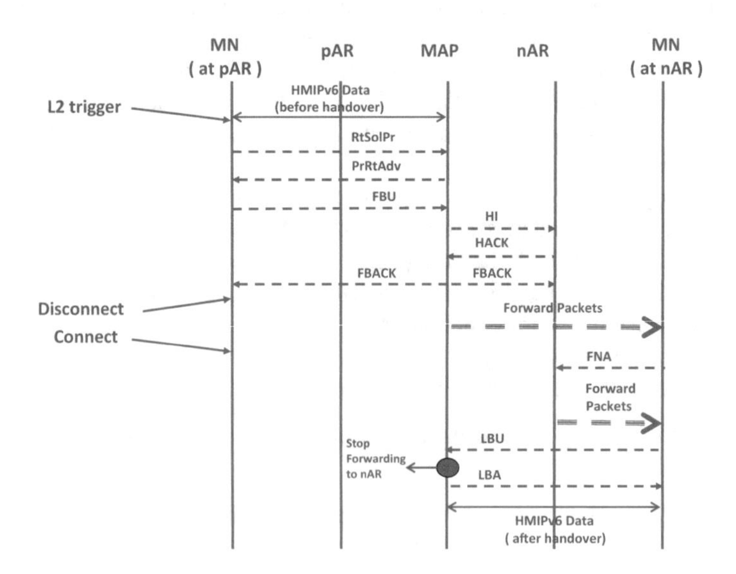

Fig. 5 FHMIPv6 flow diagram

Once the MN is in nAR region, it sends out Fast Neighbour Advertisement (FNA) to the nAR and nAR returns the FNA-ACK to the MN. Then, MN sends Local Binding Update (LBU) to MAP. Next, the HA performs Duplicate Address Detection (DAD) and updates the binding cache. Then, MAP sends a Binding Acknowledgement (BACK) to MN. After this process, MN sends binding update to it's HA and active CN's with NLCoA as its source address or HA, CN's address is set as destination address.

3.2 Wireless Mesh Network

Wireless Mesh Network (WMN) is one of the multi-hop infrastructure based network. A WMN is communication network which is made up of radio nodes organized in a mesh topology. The WMN consists of mesh client, mesh router and gateways. The mesh clients are often laptops, cell phones and other wireless devices while the mesh routers forward traffic to and from the gateway, but need not connect to the Internet [26].

The WMN primary advantages lie in its inherent fault tolerance against network failure, broadband capability and simplicity of setting up a network. Compare to cellular networks, unavailability of communication services over a large geographical area is occurred when a single base station (BS) failure [27]. The WMN is reliable and offers redundancy. When one node is failed, the rest of the nodes can still communicate with each other, directly or through one or more intermediate nodes. Besides that, the administration and maintenance costs for WMN are lower. In addition, a wireless mesh overcomes the line-of-sight issues that may occur when a space is crowded with buildings or industrial equipments.

A WMN can be seen as a special type of WANET. A WMN often has a more planned configuration, and may be deployed to provide dynamic and cost effective connectivity over a certain geographic area [28]. An ad hoc network, on the other hand, is formed ad hoc when wireless devices come within communication range of each other. The mesh routers may be mobile, and may be moved according to specific demands that is arising in the network. Often, the mesh routers are not limited in terms of resources compared to other nodes in the network. Thus, mesh routers can be exploited to perform more resource intensive functions. In this way, the WMN differs from an ad hoc network, since these nodes are often constrained by resources.

4 Simulation Design

4.1 Simulation Setup

To perform a comparison between MIPv6, HMIPv6, FMIPv6 and FHMIPv6 mobility protocols, some configurations and parameters need to be fixed to obtain the optimum results for each mobility management protocol. The environment for all Host-Based mobility management protocols are set up in Wireless Mesh Network topology environment and the data rate is fixed in 100 Mbps. Table 1 shows the type of parameters and values that are needed to fixed for the whole simulation process.

The network topology consists of MIPv6, HMIPv6, FMIPv6, FHMIPv6 and WMN. Inter network section comprises of 8 routers with 5 wired routers and 3 wireless routers that are accomplished as base stations. Intra network portion includes 9 Wireless Mesh routers which has been setup in a grid formation to maximize the coverage area. Figure 6 shown the simulation design of inter and intra networks environment for Host-Based mobility management protocol. Table 2 shows configuration details of inter and intra networks topology of each link and node.

All the Host-Based mobility management protocols are implemented in the inter network packet transmission. The inter network consists of Corresponding Node (CN), Home Agent (HA), Mobility Anchor Point (MAP), previous Access Router (pAR), new Access Router (nAR) and 3 routers namely N1, N2 and N3. The connection between CN–N1, HA–N1 and N1–MAP is set to 100 Mbps. These represent the local area network with high connectivity. MAP–N1 and MAP–N2 are set to 10 Mbps which represent 10 Mbps Ethernet network. N2–pAR and N3–nAR are set to 1 Mbps which represent WiFi networks.

Table 1 Type of parameters and value

Wireless mesh network data rate	10 Mbps
Window size (byte)	32
Transport protocol	TCP
Link delay	2 ms

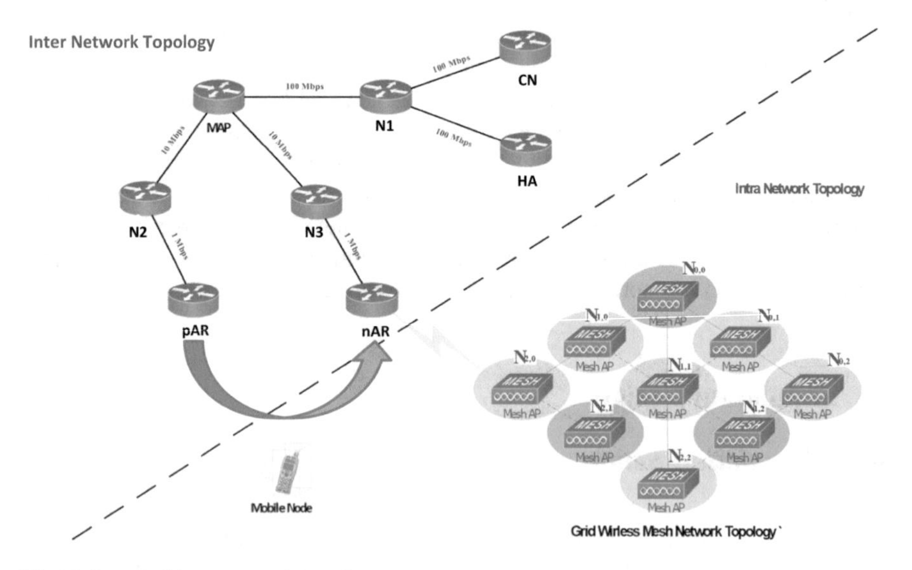

Fig. 6 Inter and intra networks environment

Table 2 Configuration details for inter and intra networks topology

Link connection	Link speed (Mbps)	Queue type
CN–N1	100	RED
HA–N1	100	RED
N1–MAP	100	RED
MAP–N2	10	RED
MAP–N3	10	RED
N2–pAR	1	DropTail
N3–nAR	1	DropTail

The network environment needs to be constructed before implementing the mobility management protocols. The Intra-network is setup by Wireless Mesh Network (WMN). The WMN topology is in grid formation to fully cover the coverage area which is either scattered or concentrated in a certain geographical terrain. The grid formation WMN topology build up by 9 mesh routers and the mobility management protocols act as the mesh client. The bandwidth for WMN is setting up at 10 Mbps and the data transfer rate among the mesh router is up to 100 Mbps. The grid WMN topology allows the data transmitted from different devices simultaneously and can withstand high load traffic. The mesh router can exchange and transfer messages directly with neighbours which are located in a vertical or horizontal position. The two mesh routers in the diagonal of a rectangle cannot generally reach each other directly, as obstacles like a tall building or mountain that highly blocks the wireless transmission. For grid formation of WMN network topology, one mesh router act as the main controller to whole topology. The main mesh router, N0 connect with the WLAN Controller which works as main controller.

For Inter network, MN is set to move into the connection area of pAR and establish connection with pAR. The pAR belongs to the corresponding home networks. The MN assigned static address from home network, which means that the MN will never change its IP address once it gets that IP address. The HA represents its home default network and it's also used to tunnel traffic when the mobile device is moving into nAR network when route optimization is not used. The CN keeps communicating with MN through the pAR.

When MN starts moving from pAR toward nAR connection area, MN sends it's binding messages while detecting motion and keeps looking for routing advertisements and solicitation messages from the neighbour router. The MN associates with the nAR. If the association is successful, then the handover process starts. If the association is not successful, the MN repeats the current steps. For MIPv6 and FMIP6, the MAP router is functioned as normal access router, while for HMIPv6 and FHMIPv6, MAP router is functioned as access router that has provide new technology, Mobility Anchor Point (MAP) service.

4.2 Performance Metrics

In order to get a full understanding about the behaviour of all Host-Based mobility management protocols performed in grid formation of WMN network topology, the following performance metrics were selected for this simulation.

i. Throughput—represents the average rate of successful packet delivery per unit time over a communication channel.
ii. Packet Delivery Ratio (PDR)—represents the ratio between the number of packets received by the receiver, and the number of packets sent by the source.
iii. Latency—represents the delay from the packets sent by the host (Computer User) to the server (Internet).

5 Results and Discussion

Simulation results are presented in this section. In Table 3, the result of each performance metric for all Host-Based mobility management protocols in Wireless Mesh Network (WMN) environment are shown. The packet sizes used for

Table 3 Performance of various types of Host-Based mobility management protocols in WMN environment

Packet size (bytes)	Latency mean (ms)	PacketDelivery ratio (%)	Throughput (bps)
Mobile Internet Protocols version 6 (MIPv6) with WMN			
256	175	74.68	69617.23
512	179	75.22	70860.98
1024	178	74.66	70215.13
2048	183	82.43	91023.06
Hierarchical Mobile Internet Protocol version 6 (HMIPv6) with WMN			
256	201	81.76	77824.00
512	204	82.22	78807.04
1024	202	84.51	80858.21
2048	237	92.65	104133.43
Fast Handover Mobile Internet Protocol version 6 (FMIIV6) with WMN			
256	195	79.87	75243.53
512	200	81.05	77672.90
1024	198	81.51	77987.84
2048	215	89.65	100761.60
Fast Handover for Hierarchical Mobile Internet Protocol version 6 (FHMIPv6) with WMN			
256	200	81.76	77824.00
512	202	82.22	78807.04
1024	201	84.51	80858.21
2048	234	92.65	104133.43

simulation are started from 256 bytes, and increase to 512 bytes, 1024 bytes, and 2048 bytes. The Host-Based mobility management protocols include MIPv6, HMIPv6, FMIPv6 and FHMIPv6.

Based on the results obtained from the simulation experiments, MIPv6 with WMN performs well in term of latency mean. The reason that MIPv6 with WMN has low latency is because the packets have been dropped that contribute to a low Packet Delivery Ratio and Throughput. The highest Packet Delivery Ratio that can be reached is only 82.43% and in the same time the Throughput only reaches 91023.06 bps. HMIPv6 and FHMIPv6 with WMN perform well in term of Throughput and Packet Delivery Ratio (PDR). This is because the Mobility Anchor Point (MAP) is implemented and the hierarchical movement pattern performs better. The FHMIPv6 with WMN performs slightly better than HMIPv6 in term of latency mean. FMIPv6 with WMN has a fair performance outcome as it is based on the anticipated handover with a slightly lower Packet Delivery Ratio and Throughput compare to HMIPv6 and FHMIPv6.

Having studied Fig. 7, it is observed that as the packet size increases, latency mean is increased. The reason is because as the packet size increases, the network needs more time to send packets over the Mobile Internet and through Wireless Mesh Network. The handoff latency is also included. However, in Wireless Mesh Network, it has minimized the time for handover process. Additionally, it is observed that MIPv6 performs better compared to HMIPv6, FMIPv6 and FHMIPv6. It is followed by FMIPv6 that performs better than HMIPv6 and FHMIPv6. Although HMIPv6 and FHMIPv6 have been implemented with MAP, but FHMIPv6 has slightly lower latency mean compared with HMIPv6. However, latency mean for Host-Based mobility management protocols do not have many differences. Thus, it can be concluded that all Host-Based mobility management protocols have not much improvement between the mechanisms with the consolidation with WMN.

As shown in Fig. 8, the packet delivery ratio of MIPv6, HMIPv6, FMIPv6 and FHMIpv6 for various packet sizes are presented. It can be observed that HMIPv6

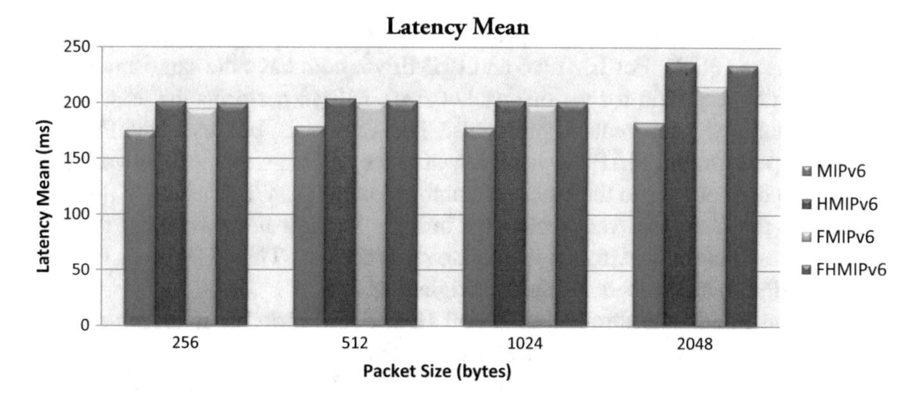

Fig. 7 Latency mean chart

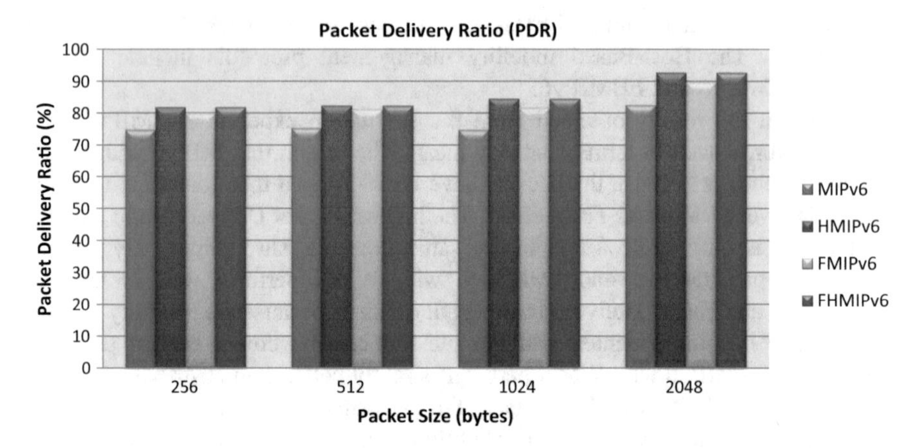

Fig. 8 Packet Delivery Ratio (PDR) chart

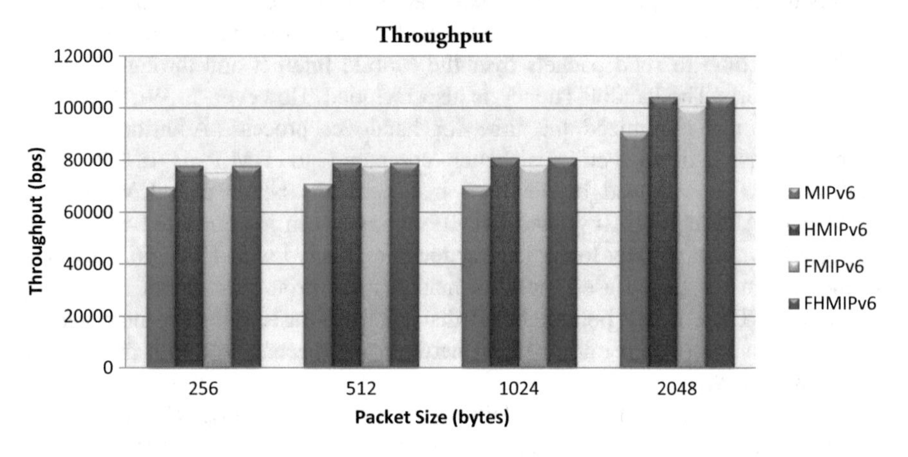

Fig. 9 Throughput chart

and FHMIPv6 perform better compared to MIPv6 and FMIPv6 in term of Packet Delivery Ratio (PDR). For HMIPv6 and FHMIPv6, both have the same amount for the packet delivery ratio for various packet sizes. MIPv6 performs the least among these 4 mobility management protocols. The reason is because HMIPv6 and FHMIPv6 perform the MAP mechanism where the same hierarchical network does not need to be sent over to the higher hierarchical network. Whereas in FMIPv6 and FHMIPv6, the fast handover mechanism informs the new network about the handover process before it performs the handover processes. Thus, HMIPv6, FMIPv6 and FHMIPv6 perform better than the original MIPv6.

Based on Fig. 9, the throughput for all Host-Based mobility management protocols has been observed. The HMIPv6 and FHMIPv6 perform better compared to the other two mechanisms. The HMIPv6 and FHMIPv6 share the same throughput for various packet sizes. For example packet size of 2048 bytes, HMIPv6 and

FHMIPv6 have the throughput of 104133.43 bps. FMIPv6 has the throughput of 100761.60 bps and MIPv6 has the throughput of 91023.06 bps. The reason of HMIPv6 and FHMIPv6 performs better than FMIPv6 and MIPv6 is explained as before where HMIPv6 and FHMIPv6 do not need to perform higher hierarchical data transmission if the nodes perform lower hierarchical network communication.

By implementing the FHMIPv6 to handle the inter network scenario, the MAP and fast handover mechanisms are activated. When an MN moves in MAP domain, MAP acts as HA of the MN locally and the MN registers only to its location information to MAP. The MAP minimize frequent BUs to its HA and CNs and the BU signaling is localized in a most time. Therefore, FHMIPv6 is smaller location update cost than MIPv6 which updates an MN's location information to the HA and CNs. For fast handover mechanism, when the mobile node senses lower signal strength, mobile node advertises to the neighbour network for the need to attach to the new higher signal strength access point. Mobile node informs the new access point of the need to change to the new access point before the process of handover. Thus, movement detection latency and new CoA configuration delay during handover are reduced. With this enhanced mobility management, it is believed that with the FHMIPv6 implementation, this can reduce latency, increase throughput and decrease distortion. With these criteria, the future aims of wireless communication while moving is made possible which eases the communication between human.

6 Conclusion

As conclusion, due to the various performance metrics results, the FHMIPv6 performs much better compares to others Host-Based mobility management protocols. By introducing this proposed expansion, it's able to reduce the signaling cost, in order can improve service quality and service range of wireless communication in areas that are affected by coverage problems. In future, the Network-Based mobility management protocols such as Proxy Mobile Internet Protocol version 6 (PMIPv6) and Fast Proxy Mobile Internet Protocol version 6 (FPMIPv6) are proposed to be consolidate and make comparisons with Host-Based mobility management protocols in Wireless Mesh Network (WMN) environment. This simulation research enables the researchers to fully understand the performance of PMIPv6 and FPMIPv6 in inter network with WMN.

Acknowledgements An acknowledgement is given to MyBrain15 for granting this research. We are deeply grateful to First EAI International Conference on Computer Science and Engineering 2016 (COMPSE 2016) that held in Penang Malaysia on 11th to 12th November 2016. The COMPSE 2016 provides an opportunity to accept and publish our paper. We thank our colleagues from School of Computer and Communication from University Malaysia Perlis who provided insight and expertise that greatly assisted the research, although they may not agree with all of the interpretations or conclusions of this paper.

References

1. Johnson, D., Perkins, C., & Arkko, J. (2004, June). Mobility support in IPv6. Internet Society, Reston, VA, IETF RFC 3775.
2. Koodli, R. (2005, July). Fast handovers for Mobile IPv6. Internet Society, Reston, VA, IETF RFC 4068.
3. Soliman, H., Castelluccia, C., ElMalki, K., & Bellier, L. (2005, August). Hierarchical Mobile IPv6 mobility management (HMIPv6). Internet Society, Reston, VA, IETF RFC 4140.
4. Jung, H., Lee, J. Y., & Soliman, H. (2005, October). Fast handover for hierarchical MIPv6 (F-HMIPv6) (pp. 1–14). IETF.
5. Hui, S., & Yeung, K. (2003, December). Challenges in the migration to 4G mobile systems. *IEEE Communications Magazine*, 54–59.
6. Zhang, Y., Luo, J., & Hu, H. (2007). Wireless mesh networking, posted at Wireless DevCenter on Jan (May 2016) (p. 507).
7. Chitedze, Z., & Tucker, W. (2012). FHMIPv6-based handover for wireless mesh networks. In *Proceedings of South Africa Telecommunication Networks and Application Conference* (pp. 1–5).
8. Shaima, Q., & Ajaz, H. M. (2014). Mobility management in next generation networks: Analysis of handover in micro and macro mobility protocols. *International Journal of Computing and Network Technology*, *3*(3).
9. Ashraf, K., Amarsinh, V., & Satish, D. (2013). Survey and analysis of mobility management protocols for handover in wireless network. In: *Conference (IACC)*, 2013 (Vol. 2, pp. 413–420).
10. Li, Y., & Chen, I. R. (2013). Dynamic agent-based hierarchical multicast for wireless mesh networks. *Ad Hoc Networks, 11*(6), 1683–1698.
11. Lee, J.-H., Bonnin, J.-M., You, I., & Chung, T.-M. (2013, March). Comparative handover performance analysis of IPv6 mobility management protocols. *IEEE Transactions on Industrial Electronics*, *60*(3).
12. Vasu, K., Mahapatra, S., & Kumar, C. S. (2012). MIPv6 protocols: A survey and comparative analysis. In *Computer Science & Information Technology (CS & IT)* (Vol. 07, pp. 73–93). © CS & IT-CSCP 2012.
13. Sun, J. H., Howie, D., & Sauvola, J. (2012). Mobility management techniques for next generation wireless networks. In *Proceedings of SPIE. Wireless and Mobile Communications* (Vol. 4586, pp. 155–166).
14. Sko, M., & Klügl, R. (2011). Analytical comparison of Mobile IPv6 handover schemes. *Electrorevue, 2*(2), 22–26.
15. Makaya, C., & Pierre, S. (2008). An analytical framework for performance evaluation of IPv6-based mobility management protocols. *IEEE TWC, 7*(3), 972–983.
16. Fu, S., & Atiquzzaman, M. (2005). Handover latency comparison of SIGMA, FMIPv6, HMIPv6, FHMIPv6. In *IEEE Global Telecommunications Conference, 2005. GLOBECOM '05*. St. Louis, MO (pp. 3809–3813).
17. Zhang, Y., & Bi, H. (2012). The simulation of hierarchical Mobile IPv6 with fast handover using NS2. *Procedia Engineering, 37,* 214–217.
18. Murtadha, M. K., Noordin, N. K., & Ali, B. M. (2015). Survey and analysis of integrating PMIPv6 and MIH wireless networks. *Wireless Personal Communications*, 1351–1376.
19. Hoh, W. S., Muthut, S., Ong, B.-L., Elshaikh, M., Nazri, M., Warip, M., & Ahmad, R. B. (2015). A survey of mobility management protocols. *ARPN Journal of Engineering and Applied Sciences*, *10*(19).
20. Khan, R. A., & Mir, A. H. (2014). Performance analysis of host based and network based ip mobility management schemes (Vol. 6, pp. 1798–1803).
21. Li, J., Zhang, P., & Sampalli, S. (2008). Improved security mechanism for Mobile IPv6. *International Journal of Network Security, 6*(3), 291–300.

22. Yadav, A., & Singh, A. (2014). Performance analysis and optimization of Hmipv6 And Fmipv6 handoff management protocols. *International Journal of Engineering Research, 5013* (3), 305–308.
23. Gelogo, Y. E., & Park, B. (2012). Reducing packet loss for Mobile IPv6 fast handover (FMIPv6). *International Journal of Software Engineering and its Applications, 6*(1), 87–92.
24. Muthut, S., Ong, B.-L., Hoh, W. S., & Badlishah Ahmad, R. (2015). Integration of fast handover and hierarchical mobile internet protocol with wireless mesh network. *Australian Journal of Basic and Applied Sciences*, *9*(25), 72–78.
25. Ortiz, J. H., & Perea, J. L. (2011). Integration of protocols FHMIPv6/MPLS in hybrid networks. *Cyber Journals: Multidisciplinary Journals in Science and Technology, Journal of Selected Areas in Telecommunications (JSAT).*
26. Ahmed, E., Shiraz, M., & Gani, A. (2013). Spectrum-aware distributed channel assignment for cognitive radio wireless mesh networks. *Malaysian Journal of Computer Science, 26*(3), 232–250.
27. Muthut, S., Ong, B., Adilah, N., Zahri, H., & Ahmad, R. B. (2015). An overview of performance enhancement of FHMIPv6 on wireless mesh network. *International Journal of Future Computer and Communication, 4*(3), 160–164.
28. Ghazisaidi, N., Kassaei, H., & Bohlooli, M. S. (2009). Integration of WiFi and WiMAX-mesh networks. In *2009 Second International Conference on Advances in Mesh Networks* (pp. 6–11).

Application of Parallel Computing Technologies for Numerical Simulation of Air Transport in the Human Nasal Cavity

Alibek Issakhov and Aizhan Abylkassymova

Abstract The use of parallel computing technologies for numerical simulation of air transport in the human nasal cavity was considered in this paper. Investigation of air flow in the human nasal cavity is of considerable interest, since breathing is done mainly through the nose. A two-dimensional numerical simulation of air transport in the model cross-sections of the nasal cavity to normal human nose based on the Navier-Stokes equations, the equations for temperature and equation for relative humidity were conducted in this study. The projection method is used for the numerical solution of this system of equations. This numerical algorithm was fully parallelized using different geometric decompositions (1D, 2D, and 3D). A preliminary theoretical analysis of the speed-up and effectiveness of various methods of decomposition of the computational domain and the real numerical experiments for this problem were made in this work. Moreover the best domain decomposition method has been determined. The obtained data transfer numerical modelling air human nasal cavity was verified with known numerical results in the form of velocity and temperature profiles.

Keywords Decomposition methods · Theoretical analysis of efficiency
Speed-up · Alveolar state · Heat transfer in the nasal cavity · Projection method
Finite volume method

1 Introduction

The current trend in the development of high-performance computers opens up new opportunities for developing highly effective methods for modelling complex problems using multi-level decomposition and hierarchical parallelization of computations. For most real physical processes with a large computational grid this approach

A. Issakhov (✉) · A. Abylkassymova
Al-Farabi Kazakh National University, Almaty, Kazakhstan
e-mail: alibek.issakhov@gmail.com

A. Abylkassymova
e-mail: abylkassymova.aizhan@gmail.com

I. Zelinka et al. (eds.), *Innovative Computing, Optimization and Its Applications*, Studies in Computational Intelligence 741,
https://doi.org/10.1007/978-3-319-66984-7_8

is the only practical way to create an adequate computing model of control objects. In addition, traditional serial computers and computational schemes have come to their technological limit. At the same time, a technological breakthrough in the field of creating means for interprocessor and intercomputer communications makes it possible to realize effective control in the distribution of computations for various components by an integrated computer, which in turn is one of the key properties of parallelism.

Through the nasal cavity there is a primary recognition of odors, through it we breathe the air, which passes into the alveolar state (it heats there to physiologically normal temperature and is completely saturated with water vapor). They serve as regulators of all air circulation, create a normal air temperature and are completely saturated with water vapor, cleaned and disinfected. Normally, the airflow passes through the nose at a speed of 6 L/min, this index can be increased to 10 L/min.

However, the nasal cavity depending on the cause of occurrence has curvatures and can be divided as:

- Physiological
- Compensatory
- Traumatic

With the aforementioned character of curvature, they negatively affect primarily the difficulty of breathing. Nasal breathing is very important, a systemic part of our body's vital activity and any of its disturbances sooner or later cause negative consequences for the human body.

The main method of eliminating the curvature of the nasal cavity is a surgical operation—septoplasty. However, it should be noted that the success of a surgical operation at the best does not exceed 80%, which leads to a repeated surgical operation. And also the surgical operation will depend on the experience and skill of the surgeon. Naturally, to increase the percentage of an operation success, it will be necessary to accurately make nasal cavity corrections. Before the surgical intervention due to X-ray images it is possible to evaluate the nature of the curvature and with the help of numerical modelling it will be possible to correct and optimize the nasal cavity in advance. Knowing the preliminary accurate correction of the nasal sinus, the surgeon can increase the percentage of success of the operation that will accordingly reduce the percentage of the reoperation.

2 Background

The nasal cavity balances the inhaled air with the internal state of the body with surprising efficiency. In the papers of Cole [1], Inglesstedt [2] and Webb [3], it was generally agreed that the inhaled air through the nasal cavity reaches up to the alveolar state (completely saturated with water vapor and at normal body temperature) by the time it reaches the pharynx. And it practically does not depend on a condition of ambient air which has arrived through nostrils. These results were also obtained in the paper of Farley and Patel [4] who collected data in natural conditions with air temperature readings along the upper respiratory tract, as well as Hannah and Scherer [5], reflecting measurements of local mass transfer coefficients on the gypsum model the human upper respiratory tract. Nevertheless, McFadden [6] noted that the conclusions are valid for quiet breathing, in some circumstances at high ventilation levels, conditioning of additional air should occur in the intrathoracic airways in order to completely determine the inhaled air in the alveolar state.

Numerous studies have been aimed at assessing the hydration and temperature regulation of the nasal cavity. However, mathematical models were based on axisymmetric tubes or occupied quasistationary flows [7]. As a rule, these studies confirmed the opinion that under normal conditions there is enough time for heating and humidifying the air in the nasal cavity. In addition, medications as well as surgical procedures are being used with increasing speed to restore the structure and functions of the nasal cavity [8]. For example, aromatic inhalations are used to improve airflow and to reduce congestion, as well as rhinoplasty procedures are used to overcome trauma or aesthetic deformities. These artificial interventions cause local changes, and can affect the efficiency of transport phenomena of air. However, precise intranasal characteristics and distribution of transport phenomena are not yet known even for a normal (or healthy) state [9–13].

Experimental examination of the nasal cavity is practically impossible, due to the complex internal structure and dimensions, i.e. The introduction of any measuring device or probe causes additional disturbance of the flow. Therefore, mathematical modelling is one of the only approaches to studying the flow of air in the nasal cavity.

3 Statement of the Physical Problem

Air flow that passes through the structure of the nasal cavity has a very difficult path. The complex structure of the nasal cavity and complete three-dimensional analysis of the steam flow, heat transfer in the inner part of the nasal mucosa requires significant computational resources that prevent a systematic analysis of the relevant factors (Fig. 1).

Having the available computational resources, a complex study of transport mechanisms was carried out in two-dimensional form, through the cross sections of the nose.

In addition, the following assumptions are made for numerical modelling:

- The walls of the nasal cavity and nasal concha are assumed to be immovably hard.
- The air flow in the nasal cavity is considered as a laminar flow, and the air as an incompressible medium (since the Reynolds and Mach numbers are very small).
- The velocities on the walls of the cavity are taken as zero ($u = 0$, $v = 0$).

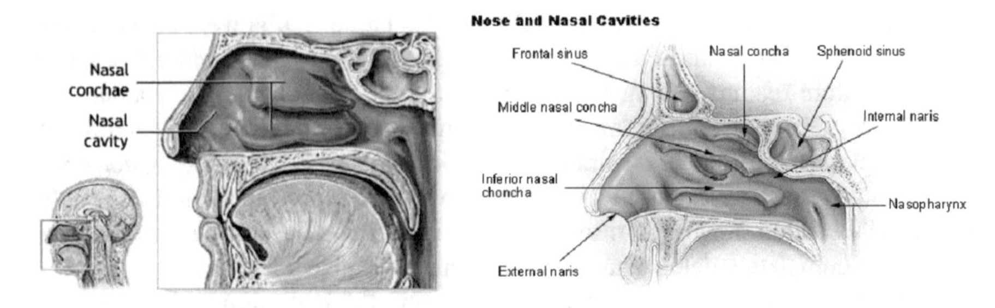

Fig. 1 Nose model with a longitudinal section

- The walls of the nasal cavity are considered fully saturated with water vapor and the temperature near the body due to the moist mucous layer reaches the vascular vessels of the nasal wall.

Thin features of the nose do not have exact dimensions, because there are differences in the structure of the nasal cavity in healthy people, so it is almost impossible to determine the exact model of a "normal nose". Thus, a simplified model of the nose is developed, where the main essential signs of the nasal cavity are revealed. The dimensions are taken from the averaged data of the human nasal cavity (Fig. 2). The physical area of the problem is the second cross-section (Fig. 2c "-2-"), which is important for the study, because it is an area where a significant proportion of the air flow takes place, and also it has a complex structure, through which the basic functions of the nasal cavity are performed.

The mathematical model is constructed on the basis of the Navier-Stokes equations, including the continuity equation, the momentum equation, the energy (temperature) equation, and also relative humidity equation are used [14–16].

$$\nabla U = 0,$$

$$\frac{\partial U}{\partial t} + (U \cdot \nabla)U = -\frac{1}{\rho}\nabla p + \nu \nabla^2 U,$$

$$\frac{\partial T}{\partial t} + (U \cdot \nabla)T = \frac{k}{\rho c_p}\nabla^2 T,$$

$$\frac{\partial C}{\partial t} + (U \cdot \nabla)C = D\nabla^2 C$$

where U—velocity vector, t—time, p—pressure, ν—kinematic viscosity, T—temperature, C—humidity, c_p—specific heat of the medium at constant pressure, k—coefficient of thermal conductivity, ρ—density, D—coefficient of molecular diffusion.

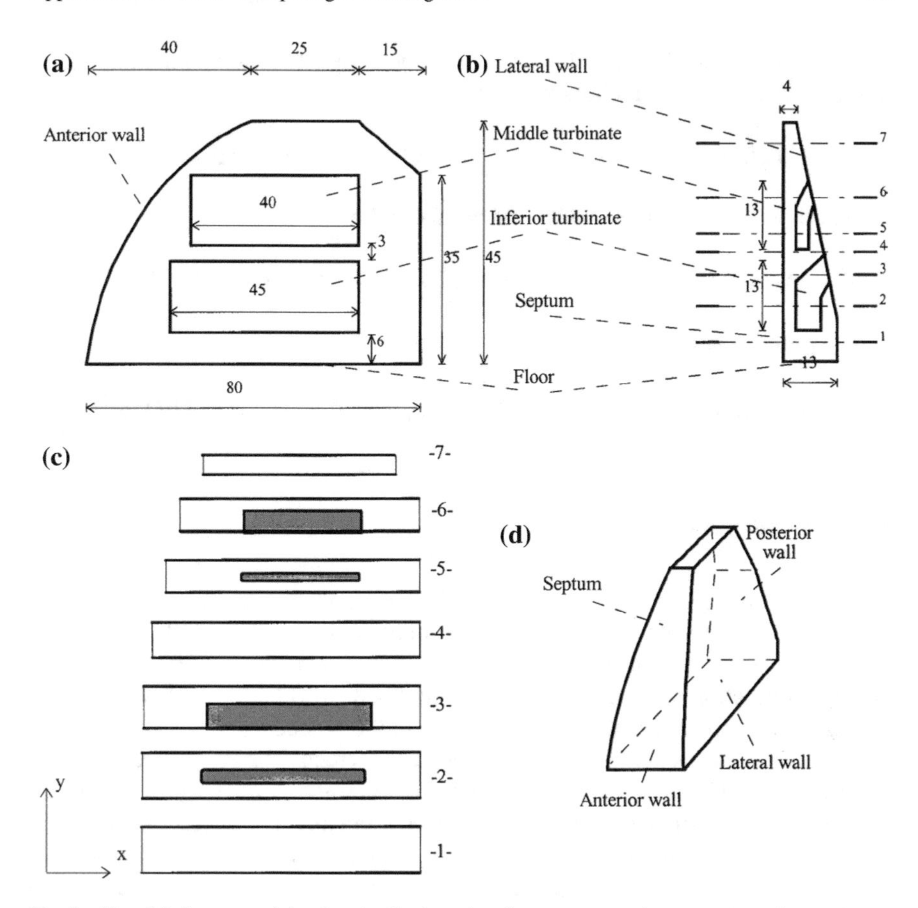

Fig. 2 Simplified nose model: **a** longitudinal section, **b** coronary section, **c** cross sections at height h = 3, 13, 17, 20, 26, 33, 40 mm from the bottom point of the nasal cavity, **d** perspective view

The instantaneous velocity at the inlet in each cross section is assumed to have a parabolic profile with a maximum velocity $(U_{in}^{M})_{\max}$ that varies during the breathing cycle. In the paper of Girardin et al. [17] measurements were made using laser anemometry in the model of the human nose and it was found that the field flow basically has layered parabolic velocity profiles in any cross section. At rest, a normal adult breathes a volume about $V_T = 0.5$ L (inhaling and exhaling) once a minute at an average flow rate of about 0.125 L/s to each nostril. Accordingly, the instantaneous velocity distribution at the input U_{in}^{M} in the direction is given in the following form:

$$u_{in}(t, x = 0, y) = (U_{in}^{M})_{\max}\left[2\sin^2\frac{\pi t}{2} - 1\right] \times \frac{(12y - y^2)}{36}$$

The input boundary conditions for the temperature and relative humidity of the external air are given in the following form:

$$T_{in}(t, x = 0, y) = 25\,°\mathrm{C}, \quad C_{in}(t, x = 0, y) = 0.0047\,\mathrm{kg\,H_2O}/m^3$$

On the walls of the nasal cavity and nasal concha:

$$u_{wall}(t, x, y) = 0, \quad v_{wall}(t, x, y) = 0, \quad T_{wall}(t, x, y) = 37\,°\mathrm{C}, \quad C_{in}(t, x, y) = 0.0438\,\mathrm{kg\,H_2O/m^3}$$

The initial conditions are given in this form:

$$u_0(t = 0) = 0, \quad T_0(t = 0) = 32\,°\mathrm{C}, \quad C_0(t = 0) = 0.0235\,\mathrm{kg\,H_2O/m^3}.$$

4 The Numerical Algorithm

The projection method is used for a numerical solution of this system of equations [18, 19]. Equations are discretized by the finite volume method [18, 20, 21]. At the first stage it is assumed that the transfer of the momentum is carried out only through convection and diffusion, and an intermediate velocity field is calculated by the fourth-step Runge-Kutta method [16]. In the second stage, according to the found intermediate velocity field, there is a pressure field. The Poisson equation for the pressure field is solved by the Jacobi method. Then at the third stage, it is assumed that the transfer is carried out only due to the pressure gradient. Further at the fourth stage, the equations for the temperature are calculated by the fourth-step Runge-Kutta method. And in finally, equations for the relative humidity are calculated, and also solved by the fourth-step Runge-Kutta method [9, 10, 16].

$$\text{(I)}\ \int_\Omega \frac{\vec{u}^* - \vec{u}^n}{\tau} d\Omega = -\oint_{\partial\Omega} (\vec{u}^n \vec{u}^n - \upsilon \nabla \vec{u}^n) n_i d\Gamma$$

$$\text{(II)}\ \oint_{\partial\Omega} (\nabla p) n_i d\Gamma = \int_\Omega \frac{\nabla \vec{u}^*}{\tau} d\Omega$$

$$\text{(III)}\ \frac{\vec{u}^{n+1} - \vec{u}^*}{\tau} = -\nabla p.$$

$$\text{(IV)}\ \int_\Omega \frac{T^{n+1} - T^n}{\tau} d\Omega = -\oint_{\partial\Omega} \left(\vec{u}^n T^n - \frac{k}{\rho c_p} \nabla T^n\right) n_i d\Gamma$$

$$\text{(V)}\ \int_\Omega \frac{C^{n+1} - C^n}{\tau} d\Omega = -\oint_{\partial\Omega} (\vec{u}^n C^n - D \nabla C^n) n_i d\Gamma.$$

5 Algorithm of Parallelization

Well known three dimensional lid-driven cavity problem is used to check various geometric decompositions method. A computational grid was constructed using the Pointwise software to carry out the numerical simulation. The problem was launched on the ITFS-MKM software complex using a high-performance cluster. A cluster system (Intel(R) Xeon(R) CPU E5645 2.40 GHz CPU, 26 nodes with two processors per node and total number of cores are 312, 624 GB RAM) is used to decrease CPU time. This numerical algorithm is completely parallelized using various geometric decompositions (1D, 2D and 3D). Geometric partitioning of the computational grid is chosen as the main approach of parallelization. In this case, there are three different ways of exchanging the values of the grid function on the computational nodes of a one-dimensional, two-dimensional, and three-dimensional grid. After the decomposition stage, when parallel algorithms are built on separate blocks, we proceed to the relations between the blocks, the calculations on which will be performed in parallel on each processor. For this purpose, a numerical solution of the equation system was used for an explicit scheme, since this scheme is very well parallelized. In order to use the decomposition method as a parallelization method, this algorithm uses the boundary nodes of each subdomain in which it is necessary to know the value of the grid function that borders on the neighboring elements of the processor. To achieve this goal, at each computational node, fictitious points store values from neighboring computational nodes, and organize the transfer of these boundary values necessary to ensure homogeneity of calculations for explicit formulas (Fig. 3).

Data transmission is performed using the procedures of the MPI library [22–30]. By doing the preliminary theoretical analysis of the various effectiveness methods of the computational domain decomposition for this problem, we will estimate the time of the parallel program as the time of the sequential program divided by the number of processors used T_{calc}, plus the transmission time $T_p = T_{calc}/p + T_{com}$. While transmissions for different decomposition methods can be approximately expressed through bandwidth:

$$T_{com}^{1D} = t_{send} 2N^2 x2$$

$$T_{com}^{2D} = t_{send} 2N^2 x4p^{1/2} \tag{1}$$

$$T_{com}^{3D} = t_{send} 2N^2 x6p^{2/3}$$

where N^2—the number of nodes in the computational grid, p—the number of processors (cores), t_{send}—the time for sending one element (number).

It should be noted that for different decomposition methods, the data transmission cost can be represented as $T_{com}^{1D} = t_{send} 2N^2 xk(p)$ in accordance with the formula (1), where $k(p)$ is the proportionality coefficient, depending on the decomposition method and the number of processing elements used.

Table 1 shows numerical values of $k(p)$. We can see that if $p > 5$ and by using 3D decomposition this algorithm is more efficient. While for $p > 11$ and by using 3D decomposition, the necessary time of sending between the processors of the value of the function $u_{i,j,k}^{n+1}$, $v_{i,j,k}^{n+1}$, $w_{i,j,k}^{n+1}$, $p_{i,j,k}^{n+1}$ in a node with a smaller number of elements, it is expected that the time spent on data transmission will be minimal.

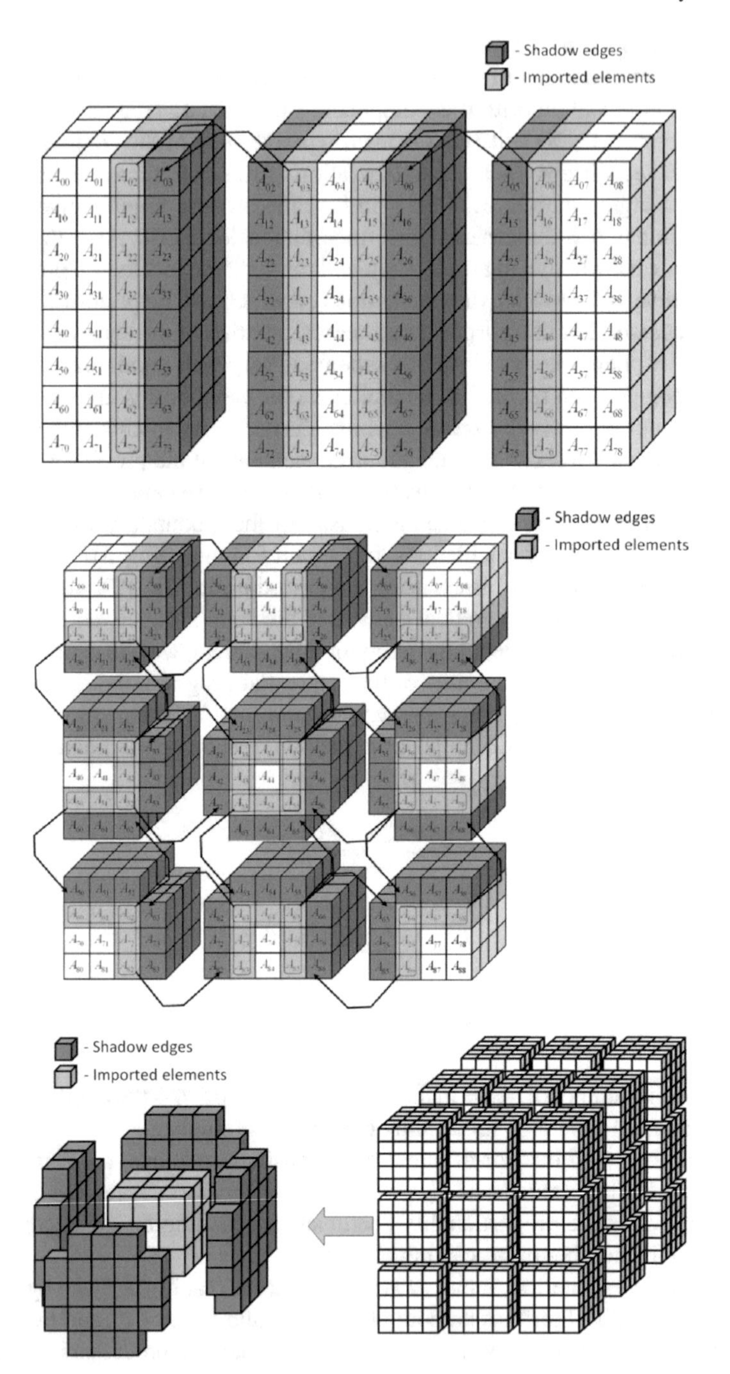

Fig. 3 Different methods of decomposition. Schemes of mechanisms for the exchange of 1D, 2D and 3D decompositions

Table 1 Dependence of the proportionality $k(p)$ coefficient on the number of processor elements and the decomposition method

Number of processes	3	4	5	6	10	11	12	16	60	120	250
1D Decomposition	2	2	2	2	2	2	2	2	2	2	2
2D Decomposition	2.31	2.00	1.79	1.63	1.26	1.20	1.15	1.00	0.51	0.36	0.25
3D Decomposition	2.88	2.38	2.05	1.82	1.29	1.21	1.14	0.94	0.39	0.24	0.15

All calculations were carried out on cluster systems T-Cluster and URSA at the Faculty of Mechanics and Mathematics, al-Farabi Kazakh National University using $128 \times 128 \times 128$ and $256 \times 256 \times 256$ computational grids. Computational experiments were conducted using up to 250 processors. The results of the computational experiment showed the presence of a good speed in solving problems of this class. They are mainly focused on additional transmissions and time calculations for various decomposition methods.

At the first stage, one common program was used, the size of the array from start to run did not change, and each element of the processor is numbered by an array of elements, starting from zero. Despite the fact that according to the theoretical analysis of 3D decomposition is the best option for parallelization (Fig. 4), computational experiments showed that the best results were achieved with 2D decomposition when the number of processes varies from 25 to 144 (Fig. 4).

Based on a preliminary theoretical analysis of the graphs, the following character can be noted. The calculation time without the cost of inter-processor communications with different decomposition methods should be approximately the similar for the equal number of processors and divided by T_{calc}/p. In fact, the calculated data (Fig. 5) show that using 2D decomposition on various computational grids gives a minimal cost for the calculation and the cost graphs are significantly higher depending on the computation time on several taken processors T_{calc}/p (Fig. 6).

To explain these results, it is necessary to pay attention to the assumptions made in the preliminary theoretical analysis of the effectiveness for this task. First, it was assumed that, regardless of the distribution of data per one processor element, the same amount of computational work was performed, which should lead to the same time expenditure. Secondly, it was assumed that the time spent on interprocessor sendings of any degree of the same amount of data does not depend on their choice of memory. In order to understand what is actually happening, the following sets of test computational calculations were carried out. For the evaluation, the sequence of the first approach was considered, when the program is executed in a version with one processor, and thus simulates various ways of geometric decomposition of data for the same amount of calculations performed by each processor.

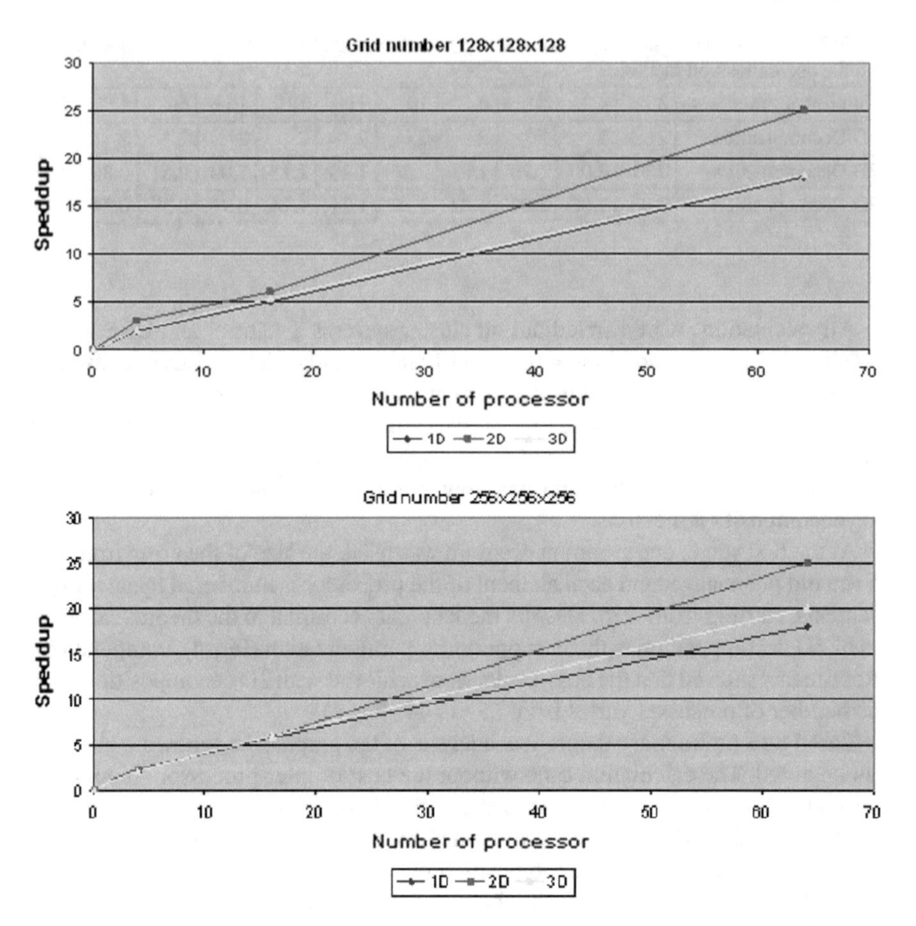

Fig. 4 Speed-up for various methods of the computational domain decomposition

6 Results of Numerical Simulation

As a result of numerical modelling of the aerodynamics of the human nasal cavity, the following data were obtained. Also, to verify this numerical algorithm, we used the calculation data from paper [7], which describes the profiles of the longitudinal component of velocity and temperature in three cross sections: at a distance $x_1 = 17$ mm, $x_2 = 49$ mm and $x_3 = 80$ mm from the entrance (Fig. 7). For the numerical simulation, the corresponding parameters for air constants were used: $\rho = 1.12$ kg/m^3, $\mu = 1.9 \times 10^{-5}$ kg/ms, $c_p = 1005.5$ J/kgK, $k = 0.0268$ W/mK, $D = 2.6 \times 10^{-5}$ m^2/s.

Figure 8 shows the comparison of profiles for x_1, x_2 and x_3 the longitudinal velocity component of the calculation results and data from the article Naftali et al. [7]. Figure 9 shows a comparison of temperature profiles for cross sections x_1, x_2 and x_3

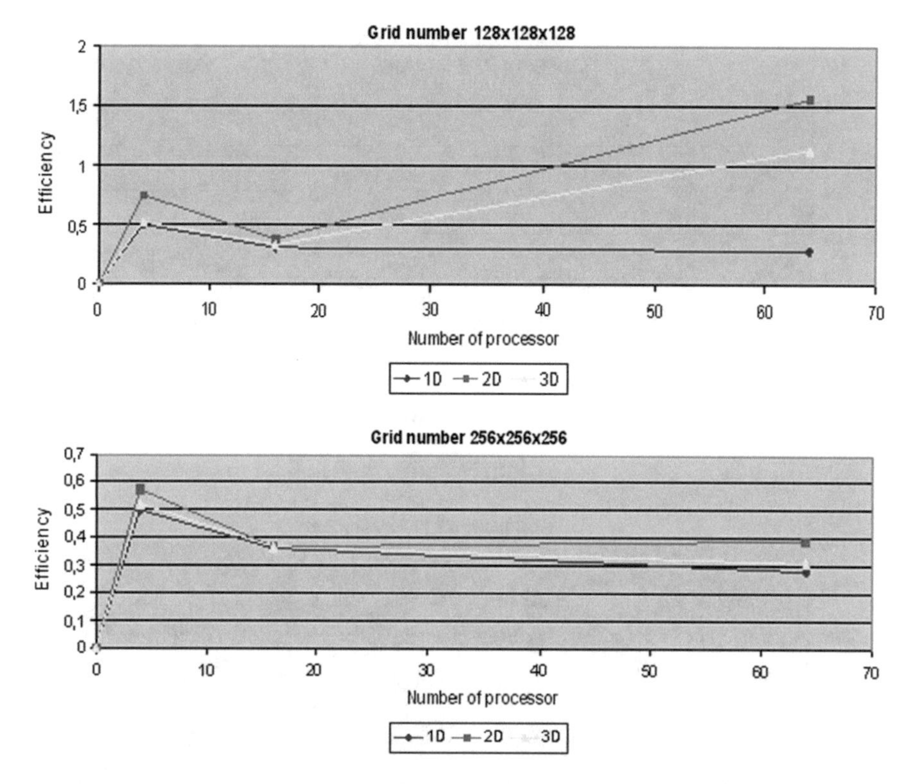

Fig. 5 Efficiency for various methods of the computational domain decomposition

with paper [7]. Figure 10 shows the relative humidity profiles for cross sections x_1, x_2 and x_3. In all the figures, numerical results were shown to be dimensionless.

It can be seen from the figures that when passing through narrow areas of the nasal cavity air is heated downstream, and relative humidity also increases. And also from Fig. 9 it can be noted that behind the nasal septum the temperature increases and the air temperature is heated to the alveolar state when reaching to the nasopharynx. And at low ambient temperatures, relative humidity plays a very important role. Figure 11 shows the longitudinal velocity component in the cross section for time $t = 1$ s. Figure 12 shows the transverse velocity component in the cross section for time $t = 1$ s. It can be seen from the figures that vortex currents appearing from the nasal conchas, which play an important role in the process of heating the air. Figures 13 and 14 show the temperature components, and the relative humidity in the calculated area for time $t = 1$ s. It can be seen from the figures that when passing through narrow areas of the nasal cavity air is heated downstream, and relative humidity also increases.

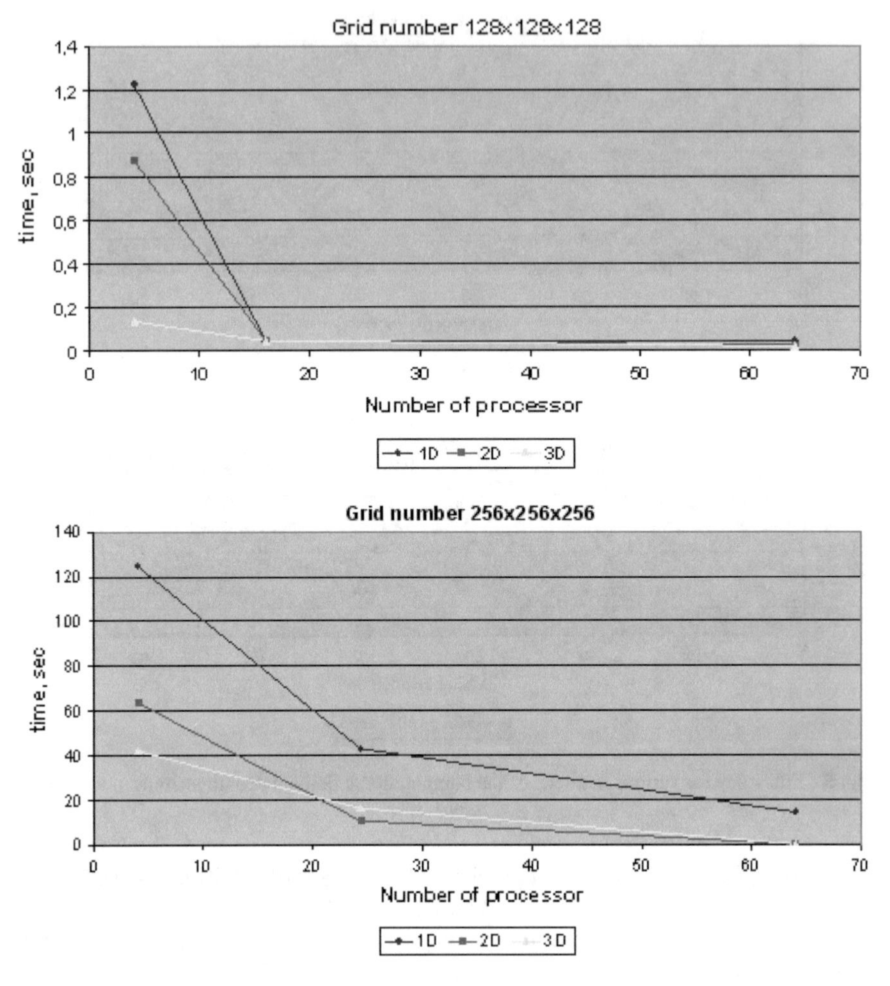

Fig. 6 Time of calculation without taking into account the cost of data transfer for different decomposition methods

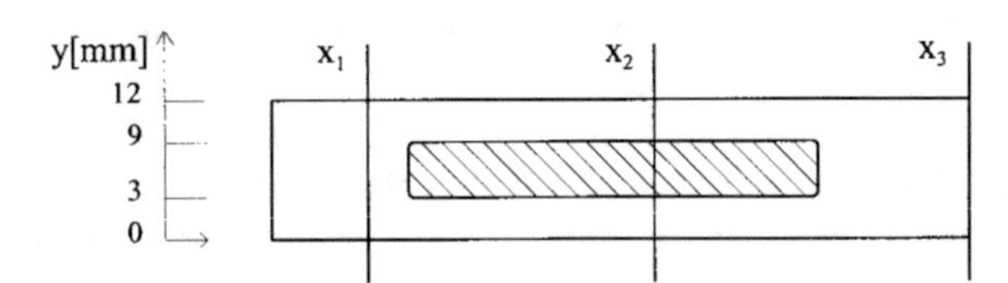

Fig. 7 Evaluation in three locations for temperature and velocity for a cross section

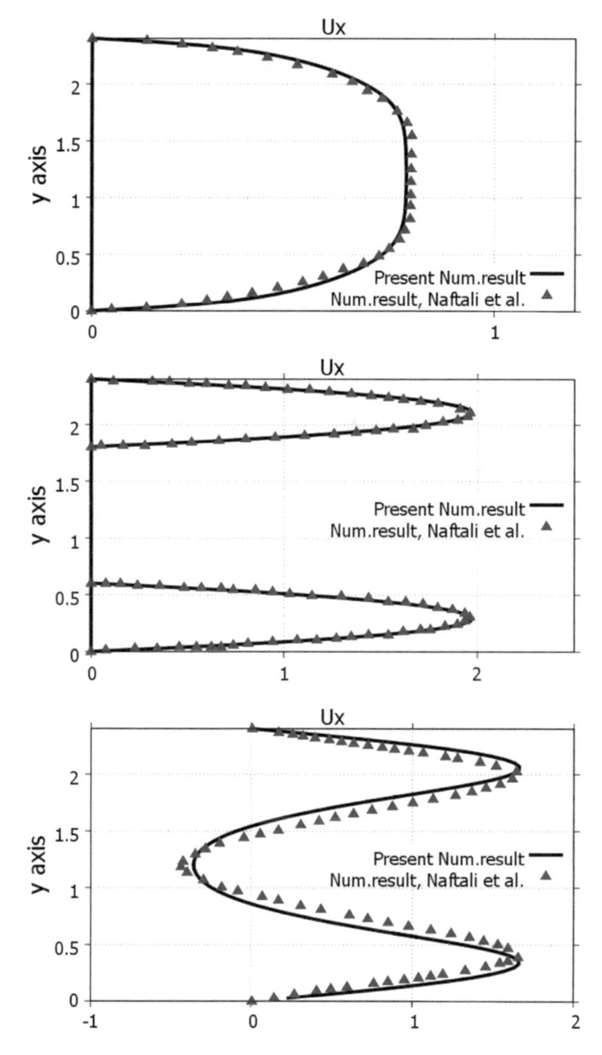

Fig. 8 Comparison of the velocity component profile for the cross sections $x_1 = 17$ mm, $x_1 = 49$ mm and $x_1 = 80$ mm with the calculation results [7]

7 Future Research Directions

Moreover, future increases in computational resources will be accompanied by continuing demands from application scientists for increased resolution and the inclusion of additional physics. Consequently, new approaches are needed in order to decrease simulation computational costs. Such techniques will be required to harness both the underlying model hierarchy and the stochastic hierarchy.

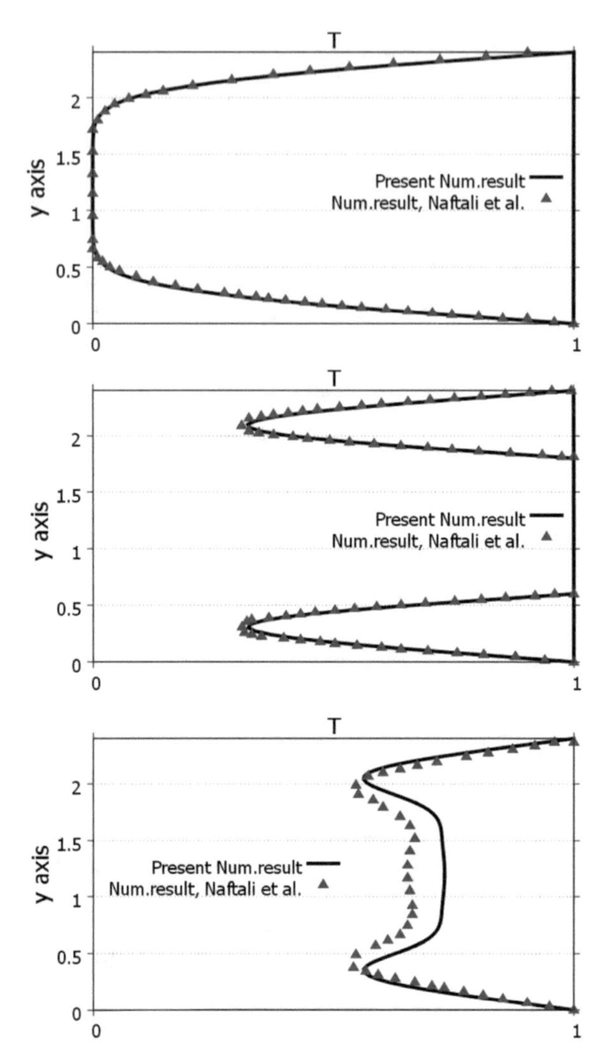

Fig. 9 Comparison of temperature profiles for cross sections $x_1 = 17$ mm, $x_1 = 49$ mm and $x_1 = 80$ mm with the calculation results [7]

Future research should continue to examine knowledge in qualitatively and quantitatively approximate the basic laws of air transport in the human nasal cavity, which will tie to real conditions. These research will give more real numerical results.

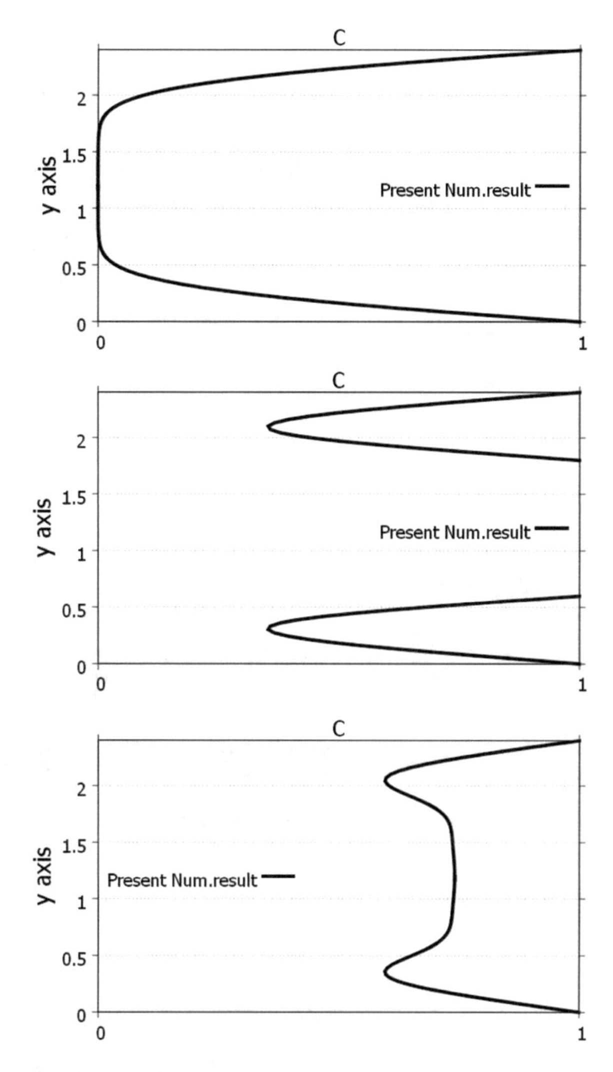

Fig. 10 Relative humidity profiles for sections $x_1 = 17$ mm, $x_1 = 49$ mm and $x_1 = 80$ mm

8 Conclusion

Thus, for explicit numerical methods for lid-driven cavity problem solved by the system of the Navier-Stokes equation with one-dimensional, two- and three-dimensional decompositions are applied. But the results of the testing programs showed that for the number of processors not exceeding 250. While 3D decomposition is not time-consuming compared to 2D decomposition, and 3D decomposition has more time-consuming software implementation and the use of 2D decomposition is sufficient for the scale of the task. And also during the examination of the nasal cavity, the

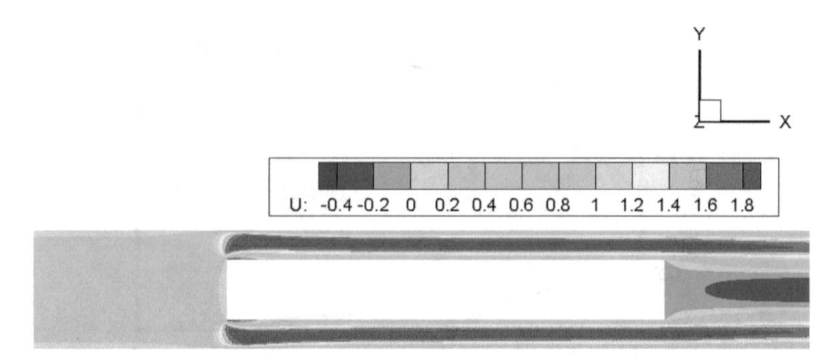

Fig. 11 Longitudinal components of the flow velocity for a cross section

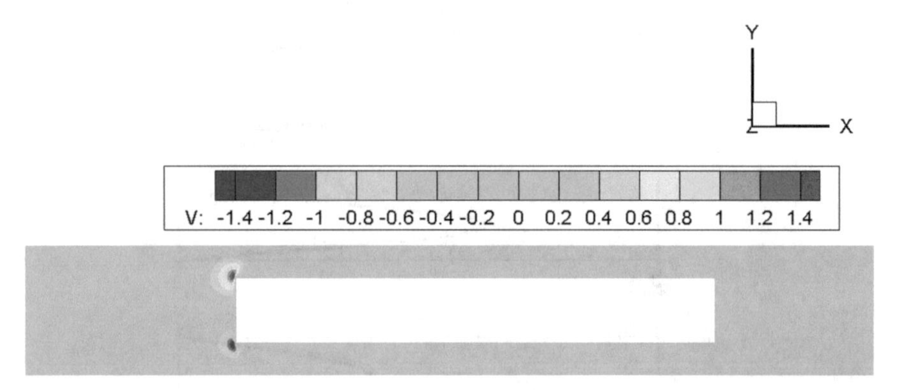

Fig. 12 Transverse velocity components flow for a cross section

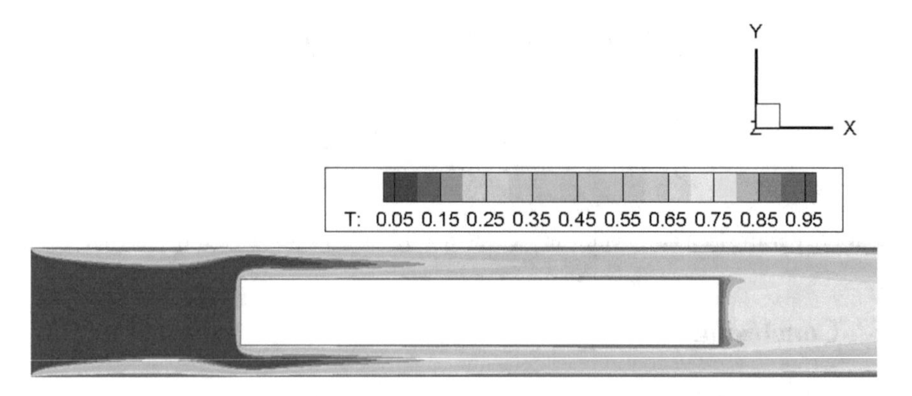

Fig. 13 Temperature flow for cross section

following conclusion can be drawn: the walls of the nasal cavity contribute to the heating of air and the appearance of vortices, which are not a small importance for the transition of air to the alveolar state, before entering the nasopharynx. Not less

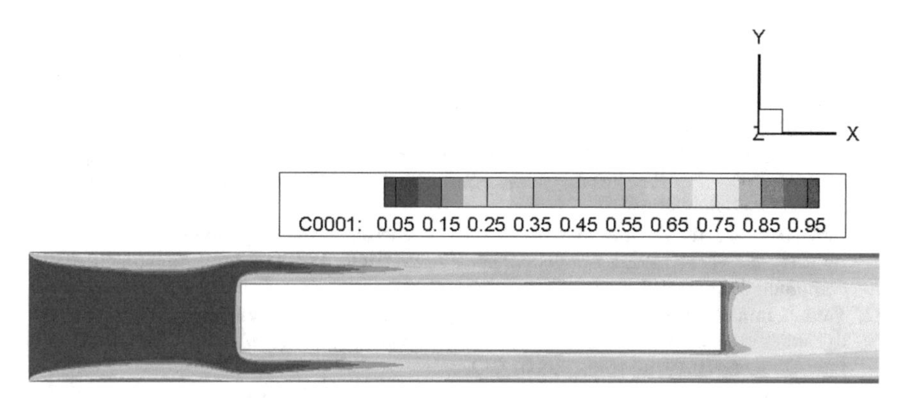

Fig. 14 Relative humidity flow for cross section

importance has the relative humidity in the nasal cavity, since at low ambient temperatures, due to moisture, the incoming air is heated. Studies of air movement in the nasal cavity are very important, as at present, for various reasons. For instance, one of them is that the number of people with nasal breathing problems is increasing. This problem is solved by surgery, where it is important to optimally operate the nose structure, so that the nasal cavity functions correctly, as normal breathing is carried out with the help of the nose.

Acknowledgements This work is supported by the grant from the Ministry of education and science of the Republic of Kazakhstan.
The authors wish to thank anonymous referees for their helpful and constructive comments on earlier versions of this paper. And would like to present our sincere thanks to the Editor in Chief and First EAI International Conference on Computer Science and Engineering (COMPSE 2016) to publish this paper.

References

1. Cole, P. (1953). Some aspects of temperature, moisture and heat relationships in the upper respiratory tract. *Journal of Laryngology and Otology*, *67*, 669–681.
2. Ingelstedt, S. (1956). Studies on conditioning of air in the respiratory tract. *Acta Oto-Laryngologica Supplementum*, *131*, 1–80.
3. Webb, P. (1951). Air temperatures in respiratory tracts of resting subjects. *Journal of Applied Physiology*, *4*, 378–382.
4. Farley, R. D., & Patel, K. R. (1988). Comparison of air warming in human airway with thermodynamic model. *Medical and Biological Engineering and Computing*, *26*, 628–632.
5. Hanna, L. M., & Scherer, P. W. (1986). Measurement of local mass transfer coefficients in a cast model of the human upper respiratory tract. *Journal of Biomechanical Engineering*, *108*, 12–18.
6. McFadden, E. R. (1983). Respiratory heat and water exchange: Physiological and clinical implications. *Journal of Applied Physiology*, *54*, 331–336.
7. Naftali, S., Schroter, R. C., Shiner, R. J., & Elad, D. (1998). Transport phenomena in the human nasal cavity: A computational model. *Annals of Biomedical Engineering*, 831–839.

8. Maran, A. G. D., & Lund, V. J. (1990). *Clinical Rhinology*. New York: Thieme Medical.
9. Issakhov, A. (2016). Mathematical modeling of the discharged heat water effect on the aquatic environment from thermal power plant under various operational capacities. *Applied Mathematical Modelling, 40*(2), 1082–1096. https://doi.org/10.1016/j.apm.2015.06.024.
10. Issakhov, A. (2011). Large eddy simulation of turbulent mixing by using 3D decomposition method. *Journal of Physics: Conference Series, 318*(4), 1282–1288. https://doi.org/10.1088/1742-6596/318/4/042051.
11. Inthavong, K., Tu, J. Y., & Heschl, C. (2011). Micron particle deposition in the nasal cavity using the v(2)-f model. *Computers and Fluids, 51*(1), 184–188. https://doi.org/10.1016/j.compfluid.2011.08.013.
12. Wen, J., Inthavong, K., Tu, J., & Wang, S. M. (2008). Numerical simulations for detailed airflow dynamics in a human nasal cavity. *Respiratory Physiology and Neurobiology, 161*(2), 125–135. https://doi.org/10.1016/j.resp.2008.01.012.
13. Zubair, M., Ahmad, K. A., Abdullah, M. Z., & Sufian, S. F. (2015). Characteristic airflow patterns during inspiration and expiration: Experimental and numerical investigation. *Journal of Medical and Biological Engineering, 35*(3), 387–394. https://doi.org/10.1007/s40846-015-0037-4.
14. Fletcher C. A. J., & Fletcher C. A. (2013). *Computational techniques for fluid dynamics, Vol. 1: Fundamental and general techniques* (401 pp.). Springer.
15. Roache, P. J. (1972). *Computational fluid dynamics* (434 pp.). Albuquerque, NM: Hermosa Publications.
16. Chung, T. J. (2002). *Computational fluid dynamics* (1034 pp.).
17. Girardin, M., Bilgen, E., & Arbour, P. (1983). Experimental study of velocity fields in a human nasal fossa by laser anemometry. *Annals of Otology, Rhinology, and Laryngology, 92*, 231–236.
18. Issakhov, A. (2015). Mathematical modeling of the discharged heat water effect on the aquatic environment from thermal power plant. *International Journal of Nonlinear Science and Numerical Simulation, 16*(5), 229–238. https://doi.org/10.1515/ijnsns-2015-0047.
19. Chorin, A. J. (1968). Numerical solution of the Navier-Stokes equations. *Mathematics of Computation, 22*, 745–762.
20. Pletcher, R. H., Tannehill, J. C., & Anderson, D. (2011). *Computational fluid mechanics and heat transfer* (3rd ed.). Series in Computational and Physical Processes in Mechanics and Thermal Sciences (774 pp.). CRC Press.
21. Ferziger, J. H., & Peric, M. (2013). *Computational methods for fluid dynamics* (3rd ed., 426 pp.). Springer.
22. Karniadakis, G. E., & Kirby II, R. M. (2000). *Parallel scientific computing in C++ and MPI: A seamless approach to parallel algorithms and their implementation* (p. 630). Cambridge University Press.
23. Pacheco, P. (1996). *Parallel programming with MPI* (p. 500). Morgan Kaufmann.
24. Issakhov, A. (2013). Mathematical modelling of the influence of thermal power plant on the aquatic environment with different meteorological condition by using parallel technologies. In *Power, control and optimization*. Lecture Notes Electrical Engineering (Vol. 239, pp. 165–179).
25. Issakhov, A. (2012). Mathematical modelling of the influence of thermal power plant to the aquatic environment by using parallel technologies. AIP Conf. Proc. *1499*, 15–18. http://dx.doi.org/10.1063/1.4768963.
26. Issakhov, A. (2016). Mathematical modelling of the thermal process in the aquatic environment with considering the hydrometeorological condition at the reservoir-cooler by using parallel technologies. In *Sustaining power resources through energy optimization and engineering*, Chapter 10 (pp. 227–243). https://doi.org/10.4018/978-1-4666-9755-3.ch010.
27. Issakhov, A. (2014). Modeling of synthetic turbulence generation in boundary layer by using zonal RANS/LES method. *International Journal of Nonlinear Sciences and Numerical Simulation, 15*(2), 115–120.

28. Issakhov, A. (2016). Numerical modelling of distribution the discharged heat water from thermal power plant on the aquatic environment. *AIP Conference Proceedings, 1738*, 480025. https://doi.org/10.1063/1.4952261.
29. Issakhov, A. (2017). Numerical modelling of the thermal effects on the aquatic environment from the thermal power plant by using two water discharge pipes. *AIP Conference Proceedings, 1863*, 560050. https://doi.org/10.1063/1.4992733.
30. Issakhov, A. (2015). Numerical modeling of the effect of discharged hot water on the aquatic environment from a thermal power plant. *International Journal of Energy for a Clean Environment, 16*(1–4), 23–28. https://doi.org/10.1615/InterJEnerCleanEnv.2016015438.

Genetic Algorithms-Based Techniques for Solving Dynamic Optimization Problems with Unknown Active Variables and Boundaries

AbdelMonaem F.M. AbdAllah, Daryl L. Essam and Ruhul A. Sarker

Abstract In this paper, we consider a class of dynamic optimization problems in which the number of active variables and their boundaries vary as time passes (DOPUAVBs). We assume that such changes in different time periods are not known to decision makers due to certain internal and external factors. Here, we propose three variants of genetic algorithm to deal with a dynamic problem class. These proposed algorithms are compared with one another, as well as with a standard genetic algorithm based on the best of feasible generations and feasibility percentage. Experimental results and statistical tests clearly show the superiority of our proposed algorithms. Moreover, the proposed algorithm, which simultaneous addresses two sub-problems of such dynamic problems, shows superiority to other algorithms in most cases.

Keywords Active · Best of feasible generations
Dynamic optimization problems · Feasibility percentage
Genetic algorithms · Mask detection

1 Introduction

Optimization is one of the essential research fields that directly relates to everyday decision making problems, such as planning, transportation and logistics. There are different classes of optimization problems. Based on the variables that affect them, optimization problems can be categorized as discrete, continuous or mixed-variables,

---

A.F.M. AbdAllah (✉) · D.L. Essam · R.A. Sarker
School of Engineering and Information Technology, University of New South Wales Canberra (UNSW Canberra@ADFA), Canberra, ACT 2600, Australia
e-mail: a.abdallah@student.adfa.edu.au; abdo_system@yahoo.com

D.L. Essam
e-mail: d.essam@adfa.edu.au

R.A. Sarker
e-mail: r.sarker@adfa.edu.au

I. Zelinka et al. (eds.), *Innovative Computing, Optimization and Its Applications*, Studies in Computational Intelligence 741,
https://doi.org/10.1007/978-3-319-66984-7_9

as discussed in the following subsections. Also, optimization problems can be categorized based on existing constraints, with constrained problems generally considered to be more complicated than unconstrained ones. Furthermore, they can be categorized as static that do not change over time [1], or dynamic, where at least one part of a problem changes over time [2]. In many real-life situations, problems change as time passes, such as the demand and the capacity at different nodes and arcs in transportation systems. In Dynamic Optimization Problems (DOPs), at least one part of the problem, such as its objective function or constraints change over time. Therefore, for DOPs solving algorithms, it is important to not only locate optimal solutions, but to also track changes as time passes [3, 4]. As a result, DOP has become a challenging topic in computer science and operations research.

In the literature, most of the research carried out in DOPs deals with changes in the objective functions and/or constraints [3, 4]. However, the CEC2009 competition presented dynamic problems which are the only attempt that considers changes in problem dimensionality. In that competition, the number of variables is simply increased or decreased by adding or eliminating a variable from the end of the problem vector. So, to the best of our knowledge, there is not a detailed study taking into consideration changes in active variables and boundaries.

In an earlier work, we defined dynamic optimization problems with unknown active variables and also proposed a type of algorithm to solve such problems [5]. Furthermore, we conducted research on an initial version of dynamic optimization problems with known changeable boundaries [6].

Here, in this paper, we introduce a DOP with unknown active variables and boundaries (DOPUAVBs), in which both the active variables and their boundaries change as time passes. Therefore, a DOPUAVB consists of two sub-problems: DOPs with unknown active variables (DOPUAVs) and DOPs with unknown dynamic boundaries (DOPUBs). To solve such a dynamic problem, we develop three variants of genetic algorithms (GAs). The first algorithm considers the activeness of variables. The second considers the changeable boundaries of the variables, and the third simultaneously considers both sub-problems. The proposed algorithms were compared with one another, as well as with a simple GA (SGA), on the basis of the average of the feasible generations and percentage.

This paper is organized as follows. In Sect. 2, dynamic optimization problems with unknown active variables and boundaries are introduced and described, and a framework is provided for generating its test problems. Section 3 introduces three proposed GA-based techniques to solve such dynamic problems, along with SGA. Section 4 includes experimental results and comparisons among all GA-based techniques. Finally, conclusions and directions for future work are presented in Sect. 5.

2 Dynamic Optimization Problems with Unknown Active Variables and Boundaries (DOPUAVBs)

In this section, we propose a new type of dynamic problem, called dynamic optimization problem with unknown active variables and boundaries (DOPUAVB). In such dynamic problems, the activeness of variables and their boundaries change as time passes. Therefore, a DOPUAVB consists of two sub-problems: a DOP with unknown active variables (DOPUAV) and a DOP with unknown dynamic boundaries (DOPUB). In a DOPUAV, active variables change, while in a DOPUB, the boundaries of variables change as time passes. Without loss of generality, this paper considers minimization problems.

To generate an instance for the DOPUAVB, its two sub-problems are considered. First, for a DOPUAV, active variables affect a decision during the time slot, while inactive variables do not. To simulate such dynamic problems, a mask with variable coefficients of 0s and 1s is randomly generated to determine inactive and active variables respectively. Let us consider a simple example of an absolute function with 5 variables: $abs(x_1 + x_2 + x_3 + x_4 + x_5)$; the minimal value for this function is when x_1: x_5 equal 0. Suppose that two of these variables are inactive; let x_2 and x_5 be chosen to be inactive (its mask value is set equal to 0) while the others are active (1). In such a case, the optimal occurs when x_1, x_3 and x_4 converge to 0s, while x_2 and x_5 have any other values. This is because the values of the inactive variables, x_2 and x_5, are ignored; to do this they are always evaluated as 0. Moreover, due to mutation, crossover and lack of selection pressure processes, x_2 and x_5 tend to diverge to different random values. Hence, the efficiency of an algorithm for solving DOPUAVs depends on how active and inactive variables are handled as time passes. In [1], to solve DOPUAVs, it is suggested that an algorithm needs to periodically detect the mask of the problem every a specified number of generations. Here in this paper, to save fitness evaluations, the solving algorithm tries to detect whether or not the problem has changed before detecting the problem mask.

Second, in DOPUB, the original/default boundaries of the variables are [−5, +5] with initial width equal to 10, and are shifted randomly inside these boundaries by a range of [−3, +3]; where "−3" and "+3" shifts the dynamic/feasible/changeable/shifted boundaries to the left and right by 3 steps respectively, while maintaining a minimum width of these dynamic/feasible boundaries being 2. Then, if any of the variable values of a solution are within current feasible boundaries, the fitness function will be assigned its actual fitness value; otherwise, a maximum value (DBL_MAX) will be assigned. This is because for such an infeasible solution, the objective function does not have any function value or information about infeasible areas. In this problem, in contrast to constrained optimization problems, in DOPUB the objective function cannot assign any constraint violation value to the infeasible solution(s). Moreover, the objective function cannot tell which variable is outside of its feasible boundaries. Note that for evaluating either feasible or infeasible solution, the number of conducted fitness evaluations is increased by 1.

3 Genetic-Based Algorithms for Solving DOPUAVBs

In this paper, four genetic algorithms (GAs) are used to solve dynamic optimization problems with unknown active variables and boundaries (DOPUAVBs). These GAs proposed to illustrate how they deal with the proposed problem under different considerations. These GA-based techniques are presented as follows:

3.1 SGA

The first GA is a simple GA (SGA), in which its operators work normally without any modifications. In other words, processes of selection, crossover and mutation deal with the original boundaries for all variables without any consideration of variables activeness and/or their current dynamic boundaries. Figure 1 shows the basic structure of SGA.

3.2 GAUAV

The second GA deals with unknown active variables (GAUAV). This GA considers the first sub-problem (DOPUAV). The GAUAV consists of two processes as follows:

Problem change detection. This process is used to detect whether or not a problem has changed. For problem change detection, some experiments were conducted. From those experiments, to reduce the probability of false problem change detection, a *nonZeroNotEqualAbsChromosome* is used. A *nonZeroNotEqualAbsChromosome* is a

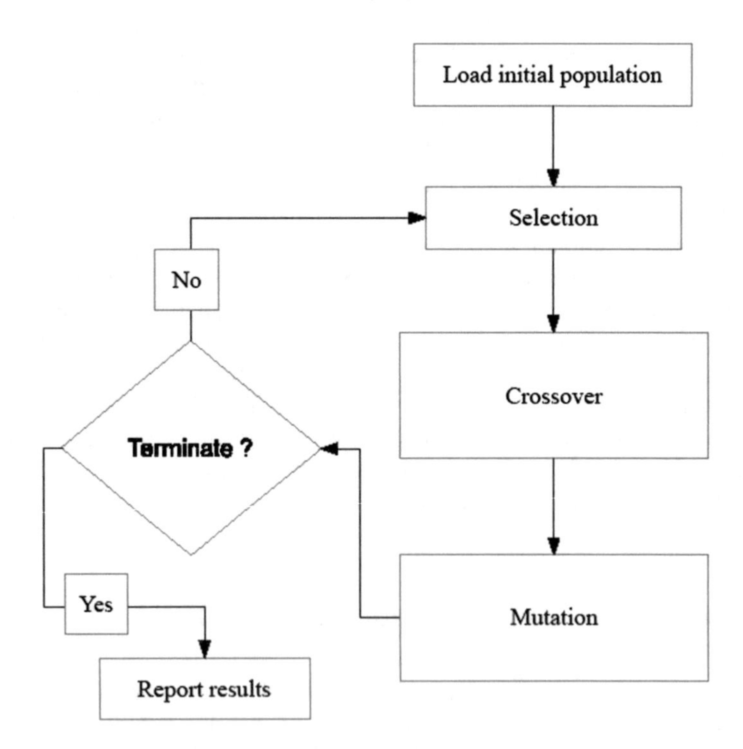

Fig. 1 The basic structure for SGA

chromosome that has non-zero, not-equal and unique absolute values, for example, if there is a chromosome with five genes, it might be (1, 2, 5, 4, 3). This chromosome is re-evaluated every generation, if its fitness is changed, then a change is detected. Once a change is detected, the GAUAV tries to detect the current problem mask using the mask detection process.

Mask detection process. This process is used to detect the mask of the inactive and active variables. Here, mask detection is done by using the single-sample mask detection procedure as follows:

- Choose a random chromosome.
- Calculate its actual fitness, let it be *F1*.
- Then for each variable, its value is replaced by a new random value within its boundary which is generated, where the absolute value of this new value is not equal to its old value:

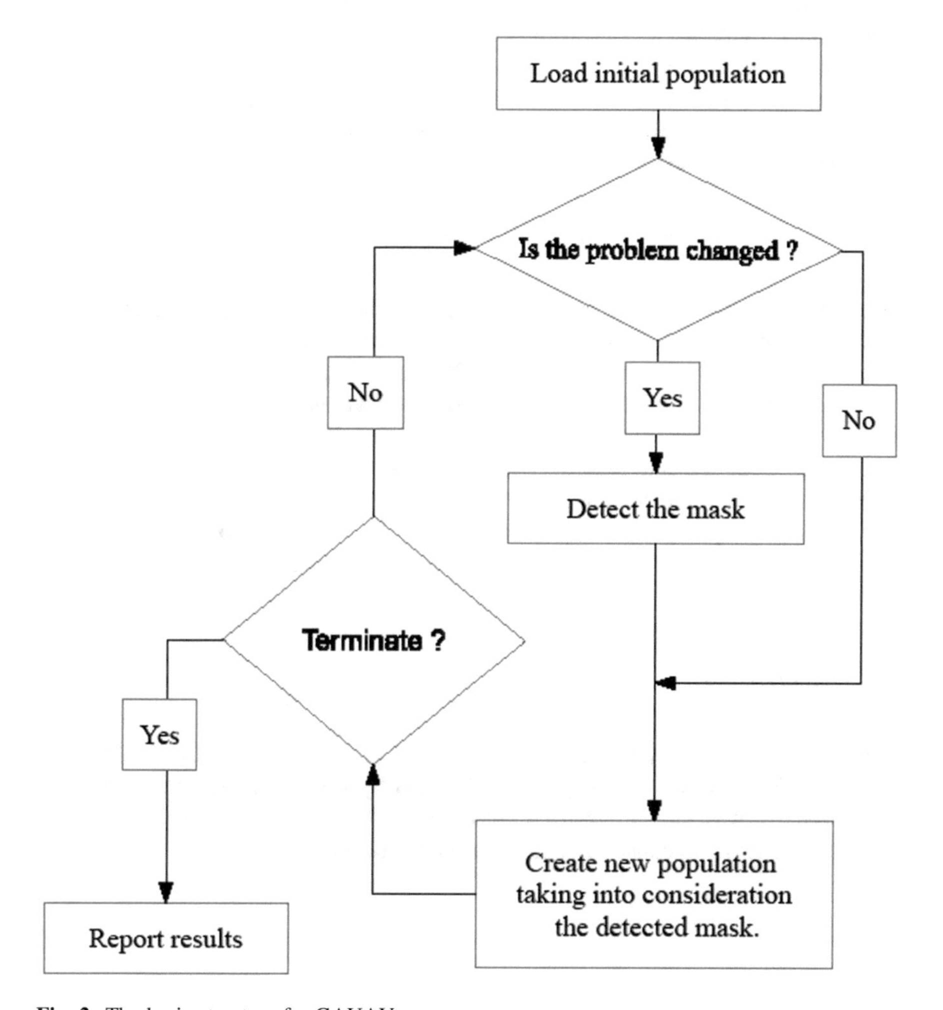

Fig. 2 The basic structure for GAUAV

- The fitness is re-calculated, let it be *F2*.
- If *F1* equals *F2*, then this variable is assumed to be inactive (its detected mask value is equal to 0); otherwise, it is assumed to be active (its detected mask value is equal to 1).

Figure 2 shows the basic structure of GAUAV.

3.3 GAUB

The third GA deals with unknown dynamic boundaries (GAUB). This GA tries to detect and use feasibility during the course of a search process. To do this, the GAUB keeps track of the feasible boundaries for feasible chromosomes, where the current lower and upper boundaries of the feasible area is the minimum and maximum variable values of feasible chromosomes respectively. Then, GAUB uses the detected feasible boundaries to evaluate infeasible chromosomes, by considering the distance between them and the centroid of the detected feasible boundaries as a degree of constraint violation. This is to guide GAUB during its selection process; it is calculated as follows:

$$X_{i(constraint\,violation)} = \sum\nolimits_{j=1}^{N} d\left(X_{ij}, feasible\,centroid_j\right), \tag{1}$$

where X_i is an infeasible chromosome, N is the number of variables, and d is the distance metric. Figure 3 shows the basic structure of GAUB.

Here an illustrative example is used to show how GAUB computes the constraint violation value of an infeasible solution. Suppose we used the Manhattan distance as the distance metric and there is a problem consists of 2 variables that have boundaries [−5, 5]. An infeasible solution (−2, 2) is exist, and the currently detected feasible boundaries are [−1, −3] and [2, 0], so the feasible centroid is (1, −1). So, using Eq. 1, the constraint violation of this infeasible solution equals (abs (−2 − (1)) + abs (2 − (−1))) = (3 + 3) = 6, where abs is the absolute function.

3.4 GAUAVB

The fourth GA is a hybrid GA of the second and the third GAs (GAUAVB). This GA tries to solve the complex DOPUAVB by simultaneously considering its sub-problems. It is shown in Fig. 4. Furthermore, it considers only active variables when calculating the constraint violation value as follows:

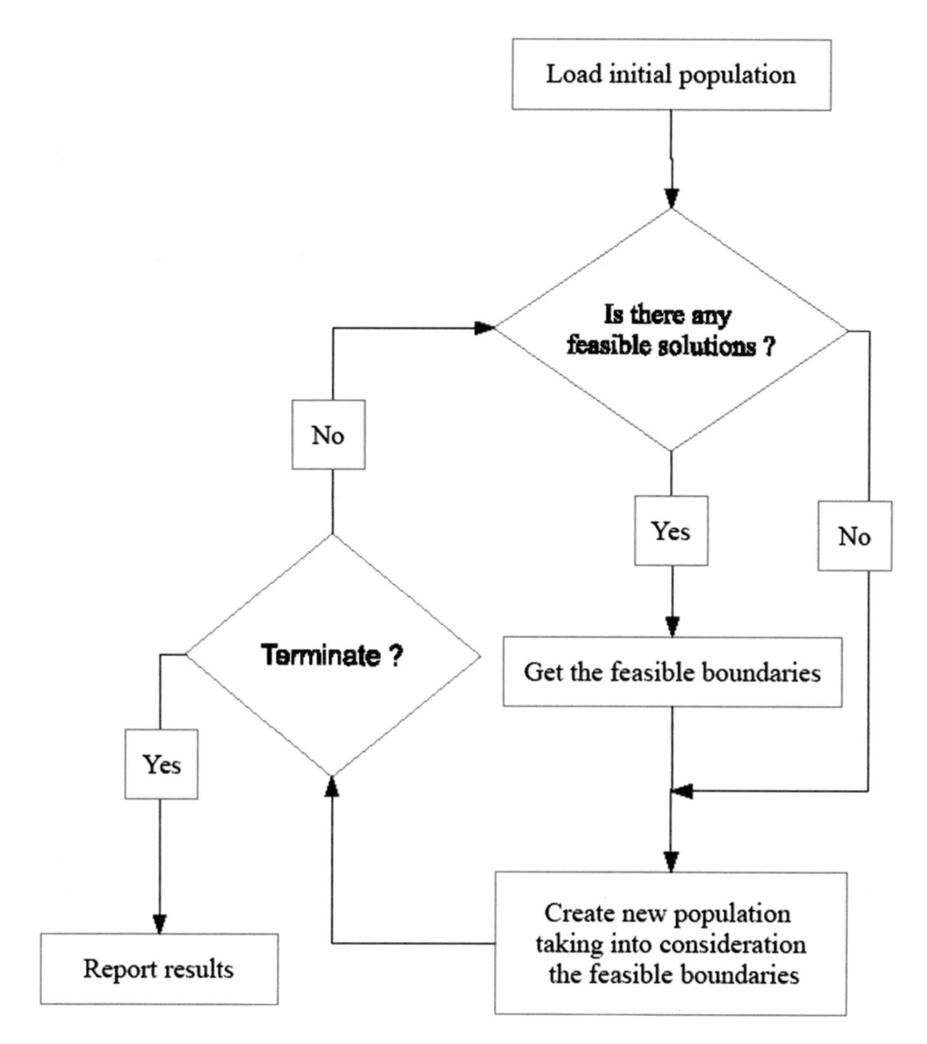

Fig. 3 The basic structure for GAUB

$$X_{i(constraint\ violation)} = \sum_{j=1}^{N} d\left(X_{ij}, feasible\ centroid_j\right),\ if\ variable\ j\ is\ active, \quad (2)$$

Using the previously used example in Sect. 3.3, suppose that the first variable is detected as an inactive variable. In this case, the distance violation of the first variable is excluded from the constraint violation calculations, So, using Eq. 2, the constraint violation of this infeasible solution equals (abs (2 − (−1))) = 3.

Note that for GAUAV and GAUAVB, when variables are detected as inactive, they are prevented from being mutated. Furthermore, the tournament selection in both GAUB and GAUAVB is adapted by using feasibility rules [2]. It works as follows:

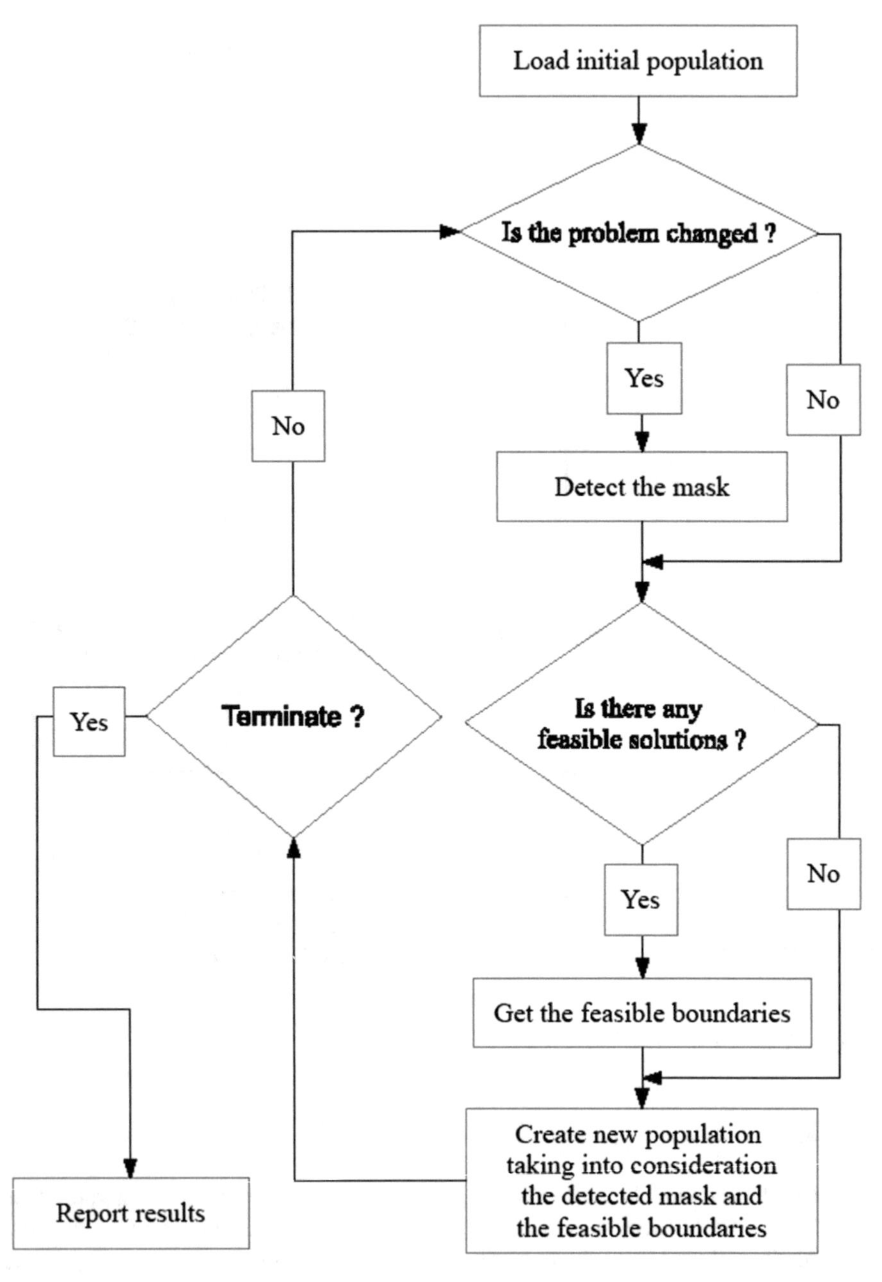

Fig. 4 The basic structure for GAUAVB

(1) If two compared solutions are both feasible, select the solution based on the minimum value of the fitness function.
(2) If one of the two compared solutions is a feasible and the other is infeasible, the feasible solution is selected.
(3) If both solutions are infeasible, select solutions based on the constraint violation (the distance between the solution and the centroid of the current feasible boundaries). The solution with the lowest violation wins.

4 Experimental Setup, Analysis and Discussion

To test the performance of the previously presented genetic algorithms (GAs)-based techniques for solving DOPUAVBs, real-coded GA-based algorithms with the same processes were implemented. The crossover is one-point, mutation is uniform and the selection is tournament. In this paper, a set of unconstrained benchmark functions, namely Sphere, Rastrigin, Weierstrass, Griewank, Ackley, Rosenbrock, Levy and NeumaierTrid are used to form these functions. Of these functions, five are completely separable problems, and three functions are non-separable. The five separable problems, namely, Sphere, Ackley, Griewank, Restrigin and Weierstrass, were used in previous test suites of dynamic problems [3], while the other three non-separable functions, namely, Levy, Rosenbrock and Trid, are taken from Surjanovic and Bingham (2015) [4, 5]. The global optimum for all these functions is F(x*) = 0 with all solution variables [x]* = 0, except Rosenbrock's where x* = 1, and Trid where the global depends on the number of dimensions. From these five separable functions, Sphere is unimodal, while Griewank's, Rastrigin's, Ackley's and Weierstrass are multimodal. Unimodal modal functions contain only one optimum and hence are easier to optimise. On the other hand, multimodal functions are complex and contain many local optima and one global optimum. Of the three non-separable functions, Trid and Rosenbrock are unimodal, and Levy is a multimodal function. Optimising non-separable functions is generally harder than optimising separable functions. Therefore, a multimodal non-separable function is more challenging than optimising a unimodal separable function. Table 1 summarises some characteristics of the used functions followed by their equations

$$F_{Sphere} = \sum_{i=1}^{N} x_i^2 \tag{3}$$

$$F_{Ackley} = -20\,\mathbf{exp}\left(-0.2\sqrt{\frac{1}{N}\sum_{i=1}^{N} x_i^2}\right) - \mathbf{exp}\left(\frac{1}{N}\sum_{i=1}^{N} \mathbf{cos}(2\pi x_i)\right) + 20 + e \tag{4}$$

$$F_{Griewank} = \sum_{i=1}^{n} \frac{x_i^2}{4000} - \prod_{i=1}^{n} cos\left(\frac{x_i}{\sqrt{i}}\right) + 1 \tag{5}$$

Table 1 Summary of used functions characteristics

Function	Separable	Multimodal	Global value and location
Sphere	Yes	No	0 where $x^* = 0$
Ackley	Yes	Yes	0 where $x^* = 0$
Griewank	Yes	Yes	0 where $x^* = 0$
Restrigin	Yes	Yes	0 where $x^* = 0$
Weierstrass	Yes	Yes	0 where $x^* = 0$
Levy	No	Yes	0 where $x^* = 0$
Rosenbrock	No	No	0 where $x^* = 1$
Trid	No	No	Depends on the number of variables

$$F_{Rastrigin} = \sum_{i=1}^{N} \left(x_i^2 - 10\,cos(2\pi x_i) + 10\right) \tag{6}$$

$$F_{Weierstrass} = \sum_{i=1}^{n} \left(\sum_{k=0}^{k_{max}} \left[a^k cos\left(2\pi b^k (x_i + 0.5)\right)\right]\right) - n \sum_{k=0}^{k_{max}} \left[a^k cos\left(\pi b^k\right)\right] \tag{7}$$

where: $a = 0.50$, $\mathrm{b} = 3$, and $\mathrm{k_{max}} = 20$.

$$\mathbf{F_{Levy}} = \mathbf{sin}^2(\pi\,\mathbf{w}_1) + \sum_{\mathbf{i}=1}^{\mathbf{n}-1} (\mathbf{w_i} - 1)^2\left[1 + 10\,\mathbf{sin}^2(\pi\,\mathbf{w_i} + 1)\right](\mathbf{w_n} - 1)^2\left[1 + \mathbf{sin}^2(2\pi\,\mathbf{w_n})\right] \tag{8}$$

where: $w_i = 1 + \frac{x_i - 1}{4}$, for all I = 1, …, n

$$F_{Rosenbrock} = \sum_{i=1}^{N-1} \left(100\left(x_i^2 - x_{i+1}\right)^2 + (x_i - 1)^2\right) \tag{9}$$

$$F_{Trid} = \sum_{i=1}^{n} (x_i - 1)^2 - \sum_{i=2}^{n} x_i x_{i-1} \tag{10}$$

The compared algorithms were tested under different settings of DOPUAVB as follows:

(1) The frequency of change (FOC), which determines how often a problem changes; was varied as 500, 2000 and 8000 fitness evaluations. This is used to test how the number of fitness evaluations might affect the algorithm performance.
(2) The number of inactive variables/the number of variables that have shifted boundaries (NOV); was varied as 1/1, 5/1, 1/5 and 5/5, where the first number represents the number of inactive variables and the second number represents the number of shifted boundaries of variables. This is used to test how the number of the active variables and changeable boundaries might affect the algorithm performance.

Experimental settings are shown in Table 2. Here, Manhattan distance [6] is used to calculate the degree of constraint violation. Note that fitness evaluations that

Table 2 Experimental settings

Parameter	Value
Population size	50
Tournament size	2
Selection pressure	0.90
Elitism percentage	2
Crossover rate	0.90
Mutation rate	0.15
Number of variables	20

are used in problem change detection and mask change detection are included in the budget of all of the algorithms. The algorithms were all coded in Microsoft C++, on a 3.4 GHz/16 GB Intel Core i7 machine running the Windows 7 operating system. Finally, for a fair comparison, all GAs had the same initial population at the beginning of each run with a total of 25 independent runs.

4.1 Comparison Based on the Quality of Solutions

To compare the quality of solutions, a variation of the Best-of-Generation measure was used, where best-of-generation values were averaged over all generations [7]. However, in DOPUAVBs, due to the change in the feasible boundaries, solving techniques might have infeasible generations, so we consider only feasible generations in these calculations. To do this, we propose a new variation of the Best-of-Generation measure, which is the average best-of-feasible-generation (ABOFG) and it is calculated as follows:

$$\overline{F}_{BOFG} = \frac{1}{F_i}\sum\nolimits_{i=1}^{G} F_{BOFG_{ij}}\left(\frac{1}{N}\sum\nolimits_{j=1}^{N} F_{BOFG_{ij}}\right), \text{where generation } i \text{ is feasible,} \quad (11)$$

where $\overline{F}_{BOFG}$ is the mean best-of-feasible-generation fitness, G is the number of generations, N is the total number of runs and $F_{BOFG_{ij}}$ is the best-of-feasible-generation fitness of generation i of run j of an algorithm on a problem [8]. As solved functions have different scales for their objective functions values, a normalized score is used so as to be able to sum obtained values of different functions to analyze the performance of compared algorithms. Note that lower values are better and the lowest are shown as bold and shaded entries. Table 3 shows the results of normalized ABOFGs for the compared techniques in regards to the number of variables (NOV).

From Table 3, it is first clearly observed that GAUAV performed better, especially when the number of unknown variables increased (5/1 and 5/5). Second, GAUB slightly performed better than SGA when the number of shifted boundaries increased (1/5 and 5/5). Third, GAUAVB outperformed all GAs in most cases.

Table 3 Normalized ABOFGs of compared algorithms in regards NOV

NOV	SGA	GAUAV	GAUB	GAUAVB
1/1	0.7444	**0.1743**	0.8111	0.2220
1/5	0.8412	0.3795	0.5727	**0.1730**
5/1	0.8898	**0.0893**	0.8226	0.1795
5/5	0.8559	0.2848	0.7030	**0.0828**
Average	0.8328	0.2320	0.7273	**0.1643**

Table 4 Normalized ABOFGs of compared techniques in regards to FOC

FOC	SGA	GAUAV	GAUB	GAUAVB
500	0.7889	**0.2540**	0.6733	0.2671
2000	0.8447	0.2264	0.7273	**0.1587**
8000	0.8648	0.2156	0.7814	**0.0672**
Average	0.8328	0.2320	0.7273	**0.1643**

Presumably, GAUAV and GAUAVB performed better as they prevented inactive variables from being mutated, as this helps GAs to effectively converge to better solutions.

From Table 4, it is clearly observed that GAUAV performed better than other GAs except GAUAVB. However, GAUAVB outperformed other GAs, especially when the frequency of changes (FOC) increased.

The Wilcoxon signed rank test [9] was also used to statistically judge the difference between paired scores, this was done because as obtained values of compared algorithms are not normally distributed, a non-parametric statistical test is used. As a null hypothesis, it is assumed that there is no significant difference between obtained values of two samples, whereas the alternative hypothesis is that there is a significant difference at the 5% significance level. Based on the obtained results, one of three signs (+, −, and ≈) is assigned when the first algorithm was significantly better, worse than, and no significance different with the second, respectively. Here, GAUAVB was paired to be compared with other GA-based variations to see how it effectively solved DOPUAVBs. In Table 5, Wilcoxon tests were applied on the total number of changes in regards to the number of variables; in this paper, there are 8 problems, and each has 10 types of changes, and each change has 3 frequency of changes, with a total of 240 values.

From Table 5, it is clear that GAUAVB was statistically better than other GAs in most cases, especially when the problem was more complex 5/5. The performance of GAUAV was statistically better for 5/1 test problems, as the number of inactive variables increases with limited changes in dynamic boundaries.

In Table 6, Wilcoxon test were again applied on the number of changes. In regards to the frequency of changes; in this paper, there are 8 problems, each has 10 changes, and each change has 4 variations of the number of variables, which gives a total of 320 values.

From Table 6, it is clear that GAUAVB was statistically better than the other GAs in most cases, especially when frequency of change increases.

Table 5 Wilcoxon signed test the compared techniques in regards to NOV

NOV	Comparison	Better	Worse	Significance
1/1	GAUAVB-to-SGA	197	43	+
	GAUAVB-to-GAUAV	123	117	≈
	GAUAVB-to-GAUB	208	32	+
1/5	GAUAVB-to-SGA	215	25	+
	GAUAVB-to-GAUAV	150	90	+
	GAUAVB-to-GAUB	198	42	+
5/1	GAUAVB-to-SGA	225	15	+
	GAUAVB-to-GAUAV	93	147	–
	GAUAVB-to-GAUB	220	20	+
5/5	GAUAVB-to-SGA	232	8	+
	GAUAVB-to-GAUAV	161	79	+
	GAUAVB-to-GAUB	228	12	+

Table 6 Wilcoxon signed test the compared techniques in regards to FOC

FOC	Comparison	Better	Worse	Significance
500	GAUAVB-to-SGA	268	52	+
	GAUAVB-to-GAUAV	149	171	≈
	GAUAVB-to-GAUB	255	65	+
2000	GAUAVB-to-SGA	288	32	+
	GAUAVB-to-GAUAV	177	143	+
	GAUAVB-to-GAUB	285	35	+
8000	GAUAVB-to-SGA	313	7	+
	GAUAVB-to-GAUAV	201	119	+
	GAUAVB-to-GAUB	314	6	+

Finally, in order to statistically compare and rank the algorithms altogether, the non-parametric Friedman test, which is similar to the ANOVA parametric, is used with a confidence level of 95% ($\alpha = 0.05$) was used [10, 11]. The null hypothesis was that there is no significant differences among compared algorithms. The computational value of the p-value was less than 0.00001. Consequently, there were significant differences among the compared algorithms. Finally, Table 7 shows Freidman test ranks; it supports above mentioned observations.

Table 7 Freidman test average ranks for compared techniques

Algorithm	SGA	GAUAV	GAUB	GAUAVB
Average rank	3.34	1.86	3.14	**1.66**

Table 8 AFPs of compared techniques in regards to the NOV

NOV	SGA (%)	GAUAV (%)	GAUB (%)	GAUAVB (%)
1/1	77.50	77.39	89.74	**90.26**
1/5	45.64	46.63	67.15	**68.83**
5/1	86.13	86.78	92.23	**93.03**
5/5	38.33	40.39	64.96	**67.31**
Average	61.90	62.80	78.52	**79.86**

Table 9 AFPs of compared techniques in regards to the FOC

FOC	SGA (%)	GAUAV (%)	GAUB (%)	GAUAVB (%)
500	58.96	60.29	71.82	**73**
2000	62.90	62.91	79.53	**80.24**
8000	63.96	64.19	83.74	**84.61**
Average	61.94	62.46	78.36	**79.28**

4.2 Comparison Based on Feasibility

In this section, the behaviors of the used algorithms are compared, based on the feasibility of the population. To do this, the average feasibility (AFP) was calculated for each algorithm. AFP indicates how an algorithm can guide its population into the changeable feasible region. Tables 8 and 9 summarize the obtained AFPs, higher values are better and the best are shown as bold and shaded entries.

From Tables 8 and 9, it is clearly observed that GAUAVB achieved higher AFPs, compared with other GAs. GAUAV also slightly achieved better AFPs than SGA when the number of shifted boundaries increased (1/5 and 5/5). It is clear that GAUB and GAUAVB achieved higher AFPs, as they guided the infeasible solution (s) towards the feasible area, by assigning a constraint violation value that guided the selection process. Also, it is clear that the founding feasible area while solving DOPUAVB is getting more complex and harder when NOV increases, especially when the number of changed boundaries increase (Table 7), and the FOC decreases (Table 8).

5 Conclusions and Future Work

Motivated by the literature [11, 12], in this paper we proposed a new type of dynamic optimization problem: single objective unconstrained dynamic optimization problems with unknown active variables and boundaries (DOPUAVBs). In such problems, both the active variables and boundaries of the variables change as time passes. Therefore, DOPUAVB consists of two sub-problems: DOP with unknown active variables (DOPUAV) and DOP with unknown boundaries (DOPUB).

Moreover, we proposed three genetic algorithms (GA)-based techniques to solve DOPUAVBs. These techniques are GAUAV (GA that deals with unknown active variables), GAUB (GA that deals with unknown changeable boundaries) and GAUAVB (GA that simultaneously deals with unknown active variables and dynamic boundaries). These techniques were compared with each another, as well as with a simple GA (SGA). Based on the quality of obtained solutions and the average of feasibility, as well as statistical tests, results showed that the proposed GAUAVB, that simultaneously considered both sub-problems, was superior to others. This is because GAUAVB had the ability to detect active variables, while also keeping track of feasible boundaries during the course of a search process. Hence it effectively solved DOPUAVBs. The advantages of the proposed technique is using the detected information of the population during the search process to solve the dynamic problem, e.g. the detected feasible boundaries and active variables. However, the disadvantage of GAUAVB is needing for existing of detected feasible boundaries and this would be difficult if the change in boundaries is rapid and has much shift rate.

There are several possible directions for future work. One direction is comparing our proposed algorithms with previously used GAs for DOPs, such as random immigration (RIGAs) and hyper-mutation (HyperM). Regarding sub-problems, we intend to solve each of them in more effective ways. For example, designing an algorithm that would implicitly detect active variables by keeping track of active variables, rather than using the mask process, as it consumes fitness evaluations. This is because the number of fitness evaluations it uses is 2 N, where N is the number of variables.

References

1. AbdAllah, A. F. M., Essam, D. L., & Sarker, R.A. (2014). Solving dynamic optimisation problem with variable dimensions. In *SEAL 2014*. Dunedin, New Zealand: Springer International Publishing.
2. Coello Coello, C. A., & Mezura Montes, E. (2002). Constraint-handling in genetic algorithms through the use of dominance-based tournament selection. *Advanced Engineering Informatics, 16*(3), 193–203.
3. Li, C., Yang, S., Nguyen, T. T., Yu, E. L., Yao, X., Jin, Y., et al. (2009). Benchmark generator for CEC' 2009 competition on dynamic optimization.
4. Surjanovic, S., & Bingham, D. (2015). Virtual library of simulation experiments: Test functions and Datasets. January 2015 [cited 2016 April 20]. Retreived from: http://www.sfu.ca/~ssurjano.
5. Adorio, E. P., & Diliman, U. P. MVF—Multivariate test functions library in C for unconstrained global optimization.
6. Padhye, N., Deb, K., & Mittal, P. An efficient and exclusively-feasible constrained handling strategy for evolutionary algorithms.

7. Morrison, R. W. (2003) Performance measurement in dynamic environments. In *GECCO workshop on evolutionary algorithms for dynamic optimization problems* (pp. 5–8).
8. Yang, S., Nguyen, T. T., & Li, C. (2013). Evolutionary dynamic optimization: Test and evaluation environments. In S. Yang & X. Yao (Eds.), *Evolutionary computation for dynamic optimization problems* (pp. 3–37). Berlin Heidelberg: Springer.
9. Corder, G. W., & Foreman, D. I. (2009) *Nonparametric statistics for non-statisticians: A step-by-step approach*. Wiley.
10. García, S., Molina, D., Lozano, M., & Herrera, F. (2009). A study on the use of non-parametric tests for analyzing the evolutionary algorithms' behaviour: A case study on the CEC' 2005 Special Session on Real Parameter Optimization. *Journal of Heuristics, 15* (6), 617–644.
11. Cruz, C., González, J. R., & Pelta, D. A. (2011). Optimization in dynamic environments: a survey on problems, methods and measures. *Soft Computing, 15*(7), 1427–1448.
12. Nguyen, T. T., Yangb, S., & Branke, J. (2012). Evolutionary dynamic optimization: A survey of the state of the art. *Swarm and Evolutionary Computation, 6,* 1–24.

Text Segmentation Techniques: A Critical Review

Irina Pak and Phoey Lee Teh

Abstract Text segmentation is a method of splitting a document into smaller parts, which is usually called segments. It is widely used in text processing. Each segment has its relevant meaning. Those segments categorized as word, sentence, topic, phrase or any information unit depending on the task of the text analysis. This study presents various reasons of usage of text segmentation for different analyzing approaches. We categorized the types of documents and languages used. The main contribution of this study includes a summarization of 50 research papers and an illustration of past decade (January 2007–January 2017)'s of research that applied text segmentation as their main approach for analysing text. Results revealed the popularity of using text segmentation in analysing different languages. Besides that, the word segment seems to be the most practical and usable segment, as it is the smaller unit than the phrase, sentence or line.

1 Introduction

Text segmentation is process of extracting coherent blocks of text [1]. The segment referred as "segment boundary" [2] or passage [3]. Another two studies referred segment as subtopic [4] and region of interest [5]. There are many reasons why splitting document can be useful for text analysis. One of the main reasons is because they are smaller and more coherent than whole documents [3]. Another reason is each segment is used as units of analysis and access [3]. Text segmentation was used to process text in emotion extraction [6], sentiment mining [7, 8], opinion mining [9, 10], topic identification [11, 12], language detection [13] and information retrieval [14]. Sentiment analyzing within the text covers wide range of

I. Pak (✉) · P.L. Teh
Department of Computing and Information Systems, Sunway University, Bandar Sunway, Malaysia
e-mail: irina.p@imail.sunway.edu.my

P.L. Teh
e-mail: phoeyleet@sunway.edu.my

I. Zelinka et al. (eds.), *Innovative Computing, Optimization and Its Applications*, Studies in Computational Intelligence 741,
https://doi.org/10.1007/978-3-319-66984-7_10

techniques, but most of them include segmentation stage in text process. For instance, Zhu et al. [15] used segmentation in thier model to identify multiple polarities and aspects within one sentence. There are studies that applied tokenization in the semantic analysis to increase the probability of obtaining the useful information by processing tokens. For example, the study of Gan et al. [16] applied tokenization in their proposed method where semantics were used to improve search results for obtaining more relevant and clear content from the search. Later, Gan and Teh [17] used technique similar to segmentation where information is organized into segments called facet and values in order to improve search algorithm. Another study of Duan et al. [18] applied text segmentation and then tokenization to determine aspects and associated features. In other words, tokenization is a similar process to text segmentation. That is splitting text into words, symbols, phrases, or any meaningful units named as a token.

This paper reviews different methods and reasons of applying text segmentation in opinion and sentiment mining and language detection. The target is to overview of text segmentation techniques with brief details. The contribution of this paper includes the categorizations of recent articles and visualization of the recent trend of research in the opinion mining and related areas, such as sentiment analysis and emotion detection. Also in pattern recognition, language processing, and information retrieval. Next section of this paper explains the scope and method used to review the past studies. Section 3 discusses the results of summarized articles. Section 4 contains a discussion on results. Lastly, Sect. 5 concludes this paper.

2 Review Method

The review process of the articles includes publications from the past 10 years. Fifty journals and conferences in total were evaluated. These articles implemented text segmentation in their main approaches. Fifty articles are summarized in Table 1 in this section.

In order to make it clear content of the Table 1, here is breakdown the column by column. The first column of the table refers to the year of the article. Next column includes the references of study. In the following column, there is a brief description of the study. There are different types of segments used in their study, it includes: (1) topic, (2) word, (3) sentence etc. Type of segment commonly selected based on their analyzing targets and specifications. The fourth column listed the type of segment used in the assessed study. The evaluation of each study has applied different sets of data or documentations. They can be categorised as: (1) corpus, (2) news, (3) articles, (4) reviews, (5) datasets etc. We listed it in the fifth column. The sixth column describes the reason for applying the text segmentation study. The last column specifies those language(s) used in those sets of documentations(s).

Table 1 Summarisation of the fifty articles applied text segmentation in their studies

No	Year	Reference	Description	Segment	Data/Document	Reason	Language
1	2007	[10]	Opinion search in web blogs (logs)	Topic	Web blog	To identify opinion in web blogs (logs)	English
2	2007	[12]	Unsupervised methods of topical text segmentation for Polish	Topic	Dataset of e-mail newsletter, artificial documents, streams, and individual documents	To detect boundaries between news items or consecutive stories	Polish
3	2007	[19]	Word segmentation of handwritten text using supervised classification techniques	Word	Dataset	To identify words within handwritten text	English
4	2007	[20]	Clustering based text segmentation	Sentence	Corpus of articles	To understand the sentences relations with consideration the similarities in a group rather than individually	English
5	2007	[6]	Comprehensive information based semantic orientation identification	Word	Comments	To identify polarity orientation in text	Chinese
6	2007	[21]	Unconventional word segmentation in Brazilian Children's early text production	Word	Handwritten texts	To define word boundaries	English
7	2007	[22]	Comparative analysis of different text segmentation algorithms on Arabic news stories	Sentence	Dataset of news	To identify boundaries within the text and measure lexical cohesion between textural units	Arabic
8	2008	[23]	Automatic story segmentation in chinese broadcast news	Topic	News	To identify story boundary	Chinese
9	2009	[15]	Aspect-based sentence segmentation for sentiment summarization	Sentence	Reviews	To extract aspect-based sentiment summary	Chinese
10	2009	[24]	Chinese text sentiment classification based on granule network Zhang	Word	Corpus	To classify text based on sentiment value	Chinese

(continued)

Table 1 (continued)

No	Year	Reference	Description	Segment	Data/Document	Reason	Language
11	2009	[25]	Automatic extraction of new words based on Google news corpora for supporting lexicon-based Chinese word segmentation systems	Word	Corpus of Google news	To identify word and sentence from Chinese language texts in real-word applications	Chinese
12	2010	[26]	An information-extraction system for Urdu-a resource-poor language	Word	Blogs, comments, news articles	To identify social and human behavior within text	Urdu
13	2010	[27]	Chinese text segmentation: a hybrid approach using transductive learning moreover, statistical association measures	Word	Corpus	To build system for cross-language information	Chinese
14	2010	[7]	Sentiment classification for stock news	Word	Chinese stock news	To classify news by sentiment orientation	Chinese
15	2010	[8]	Sentiment text classification of customers reviews on the web based on SVM	Word	Comments	To improve accuracy of sentiment classification	Chinese
16	2010	[28]	The application of text mining technology in monitoring the network education public sentiment	Word	Web documents	To analyze sentiment in text to monitor public network	Chinese
17	2010	[29]	Using text mining and sentiment analysis for online forums hotspot detection and forecast	Word	Forums	To design a text sentiment approach	Chinese
18	2011	[30]	A topic modeling perspective for text segmentation.	Topic	Corpus	To design an enhance topic extraction approach	English
19	2011	[4]	Iterative approach to text segmentation	Topic subtopic	Articles	To identify topic and subtopic boundaries within content	English
20	2011	[31]	Text segmentation of consumer magazines in PDF format	Text blocks	Articles in PDF documents	To process PDF documents	English
21	2011	[32]	Rule-based Malay text segmentation tool	Sentence	Articles	To design Malay sentence splitter and tokenizer	Malay English

(continued)

Table 1 (continued)

No	Year	Reference	Description	Segment	Data/Document	Reason	Language
22	2011	[33]	A novel evaluation method for morphological segmentation	Word	Corpus	To improve word segmentation with consideration true morphological ambiguity	English
23	2011	[34]	Usage of text segmentation and inter-passage similarities	Passage	Articles	To improve text document clustering	English
24	2012	[35]	Using a boosted tree classifier for text segmentation in hand-annotated documents	Word	Dataset of handwritten documents	To convert handwritten document into digital text	English
25	2012	[1]	Two-part segmentation of text documents	Word sentence	Corpus	To process problem and solution documents	English
26	2012	[36]	Enhancing lexical cohesion measure with confidence measures, semantic relations and language model interpolation for multimedia spoken content topic segmentation	Topic	Corpus of news	To split into segments that deal with a single topic	English
27	2012	[37]	Text line segmentation in historical documents	Line	Historical documents	To combine the strengths of top-down and bottom-up approaches	Ancient
28	2012	[38]	Topic segmentation of Chinese text based on lexical chain	Topic	News	To improve method of processing text	Chinese
29	2013	[39]	Semantic-based text block segmentation	Text blocks	Web page	To retrieve image based on text around it	English
30	2013	[40]	The first steps in developing machine translation of patents	Word	Dataset	To archive the translation from English to Russian	Russian English
31	2013	[41]	Segmentation system based on the sentiments expressed in the text.	Tag Word	Reviews	To design system which identifies a sentiment expressed in text	English

(continued)

Table 1 (continued)

No	Year	Reference	Description	Segment	Data/Document	Reason	Language
32	2013	[42]	Probabilistic Chinese word segmentation with non-local information and stochastic training	Word character	Corpus	To reduce computational complexity of learning non-local information	Chinese
33	2013	[43]	Unknown Chinese word extraction based on variety of overlapping strings	Word	Corpus	To extract words from a sentence and improve extraction of unknown words	Chinese
34	2013	[44]	Text segmentation for language identification in Greek forums	Sentence topic	Forums	To identify language	Greek English
35	2014	[13]	Recognition-based segmentation of online Arabic text recognition	Word	Dataset	To recognize Arabic text within handwriting	Arabic
36	2014	[45]	Word segmentation of overlapping ambiguous strings during Chinese reading	Character	Collected dataset	To detect word boundaries and recognize the word	Chinese
37	2015	[46]	Chinese text sentiment orientation identification	Character	Corpus	To identify sentiment in the text	Chinese
38	2015	[47]	Text segmentation based on semantic word embedding	Word	Articles	To enhance semantic word embedding approach	English
39	2015	[48]	Dynamic non-parametric joint sentiment topic mixture model	Word	Dataset of forums	To filter out low/high-frequency words, single words, improper characters and stop words	Chinese
40	2015	[49]	A multi-label classification based approach for sentiment classification	Word	Microblogs	To support a sentiment classification approach	Chinese
41	2015	[50]	Topic segmentation of TV-streams by watershed transform and vectorization	Topic	Dataset of TV streams	To enhance and a topic extraction	French
42	2015	[51]	A supervised fine-grained sentiment analysis system for online reviews	Word	Corpus of hotel reviews	To pre-process data for sentiment analysis	Chinese

(continued)

Table 1 (continued)

No	Year	Reference	Description	Segment	Data/Document	Reason	Language
43	2016	[52]	Vietnamese word segmentation	Word	Dictionaries	To check how dictionary size affects word segmentation	Vietnamese
44	2016	[53]	Text segmentation of digitized clinical texts	Line	Corpus	To identify column separator	French
45	2016	[54]	Phrase-level segmentation and labeling	Phrase	Dataset	To balance between the word and sentence levels	English
46	2016	[55]	Akkadian word segmentation	Word	Corpus	To improve the language processing in cuneiform	Ancient Akkadian
47	2016	[56]	Document segmentation and classification into musical scores and text	Word	Dataset	To detect bounding boxes containing the musical score and text	English
48	2016	[57]	Candidate document retrieval for cross-lingual plagiarism detection	Topic	Corpus	To convert the suspicious document to a set of related passages	English German Spanish
49	2016	[58]	Effects of text segmentation on silent reading of Chinese regulated poems	Character	Chinese regulated poems	To monitor eye movements of native participants in reading Chinese regulated poems	Chinese
50	2017	[59]	A new watershed model-based system for character segmentation in degraded text lines	Character	Dataset of historical document	To understand the content in historical documents	Indian English South Indian

3 Results

Following, we present the summarized review in Table 1.

Table 1 presents the summarization for the review of past years' studies where different types of segmentation are specified. As well as the reason for applying text segmentation in their methods. Doubtless, the main reason of the researchers using any text segmentation in text processing is splitting documents into segments. After the document is split, segments proceed to the different phase depending on the text analysis approach.

In order to summarize and evaluate Table 1 in more details, pie charts and graphs are presented below. Each of the charts help to look at presented papers from different perspectives. Figure 1 concentrates on types of segment used in the studies. It is essential to evaluate segments' categorization as it is one of the main aim of this paper to consider segmentation features. Figure 2 presents a pie chart which shows variety of different languages used in evaluated papers. Lastly, Fig.3

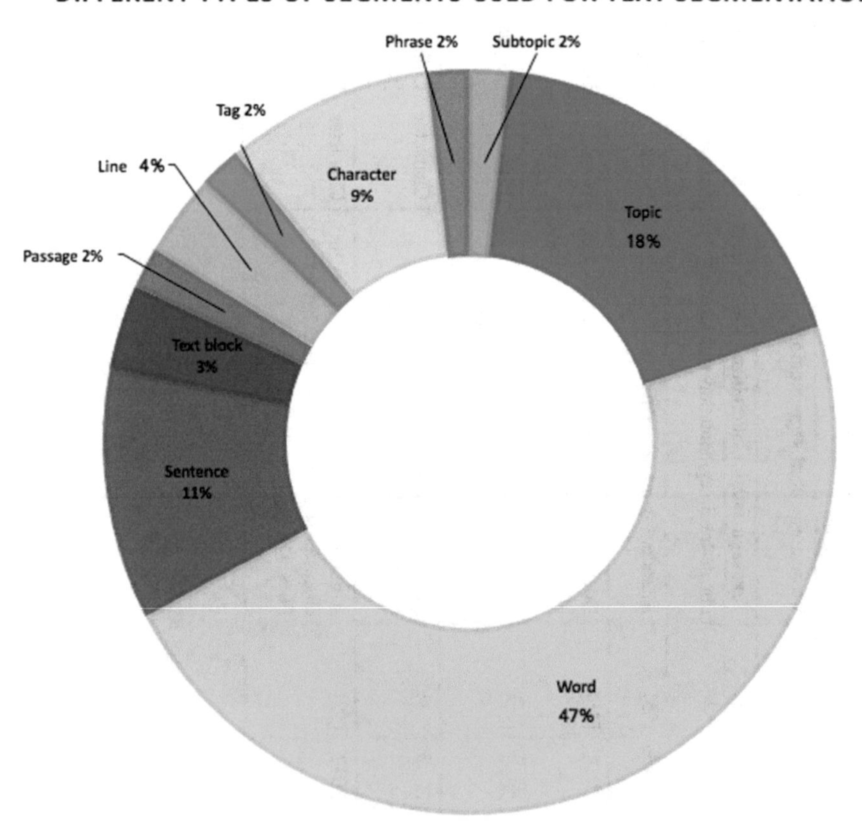

Fig. 1 Percentage of different used for text segmentation

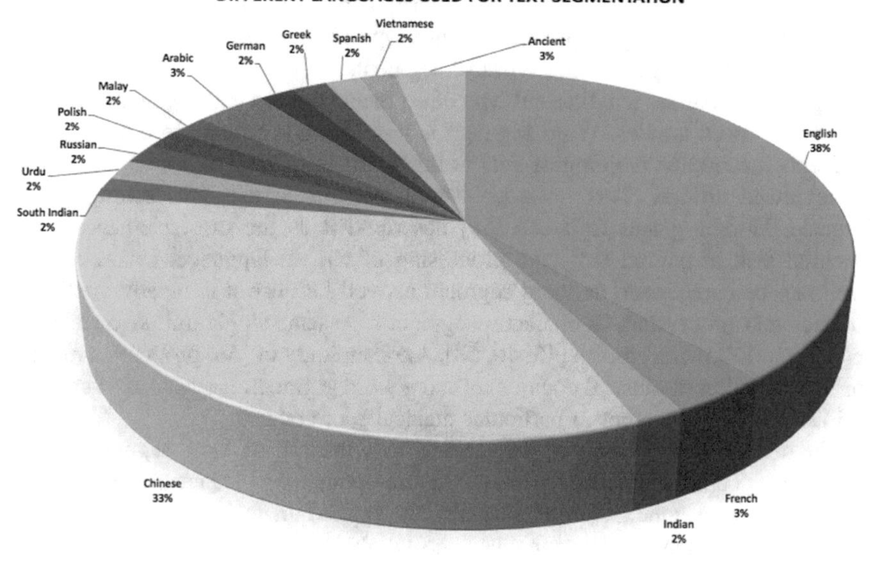

Fig. 2 Percentage of different languages used for text segmentation

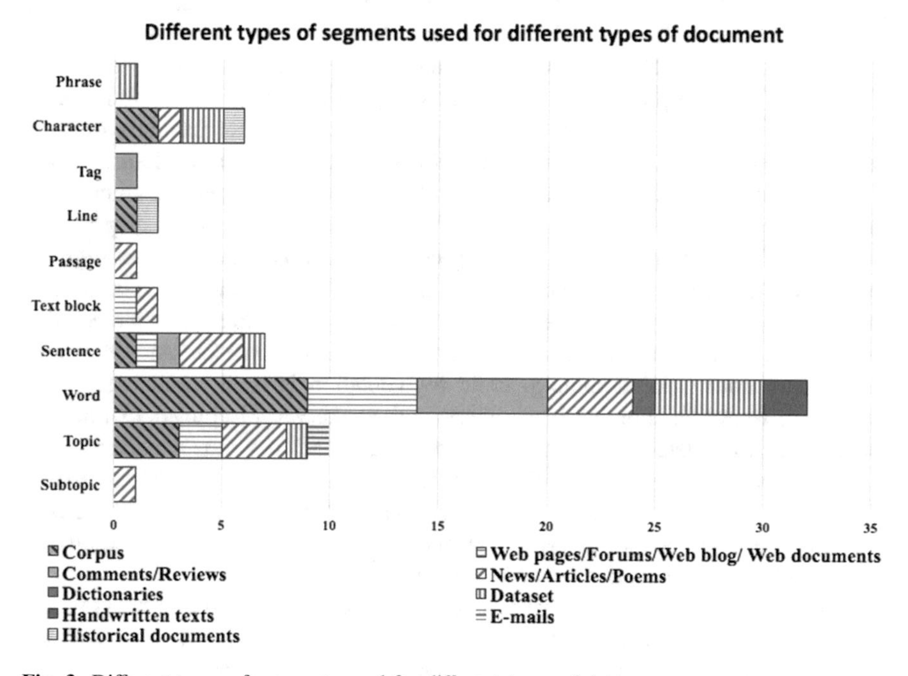

Fig. 3 Different types of segments used for different types of data

presents a graph which indicates types of segments used in different documents' types. That can help to get a picture of which type of documents are highly used and see the another trend regarding types of segment.

Figure 1 visualizes ten different type of segments used in text segmentation from the fifty reviwed articles. Word segment is highly used compare to other types of segment. It occupied the biggest area of the chart, equivalent to a total of 26 from 50 reviewed articles. There can be several reasons to use this technique. For instance, Homburg and Chiarcos [55] described it as the most elementary and essential task in natural language processing of written language. Character segment can be categorized as word segment as well because it is mostly applied in Chinese text processing, as character segment represents single unit same as word segment in following studies [45, 46, 58]. Another study of [56] proposes a method for segmentation of musical documents using word segment. It detects the segments and assigns each segment to particular musical score or text.

The second biggest area of the chart covers the use of topic segment in text processing. Topic segmentation plays an important role in data processing. For example, topic segmentation is successfully applied in tackling the problem of information overload that occur when the whole document is presented at once. Misra et al. [30] stated the reason behind splitting document could be reasonable to present only the relevant part(s) of a document. The reason they stated is that presenting the whole document without segmentation may result in previously discussed information overload. Paliwal and Pudi [34] also addressed the same problem, which led them to propose a clustering approach based on topic segmentation. Topic segmentation is popular in opinion mining area. For instance, studies of [10, 30, 38] used the topic as a segment. Another example of topic segmentation applied in opinion mining is a study of Claveau and Lefevre [50]. They proposed a new technique to compute similarities between candidate segments. Two corpora of TV broadcast in French used in evaluating of proposed technique. Furthermore, a study by Song et al. [4] has used the topic as segment too; they proposed a novel method that includes hierarchical organization and language modeling to split the text into parts. The result of that study showed that proposed method is effective in identifying the topics in evaluated dataset of articles. Apart from opinion mining, topic segmentation is used in cross-language plagiarism detection. For instance, Ehsan and Shakery [57] applied to find and examine the candidate retrieval, where proposed approach converts the document to a set of related passages. After that, it uses a proximity-based model to retrieve documents with the best matching passages. The third most applied type of segment is a sentence. The common issue in sentence segmentation among assessed articles is identifying the boundaries between sentences [20, 22].

From Table 1, it is obvious that text segmentation was applied to process text in a variety of languages including English. The pie chart on Fig. 2 illustrates the numbers of percentage for each language used including English from Table 1, in order to see trend of languages used. Leading language among evaluated articles is English with the result of 38%.

Besides English, the Chinese language is the most widely evaluated. This result shows the highest use of word segmentation technique in the process. In short, studies of [6–8, 24, [27–29], 43, 46, 48, 49, 58] used word or character segmentation, and they are all analyzing Chinese text. One of them is study of Xia et al. [24] which presents a new approach for Chinese sentiment classification based on granule network. They applied word segmentation to split sentences and later select sentimental candidate words.

Another study of Lan et al. [46] came up with another way of analyzing Chinese text in the same area of sentiment classification. Instead of using word text segmentation, they extract sentiment value based on each character, claiming that each character can contain rich sentiment information. Another study by Hog et al. [25] applied Chinese word segmentation in information retrieval process. They built an automatic statistics-based scheme for extracting news word based on the corpora. One of the main useful features of the proposed scheme is automatic and enhanced word identification. Beside the Chinese language, there are studies which applied word segmentation for Urdu [26], Russian [40], Arabic [13], Vietnamese [52], Akkadian [55], Indian [59] and South Indian [59]. However, other studies [22, 32, 44] applied sentence segmentation to analyze Arabic, Greek, and Malay languages accordingly. The topic segmentation technique is used for processing Polish [10], Chinese [38], French [50], German [57] and Spanish [57] languages. Besides that, "line" segmentation is also applied for processing Ancient [37] and French [53] languages. As a summary, it can be seen that there is a trend to apply text segmentation in the analyzing text in different languages.

We further identify if the different types of the document and datasets have any relationship on the selection of different type of segmentation technique. Figure 3 presents the number of percentage of each type of segments used for different types of document derived from Table 1. In this study, we categorize web pages, web blog, and web documents as the same group of documentations. After all, it is the most used type of document among all. The second widely used type of document is comments and reviews. Segments can be classified into the tag, word, sentence, topic, phrase and any information unit.

Figure 3 concludes that word is the most used type of segment used for corpus. Furthermore, the word segment is also highly used for comments and reviews. Web documents are utilized under word, topic and sentence segments that can indicate that web documents including pages, forums and blogs take a big part in text analyzing.

4 Discussion

As a result, we noticed the trend of applying text and sentence segmentation in processing and analyzing different languages such as Chinese, Vietnamese, Urdu, Arabic, and Ancient languages. Besides applying text segmentation for different languages, text segmentation successfully applied in opinion mining for news, blog and stock market. Finally, after comparison between evaluated types of segments, word segment is the most used compare to another types of the segment. It can be due to the smallest size of the segment which allows more detailed and deeper analysis.

5 Conclusion

This paper presents a critical review of the text segmentation methods and reasons in text processing and analyzing languages, sentiment, opinions and fifty published articles for the past decade were categorized and summarized. Those articles give contributions to text processing in information retrieval, emotion extraction, sentiment and opinion mining and language detection.

Results of this study show that word as the segment is the most used compare to other types of the segment. It means that processing smaller segments can be more useful and meaningful for more detailed and deeper analyzing of the text. Different types of document are used as a dataset for the experiment. The most popular are web pages, web blog and web document following by comments and reviews. That indicates that information from the online users and consumers plays an important role in expressing people's emotions, opinions, and feelings.

Considering the findings of this paper, the future study can include implementation of text analysis approach using text segmentation with word segment.

Acknowledgements We would like to thank First EAI International Conference on Computer Science and Engineering for the opportunity to present our paper and further extend it. This research paper was partially supported by Sunway University Internal Research Grant No. INT-FST-IS-0114-07 and Sunway-Lancaster Grant SGSSL-FST-DCIS-0115-11.

References

1. Visweswariah, P.D, Wiratunga, K., Sani N.S. (2012). Two-part segmentation of text documents. In: *Proceedings 21st ACM International Conference on Information Knowledge Management—CIKM'12* (p 793). ACM, New York: Maui.
2. Scaiano, M., Inkpen, D., Laganière, R., & Reinhartz, A. (2010). Automatic text segmentation for movie subtitles. In: Lecturer Notes Computer Science (pp. 295–298). Springer.
3. Oh, H., Myaeng, S. H., & Jang, M.-G. (2007). Semantic passage segmentation based on sentence topics for question answering. *Information Science (Ny), 177,* 3696–3717.
4. Song, F., Darling, W. M., Duric, A., & Kroon, F. W. (2011). An iterative approach to text segmentation. In: *33rd Eurobean Conference on IR Resources ECIR 2011, Dublin* (pp. 629–640). Berlin, Heidelberg: Springer.
5. Oyedotun, O. K., & Khashman, A. (2016). Document segmentation using textural features summarization and feedforward neural network. *Applied Intelligence, 45,* 1–15.
6. Wu, Y., Zhang, Y., Luo, S. M., & Wang, X. J. (2007). Comprehensive information based semantic orientation identification. *IEEE NLP-KE 2007 - Proc* (pp. 274–279). Beijing: Int. Conf. Nat. Lang. Process. Knowl. Eng. IEEE.

7. Gao, Y., Zhou, L., Zhang, Y., et al (2010). Sentiment classification for stock news. In: *ICPCA10—5th International Conference on Pervasive Computer Application* (pp. 99–104). Maribor: IEEE.
8. Xia, H., Tao, M., & Wang, Y. (2010). Sentiment text classification of customers reviews on the Web based on SVM. In: *Proceedings–2010 6th International Conference on National Computing* (pp. 3633–3637). ICNC.
9. Liu, C., Wang, Y., & Zheng, F. (2006). Automatic text summarization for dialogue style. In: *Proceedings IEEE ICIA 2006—2006 IEEE International Conference on Information Acquistics* (pp. 274–278). Weihai: IEEE.
10. Osman, D. J., & Yearwood, J. L. (2007). Opinion search in web logs In: *Conferences in Research and Practice Information Technology Service, 63,* 133–139.
11. Brants, T., Chen, F., & Tsochantaridis, I. (2002). Topic-based document segmentation with probabilistic latent semantic analysis. *CIKM'02* (pp. 211–218). Virginia: ACM.
12. Flejter, D., Wieloch, K., & Abramowicz, W. (2007). Unsupervised methods of topical text segmentation for polish. *SIGIR'13* (pp. 51–58). Dublin: ACM.
13. Potrus, M. Y., Ngah, U. K., & Ahmed, B. S. (2014). An evolutionary harmony search algorithm with dominant point detection for recognition-based segmentation of online Arabic text recognition. *Ain Shams Engineering Journal, 5,* 1129–1139.
14. Huang, X., Peng, F., Schuurmans, D., et al. (2003). Applying machine learning to text segmentation. *Information Retrieval Journal, 6,* 333–362.
15. Zhu J, Zhu M, Wang H, Tsou BK (2009) Aspect-based sentence segmentation for sentiment summarization. In: Proceeding 1st International CIKM Worshop. Top Analysis mass Open.—TSA'09 (pp. 65–72). Hong Kong: ACM New York, NY, USA ©2009.
16. Gan, K. H., Phang, K. K., & Tang, E. K. (2007). A semantic learning approach for mapping unstructured query to web resources. In: *Proceedings—2006 IEEE/WIC/ACM International Conference on Web Intelligent (WI 2006 Main Confernce Proceedings), WI'06* (pp. 494–497). Hong Kong: IEEE.
17. Hoon, G. K., Wei, & T. C. (2016). Flexible facets generation for faceted search. In: *First EAI International Conference on Computer Science Eng*ineering EAI (pp. 1–3). Penang: Malaysia.
18. Duan, D., Qian, W., Pan, S., et al (2012). VISA: A visual sentiment analysis system. In: *Proceedings 5th International Symposium Visa Information Communicate Interaction—VINCI'12*. (pp. 22–28). ACM: Hangzhou.
19. Sun, Y., Butler, T. S., Shafarenko, A., et al. (2007). Word segmentation of handwritten text using supervised classification techniques. *Applied Software Computing, 7,* 71–88.
20. Lamprier, S., Amghar, T., Levrat, B., & Saubion, F. (2007). ClassStruggle: A clustering based text segmentation. In: *Proceedings SAC'07*. (pp. 600–604). ACM: Seoul.
21. Correa, J., & Dockrell, J. E. (2007). Unconventional word segmentation in Brazilian children's early text production. *Reading and Writing, 20,* 815–831.
22. El-Shayeb, M. A., El-Beltagy, S. R, & Rafea, A. (2007). Comparative analysis of different text segmentation algorithms on arabic news stories. In: *IEEE International Conference on Information Reuse and Integration*, Las Vegas (pp. 441–446).
23. Xie, L., Zeng, J., & Feng, W. (2008). Multi-scale texttiling for automatic story segmentation in Chinese broadcast news. In: *4th Asia Information Retrieval Symposium,* Harbin (pp. 345–355). Berlin, Heidelberg: Springer.
24. Xia, Z., Suzhen, W., Mingzhu, X., & Yixin, Y. (2009). Chinese text sentiment classification based on granule network. In: *2009 IEEE International Conference on Granular Computing GRC 2009* (pp. 775–778). Nanchang: IEEE.
25. Hong, C. M., Chen, C. M., & Chiu, C. Y. (2009). Automatic extraction of new words based on Google News corpora for supporting lexicon-based Chinese word segmentation systems. *Expert Systems with Applications, 36,* 3641–3651.
26. Mukund, S., Srihari, R., & Peterson, E. (2010). An information-extraction system for Urdu-a resource-poor language. *ACM Transcations on Asian Language Information Processing, 9,* 1–43.

27. Tsai, R. T.-H. (2010). Chinese text segmentation: A hybrid approach using transductive learning and statistical association measures. *Expert Systems with Applications, 37,* 3553–3560.
28. Liu, X., Zuo, M., & Chen, L. (2010). The application of text mining technology in monitoring the network education public sentiment. In: *2010 International Conference on Computing Intelligence and Software Engineering* (pp. 1–4). Wuhan: IEEE.
29. Li, N., & Wu, D. D. (2010). Using text mining and sentiment analysis for online forums hotspot detection and forecast. *Decision Support Systems, 48,* 354–368.
30. Misra, H., Yvon, F., Cappé, O., & Jose, J. (2011). Text segmentation: A topic modeling perspective. *Information Process Management, 47,* 528–544.
31. Fan, J. (2011). Text segmentation of consumer magazines in PDF format. In: *International Conference on Document Analysis and Recognition (ICDAR)* (pp. 794–798).
32. Ranaivo-Malançon, B. (2011). Building a rule-based Malay text segmentation tool. In: *2011 International Conference on Asian Language Processing IALP 2011* (pp. 276–279). Penang: IEEE.
33. Nouri, J., & Yangarber, R. (2011). A novel evaluation method for morphological segmentation. In: *Proceedings Tenth International Conference on Language Resources Evaluation (LREC 2016)* (pp. 3102–3109). Portoroz: European Language Resources Association (ELRA).
34. Paliwal, S., & Pudi, V. (2012). Investigating usage of text segmentation and inter-passage similarities. In: Machine Learning and Data Mining Pattern Recognition (pp. 555–565). Berlin, Heidelberg: Springer.
35. Peng, X., Setlur, S., Govindaraju, V., & Ramachandrula, S. (2012). Using a boosted tree classifier for text segmentation in hand-annotated documents. *Pattern Recognition Letters, 33,* 943–950.
36. Guinaudeau, C., Gravier, G.S & Billot, P. (2012). Enhancing lexical cohesion measure with confidence measures, semantic relations and language model interpolation for multimedia spoken content topic segmentation. *Computer Speech Language. 26*, 90–104.
37. Clausner, C., Antonacopoulos, A., & Pletschacher, S. (2012). A robust hybrid approach for text line segmentation. In: *21st International Conference on pattern Recognition* (pp. 335–338). Tsukuba: IEEE.
38. Ye, F.Y., Chen, Y., Luo, X., et al (2012). Research on topic segmentation of Chinese text based on lexical chain. In: *12th International Conference on Computer and Information Technology CIT 2012* (pp. 1131–1136) .Chengdu: IEEE.
39. Myint, N., Aung, M., & Maung, S. S. (2013). Semantic based text block segmentation using wordnet. *International Journal of Computer Communication and Engineering, 2,* 601–604.
40. Kravets, L. G. (2013). The first steps in developing machine translation of patents. *World Patent Information, 35,* 183–186.
41. Chiru, C., & Teka, A. (2013). Sentiment-based text segmentation. In: *2nd International. Conference on Systems Computer Science* (pp. 234–239). Villeneuve d'Ascq: France, IEEE.
42. Sun, X., Zhang, Y., Matsuzaki, T., et al. (2013). Probabilistic Chinese word segmentation with non-local information and stochastic training. *Information Processing Management, 49,* 626–636.
43. Ye, Y., Wu, Q., Li, Y., et al. (2013). Unknown Chinese word extraction based on variety of overlapping strings. *Information Processing Management, 49,* 497–512.
44. Fragkou, P. (2013). Text segmentation for language identification in Greek forums. In: *Proceedings of Adaptation of Language Resources and Tools for Closely Related Languages and Language Variants* (pp. 23–29). Hissar: Elsevier B.V.
45. Ma, G., Li, X., & Rayner, K. (2014). Word segmentation of overlapping ambiguous strings during Chinese reading. *Journal of Experimental Psychology: Human Perception and Performance, 40,* 1046–1059.
46. Lan, Q., Li, W., & Liu, W. (2015). Chinese text sentiment orientation identificat.ion based on chinese-characters. In: *International Conference on IEEE 2015 12th Fuzzy Systems and Knowledge Discovery (FSKD)* (pp. 663–668). Zhangjiajie.

47. Alemi, A. A., & Ginsparg, P. (2015). Text segmentation based on semantic word embeddings. *KDD2015* (pp. 1–10). Sydney, Australia: ACM.
48. Fu, X., Yang, K., Huang, J. Z., & Cui, L. (2015). Dynamic non-parametric joint sentiment topic mixture model. *Knowledge-Based Systems, 82,* 102–114.
49. Liu, S. M., & Chen, J.-H. (2015). A multi-label classification based approach for sentiment classification. *Expert Systems with Applications, 42,* 1083–1093.
50. Claveau, V., & Lefevre, S. (2015). Topic segmentation of TV-streams by watershed transform and vectorization. *Computer Speech and Language, 29,* 63–80.
51. Shi, H., Zhan, W., & Li, X. (2015). A supervised fine-grained sentiment analysis system for online reviews. *Intelligent Automation and Soft Computing, 21,* 589–605.
52. Liu, W., & Wang, L. (2016). How does dictionary size influence performance of Vietnamese word segmentation? In: *Proceedings Tenth International Conference on Language Resources Evaluation (LREC 2016)* (pp. 1079−1083). European Language Resources Association (ELRA), Portorož: Slovenia.
53. Grouin, C. (2016). Text segmentation of digitized clinical texts. In: *Proceedings Tenth International Conference on Language Resource Evaluation (LREC 2016)* (pp. 3592−3599). European Language Resources Association (ELRA), Portorož: Slovenia.
54. Logacheva, V., & Specia, L. (2016). Phrase-level segmentation and labelling of machine translation errors. In: *Tenth International Conference on Language Resource Evaluation (LREC 2016)* (pp. 2240–2245). European Language Resources Association (ELRA), Portorož: Slovenia.
55. Homburg, T., & Chiarcos, C. (2016). Akkadian word segmentation. In: *Proceedings Tenth International Conference on Language Resource Evaluation. (LREC 2016)* (pp. 4067−4074). European Language Resources Association (ELRA), Portorož: Slovenia.
56. Pedersoli, F., & Tzanetakis, G. (2016). Document segmentation and classification into musical scores and text. *International Journal Document Analysis and Recognition, 19,* 289–304.
57. Ehsan, N., & Shakery, A. (2016). Candidate document retrieval for cross-lingual plagiarism detection using two-level proximity information. *Information Processing and Management, 52,* 1004–1017.
58. Qingrong, C., Wentao, G., Scheepers, C., et al. (2017). Effects of text segmentation on silent reading of Chinese regulated poems: Evidence from eye movements. *44*, 265–286.
59. Kavitha, A. S., Shivakumara, P., Kumar, G. H., & Lu, T. (2017). A new watershed model based system for character segmentation in degraded text lines. *AEU—International Journal of Electronics and Communications, 71,* 45–52.

On-Line Power Systems Security Assessment Using Data Stream Random Forest Algorithm Modification

Aleksei Zhukov, Nikita Tomin, Denis Sidorov, Victor Kurbatsky and Daniil Panasetsky

Abstract Voltage instability is among the main factors causing large-scale blackouts. One of the major objectives of the Control centers is a prompt assessment of voltage stability and possibly self-healing control of electric power systems. The standing alone solutions based on classical approximation methods are known to be redundant and suffer with limited efficiency. Therefore, the state-of-the-art machine learning algorithms have been adapted for security assessment problem over the last years. This chapter presents an automatic intelligent system for on-line voltage security control based on the Proximity Driven Streaming Random Forest (PDSRF) model using decision trees. The PDSRF combined with capabilities of L-index as a target vector makes it possible to provide the functions of dispatcher warning and "critical" nodes localization. These functions enable self-healing control as part of the security automation systems. The generic classifier processes the voltage stability indices in order to detect dangerous pre-fault states and predict emergency situations. Proposed approach enjoy high efficiency for various scenarios of modified IEEE 118-Bus Test System enabling robust identification of dangerous states.

A. Zhukov · N. Tomin · D. Sidorov (✉) · V. Kurbatsky · D. Panasetsky
ESI SB RAS, Lermontov Str. 130, Irkutsk, Russia
e-mail: dsidorov@isem.irk.ru

A. Zhukov
e-mail: zhukovalex13@gmail.com

N. Tomin
e-mail: tomin.nv@gmail.com

V. Kurbatsky
e-mail: kurbatsky@isem.irk.ru

D. Panasetsky
e-mail: panasetsky@gmail.com

N. Tomin · D. Sidorov · D. Panasetsky
INRTU, Lermontov Str. 83, Irkutsk, Russia

D. Sidorov
Hunan University, 2 Lushan S Rd, Changsha, China

I. Zelinka et al. (eds.), *Innovative Computing, Optimization and Its Applications*, Studies in Computational Intelligence 741,
https://doi.org/10.1007/978-3-319-66984-7_11

1 Introduction

In recent years electric power systems (EPS) are being operated closer and closer to their limits due to liberalization. At the same time, EPS have increased in size and complexity. Both factors increase the risk of major power outages and blackouts.

Automatic emergency and operational control systems are required to prevent cascading emergencies and blackouts. Unfortunately, in many cases, the current generation of these systems is ineffective and unreliable. A low level fault-tolerant of elements and lack of local devices coordination are the main disadvantages of modern emergency control systems causing large-scale blackouts. Greater centralization of control systems and increasing redundancy level ($N-2, N-3$) could help to handle these significant problems. However, excessive redundancy and centralization can lead to a cost increase and reliability decrease of emergency control due to the structural complexity improvement of systems elements. As for the operational control, practical experience demonstrates that power system operator becomes difficult to formulate the correct decision when control actions must be taken instantly under emergencies.

An essential tool for monitoring the power system is security assessment. The EPS security is related to the ability of a power system to continue normal operation despite unplanned casualties to operating equipment, known as contingencies. A failure of security can cause equipment damage, low frequency or low voltages, and localized loss of power. But the most severe, spectacular, costly and therefore most interesting security failures result in blackouts. Analysis of data streams characterizing the behavior of non-stationary dynamic systems is an essential task contributing the EPS security monitoring. The data stream term here refers to data records collected evenly or not evenly. These include telemetry or vector data records using SCADA systems or PMUs. For bibliography on online voltage security assessment using WAMS readers may refer to [1].

Conventional methods for the power system security evaluation require substantial computation time due to due the combinatorial nature of the problem. That is the reason why such methods are not suitable for real-time applications. A more promising solution is to employ the decentralized adaptive self-healing control systems with simultaneous increase of the intelligent level of the local devices.

In recent years it has been suggested to estimate the EPS state using the machine learning algorithms such as artificial neural networks, decision trees, and deep learning model and other the state of the art methods, see bibliography in [2–7]. Machine learning algorithms are efficient due to their higher ability to learn and generalize. Moreover, they enable high speed identification of the boundaries of the instability that allows to provide an efficient solution in real time.

The reminder of the chapter is organized as follows. Section 2 introduces the concept of electric power systems states classification. Section 3 deals mainly with architecture of an automatic intelligent system based on proximity driven streaming random forest classification algorithm. Experimental evaluation of designed intelligent system on IEEE 118-Bus test system is given in Sect. 5 followed by conclusions.

2 Problem Formulation

To monitor the power system within its limit, the SCADA systems are the primary measurement tools and a state estimator [8] is conventional post processing tool. The ENTSO-E (European Network of Transmission System Operators for Electricity) network code on operational security requires each transmission system operator (TSO) to classify its system according to the system operating states. Figure 1 shows the different operating states of a power system as identifed by Dy Liacco and adopted by authors of this chapter.

Most of the existing emergency control ideologies use two approaches: control in case of accidental disturbance and control in case of a parameter deviation. Thus, taking into account the existing ideologies, emergency control automation does not “predict” the possible development of the current situation, but acts only in case of emergency or deviation occurrence. Using such approach it is not always possible to implement the necessary and timely control actions. In addition, low fault tolerance and lack of coordination of different local devices may cause a cascade failure. Apparently, the current generation of emergency control systems faced with new challenges, recent examples of large-scale blackouts in North America in 2003, Moscow in 2005, Europe in 2003 and 2006, and India in 2012 testify to this.

The operational control (including the TSO participation) deals with the concept of emergency “prediction” (the probability of its occurrence). Functions of “prediction” can be implemented in a manual (system operator), automatic or semi-automatic manner, using different auxiliary software systems, such as expert/advice software. However, in a critical situation a radical solution, called to save the EPS, is implemented by the operator who, for various reasons is not always able to take fast and effective measure that would prevent further development of the emergency. Thus, in the existing control ideology there is a lack of procedures that would provide a smooth transition from operational to the effective emergency control, as well as a lack of algorithms that would increase the fault tolerance and adaptability of the local control.

The basic idea of our approach is to build the generic classifier of the EPS states. It is supposed to process the special security indices of the system in order to detect dangerous pre-fault states and predict emergency situations as shown in Fig. 1. One of the most common approaches to classification problem employes algorithms' combination which called *ensembles*. Ensembles employ the set of different algorithms to achieve better accuracy comparing with each of these algorithms individual accuracy. Ensembles are widely used in various studies, including the power system state monitoring. One of the most effective ensembles uses decision trees as part of the Random Forest approach [9]. Periodically updated decision trees are widely used for voltage security assessment for large-scale power outages prevention [2]. This family of algorithms uses bagging and random subspaces to create compositions of highly decorrelated decision trees to achieve a higher accuracy and stability to presence of noise in the data streams. The presence of noise in the data streams may severely affect the intrinsic characteristics of a classification problem.

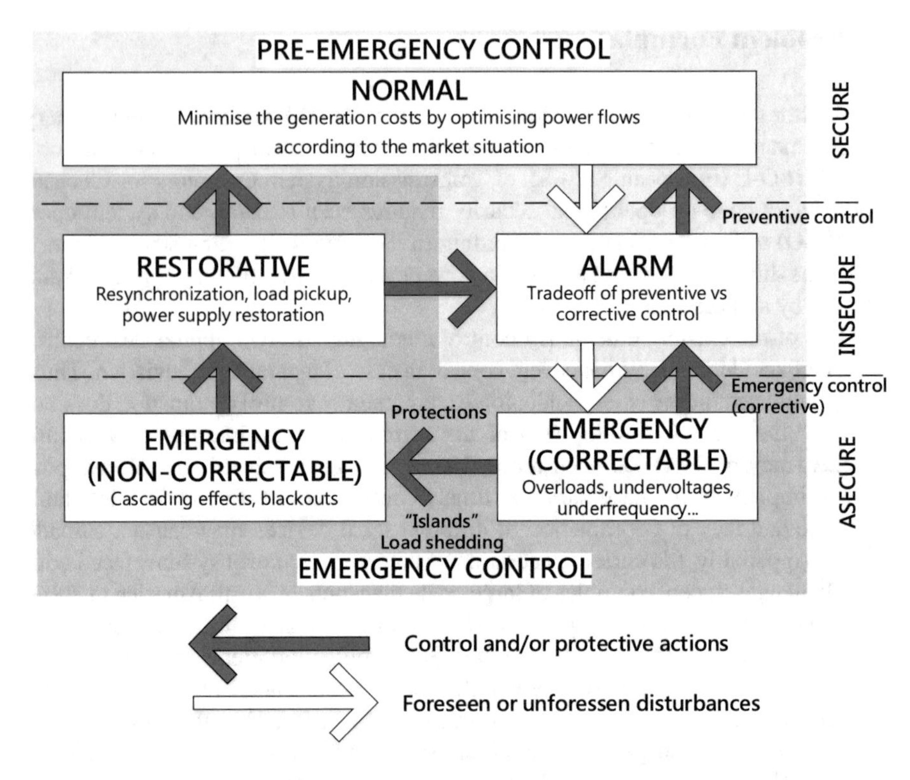

Fig. 1 Data flow diagram of the security assessment system

Wang et al. [10] have demonstrated that a classifiers' ensemble can outperform a single classifier in the presence of concept drifts when the base classifiers of the ensemble are adherence weighted to the error similar to current testing samples. We employ another approach which exploits Random Forest properties.

2.1 *Existing Approaches to Emergency and Operational Control*

In the existing control structure, it is necessary to identify reactive power coordination problems solved by emergency and operational measures. First, we concentrate here on emergency control of voltage and reactive power. Then, we consider operational control of voltage and reactive power.

Emergency control of voltage and reactive power is traditionally carried out by the following local automations:

- Local automations that provide sharp reactive power increase in case of emergency (excitation forcing of AVR and FACTS devices);

- Local automations against undervoltage that switch on and off reactive power sources or trip the load in case of unacceptable voltage decrease (undervoltage load shedding function);
- Local automated load tap changers that provide smooth secondary voltage regulation.

Let us discuss some shortcomings of the described local reactive control systems. Different local automations against undervoltage is an effective measure against voltage collapse. However, it is possible to note some disadvantages of these devices. The main drawback is a low intellectual level, undervoltage automation implements control actions using only local information (local measurements of voltage, current, etc.), and it is usually leads to excess control actions. Future emergency control should perform load shedding only if it is necessary in case if there are no reactive power reserves remained in the system. Another disadvantage of these types of automation is a large time delay, which is needed to exclude any tripping during the short circuit. Local automated load tap changers are extremely important to maintain the required secondary voltage, but lack of coordination of their control actions may cause cascading failure.

In recent decades, the *operational control of voltage and reactive power* has become increasingly important due to the general trend of power systems to work near stability limits. In general, existing and prospective reactive power control systems are built using the following hierarchical approach [11]:

- Primary voltage regulation is carried out by primary voltage controllers (local automatic voltage regulators) of synchronous generators, different FACTS devices, etc. This regulation is performed using only local information; the reaction rate is a fraction of seconds.
- Secondary voltage regulation is the coordination of primary voltage controllers. Most of secondary algorithms provide voltage control not for all but only for the key nodes ("pilot" nodes in case of coordinated algorithms [12]). This regulation can be manual, semi-manual or automatic.
- Tertiary voltage regulation related to the solution of the optimal control problem. Primary and secondary regulation reserves are recovered within tertiary regulation.

It is secondary regulation that provides resistance to the spreading of the overload through the network, which usually occurs before the cascading failure. Literature analysis showed that there are two basics approaches to the secondary voltage regulation implementation: traditional and coordinated. The traditional approach to secondary voltage regulation is simpler in terms of implementation, but provides a lower automation level. It is to install a number of reactive power sources in transmission network that are used to maintain voltage during load peak hours or in case of emergency. The realization of control actions can be both automatic and manual. The coordinated approach to secondary voltage regulation is based on preliminary separation of power system on control areas, followed by the introduction of reactive power coordination means in each area. The need of division is due to complexity of the problem, as well as due to local character of reactive power production. Secondary regulation in every subsystem is provided by selecting the "pilot" node which voltage is considered as the average index of the current state of the controlled area. A coordinated approaches to the secondary and tertiary regulation were first investigated and implemented in Europe [13–16]. Some of the developed systems have been used at the national level. In particular, their introduction into Italian EPS was widely reported [17]. As for the technical and economic comparison of traditional and coordinated approaches, there is no single opinion [18]. Apparently, each approach has its advantages and disadvantages.

3 Proposed Approach

Methodologically ensemble approaches allow concept-drift to be handled in the following ways: base classifier adaptation, changing in training dataset (such as Bootstrap [19] or RSM [20]), ensemble aggregation rule changing or changing in structure of an ensemble (pruning or growing). In this chapter combinations of these approaches are exploited as part of the Proximity Driven Streaming Random Forest (PDSRF). Besides some methods are already incorporated to the original Random Forest. Contrary to conventional algorithms we use weighted majority voting as an aggregation rule of ensemble. This allows us to adapt the entire classifier by changing the weights of the base learners. In order to obtain the classifiers weight estimation we should store samples. For this purpose we use a sliding windows approach which is used in the periodically updated Random Forest [6]. The length of this window is fixed and can be estimated by cross-validation. For the sake of time and memory optimization Extremely randomized trees [21] is used as a base learner instead of original randomized trees.

Random Forest [9] uses unpruned CART [22] trees and for N instances and $Mtry$ attributes chosen at each decision tree node average computational cost of ensemble building is $O(T\, Mtry\, N \log^2 N)$, where T is a number of trees. It can be insufficient for online applications. To reduce the complexity we use the randomization approach proposed in Extremely Randomized Trees [21]. In our implementation the split set consists of randomly generated splits and the best one is chosen by minimizing the Gini-index measure. So that $O(T\, Mtry\, N \log N)$ cost complexity can be achieved.

3.1 Base Classifier Weighting Function

We employ the assumption that the base classifiers make similar errors on similar samples even under concept-drift. The conventional Random Forest employes the special metrics called *proximity* measure. It uses a tree structure to obtain similarity in the following way. If two different sample are in the same terminal node, their proximity will be increased by one. At the end, proximities are normalised by dividing by the number of trees [9].

Following the AWE approach proposed in [10] we use an error rate to produce weights of classifiers as follows

$$w_i = 1/(E^2 + \varepsilon).$$

Here E is an new block testing error for ith classifier, and ε is a small parameter.

3.2 Forgetting Strategy

One of the main problems in concept-drifting learning is to select the proper forgetting strategy and forgetting rate [23]. The classifier should be adaptive enough to handle changes. In this case, different strategies can be more appropriate to different types of drift (for example, sudden and gradual drifts). In this paper we focus on gradual changes only.

In this paper we propose to employ ensemble pruning technique to handle the concept-drift in EPS security assessment problem. This technique uses the classic *replace-the-looser* approach [23] to discard trees with high error on new block samples.

3.3 Proximity Driven Streaming Random Forest

In this paper we demonstrate that when it comes to the real time security assessment, the most effective algorithm in terms of accuracy, robustness, online adaptability and versatility, is the algorithm of Proximity Driven Streaming Random Forest (PDSRF) proposed in [24]. Therefore, it is the PDSRF algorithm implemented in the programming language C++, which was used as a basis for the developed automatic intelligent system for a real time security assessment of a power system. To predict the sample we employ the PDSRF algorithm. The data flow diagram of PDSRF is shown in Fig. 2.

First, a stored window is used to find similar items using the specified similarity metric. Second, we evaluate our current ensemble on similar examples. Then we compute weights adherence to errors on k similar samples. Thus, to predict the sample we employ the following algorithm:

(1) A stored window is used to find similar items using the specified similarity metric;
(2) Current ensemble on similar examples is evaluated;
(3) The weights adherence to errors on k similar samples are computed.

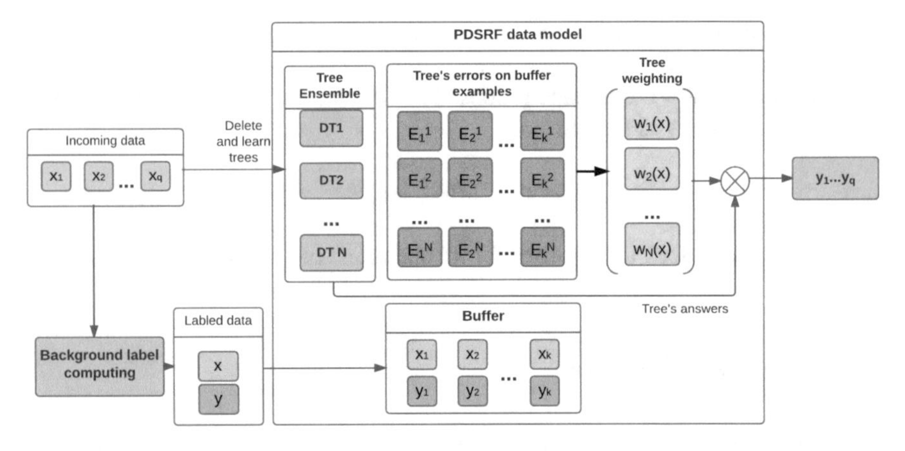

Fig. 2 PDSRF data flow diagram

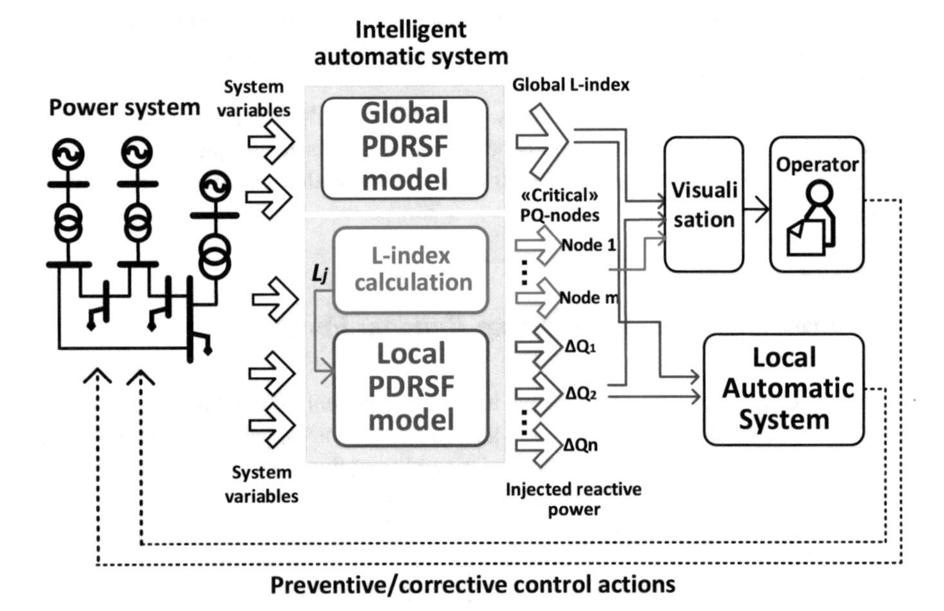

Fig. 3 Structural scheme of an automatic intelligent system for voltage security control

On every chunk the algorithm tests all the trees to choose the poorest base learner and replace it with new one trained on new block data. This process is iterative while the ensemble error on new block samples is higher than a specified threshold.

4 Structure of an Automatic Intelligent System

The general structure of the developed system is presented in Fig. 3. The scheme shows that the proposed system consists of two main models on the basis of the PDSRF algorithm: local and global.

Global model is trained to correctly identify the global L-index on the basis of system variables of a power system such as voltage at nodes, loads, power flows, etc. Here an output value of the L-index is interpreted as a security signal (indicator) of the entire power system. The local model on the basis of the PDSRF algorithm, in turn, is trained to determine the required reactive power injections ΔQ for load nodes. The inputs for the local model are represented by the same operating parameters and calculated values of the local L-index at a current time instant. Training of the local PDSRF model is based on the expression (3) as shown below. As a result, the trained model is able to determine the values ΔQ to perform corrective and/or preventive control actions on-line. Additionally, the security intelligent system can provide local signals in the form of local L-indices for each load at load node.

In the end, such a structure makes it possible to implement the above functions of on-line alarming, localization and interaction with automatic systems. For example, the system signal on the basis of L-index, which is delivered to operator through the visualization block, informs the dispatcher on the general level of security in the analyzed power system ("high", "low", "emergency"), and allows the operator to predict (estimate) the extent to which the current state of the electric power system is dangerous in terms of its closeness to voltage collapse.

In the case of a dangerous state identification ("low security") the local signals on the basis of local L-index, formed by local model, enable us to localize "critical load nodes" at which the system is at its closest to the stability loss. The corrective and/or preventive control actions can be implemented on-line on the basis of the injections ΔQ generated by the local model based on the PDSRF algorithm for load nodes. Such control actions can both adjust the operating conditions in terms of their optimality according to some economic criteria, or, in the case of a decrease in security, keep the conditions away from the instability boundaries.

It is important to note that the output signals of the alarming system are delivered both to the operator and directly to the operating automation. Interactions with the automation allow us to know where need to correct the actions of agents, since agents control the reactive power sources to regulate voltage in order to prevent the development of an emergency process. Operator, in turn, using the recommendations of the intelligent system (in case of a security decrease), can adjust the protective relay settings by decreasing the settings with respect to time, increasing sensitivity of startup signals of the emergency control functions through the selection of an appropriate group of settings, etc. Despite the fact that the proposed structure suggests a certain interaction with dispatcher to control the power system security, we see the main mechanism of the developed intelligent system operation to be mostly automatic, where many control actions are generated with the minimum involvement of the operator.

5 Experimental Evaluation

5.1 L-index

This paper employ the PDSRF model for voltage security assessment based on the L-index as security label. The L-index is proposed by Kessel and Glavitsch in [25] as an indicator of impeding voltage stability. The authors developed a voltage stability index based on the solution to power flow equations starting from the subsequent analysis of a power line equivalent model. The L index is a quantitative measure for the estimation of the distance of the actual state of the system to the stability limitation. This voltage stability index based on fundamental Kirchoff-Laws and can reflect the weak point where to locate the vulnerable locations and can predict collapse point of the system [26].

In the traditional statement, the application of such stability indices implies a purely algorithmic approach where the specified equations are directly employed to calculate numerical values of these indices for each current state of power system. However, as the practice of operation shows, such an approach has a number of significant downsides such as low robustness to erroneous inputs, computational complexity and erroneous identification of states.

The L-index describes the stability of the entire system with the expression:

$$L = \max_{j \in \alpha_L} \left(L_j \right), \tag{1}$$

where α_L is a set of load nodes. L_j is a local indicator that determines the buses which can be sources of collapse. The L-index varies in a range between 0 (no load) and 1 (voltage collapse) and used to provide meaningful voltage instability information during dynamic disturbances in the system.

The local indicator L_j was formulated in terms of the power by Kessel and Glavitsch [25] as follows:

$$L = \left| 1 + \frac{\dot{U}_{0j}}{\dot{U}_j} \right| = \left| \frac{\dot{S}_j^+}{\dot{Y}_{jj}^{+*} U_j^2} \right| = \frac{S_j^+}{Y_{jj}^+ U_j^2}, \tag{2}$$

where Y_{jj}^+ is transformed admittance; U_j is voltage of the load bus j; S_j^+ is transformed complex power, which can be calculated as

$$\dot{S}_J^+ = \dot{S}_j + \left(\sum_{j \in \alpha_L, i \neq j} \frac{\dot{Z}_{ji}^* \dot{S}_i}{\dot{Z}_{jj}^* \dot{U}_i} \right) \dot{U}_j,$$

where $\dot{Z}_{ji}^*, \dot{Z}_{jj}^*$ are off-diagonal elements and leading elements of impedance matrix.

The voltage stability indicators are known to be not only an effective method to assess the system stability but also they underlie the control of electric power system security.

According to the basic differential property of the L-index we suggest a common analytical algorithm for reactive power optimization. The algorithm can be used to determine the required reactive power injection for the load node, ΔQ. Based on the applied methodology, a large-scale power system will operate in an optimal steady state under the minimum

$$L_{\text{sum}} = L_1 + L_2 + \cdots + L_m,$$

which represents a sum of local indices L_j. In this case the function of the first partial derivative is defined as follows

$$\frac{\partial L_{sum}}{\partial \Delta Q} = \begin{bmatrix} \frac{\partial L_{sum}}{\partial \Delta Q_1} \\ \vdots \\ \frac{\partial L_{sum}}{\partial \Delta Q_m} \end{bmatrix} = \begin{bmatrix} \sum\limits_{i \in \alpha_L, i \neq j} -\frac{1}{U_j} \frac{X_{j1}}{U_1} \frac{\mu}{\sqrt{\mu^2+\gamma^2}} \\ \vdots \\ \sum\limits_{i \in \alpha_L, i \neq j} -\frac{1}{U_j} \frac{X_{jm}}{U_m} \frac{\mu}{\sqrt{\mu^2+\gamma^2}} \end{bmatrix}, \tag{3}$$

where $\mu = \sum\limits_{i \in \alpha_L, i \neq j} \frac{(Q_i + \Delta Q_i) X_{ji}}{U_i}$; $\gamma = \sum\limits_{i \in \alpha_L, i \neq j} \frac{-P_i X_{ji}}{U_i}$.

Gong et al. in [27] have demonstrated that such approach improves the voltage stability by reactive power injections at load nodes on different IEEE test systems. The injections are calculated from the minimization conditions based on L-index, and keep a system under heavy load conditions away from instability boundaries. It is to be noted, the authors state that despite the relative simplicity of the calculation, this method requires considerable computational efforts and its application in the real time problems can be complicated.

The bus with the highest L-index value will be the most vulnerable bus in the system and hence this method helps in identifying the weak areas in the system which need critical reactive power support. Among the different indices for voltage stability and voltage collapse prediction the L-index gives fairly consistent results. In next chapter we consider failure scenarios and discuss results on IEEE 118-Bus test system.

5.2 *Failure Scenarios and Results on IEEE 118-Bus Test System*

In this experiment, we evaluate our algorithm on IEEE 118-Bus Test System and compared it to the most popular classifiers of EPS states. Our algorithm was implemented natively in C++ according to the same testing methodology. With adherence to this methodology classification accuracy was calculated using the *data block eval-*

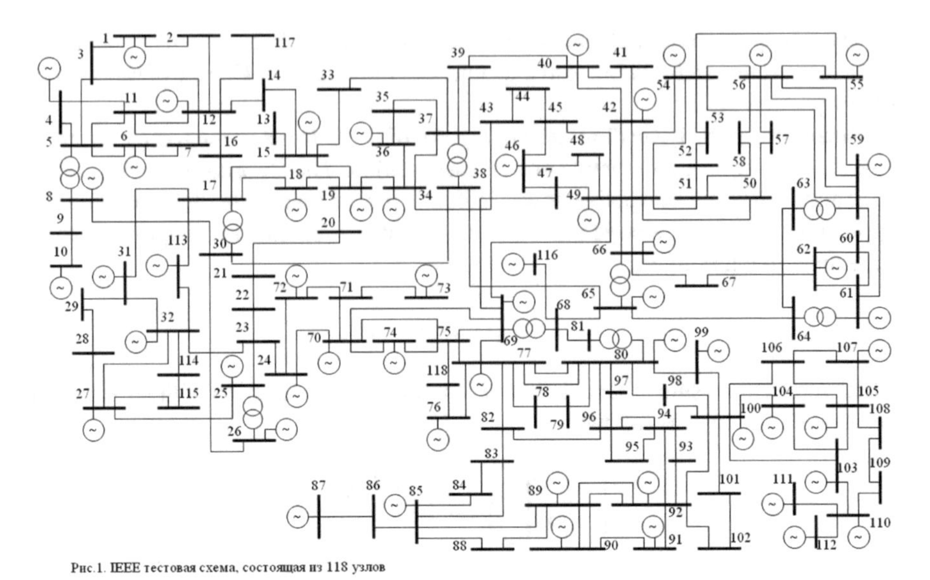

Рис.1. IEEE тестовая схема, состоящая из 118 узлов

Fig. 4 IEEE 118 Bus Test System

uation method, which exploits the *test-then-train* approach. The data block evaluation method reads incoming samples without processing them, until they form a data block of size *d*. Each new data block is first used to test the existing classifier, then it updates the classifier [28] (Fig. 4).

The efficiency of the proposed intelligent system was tested on IEEE 118 test system. The database of possible states of the test system for model training was generated by quasi-dynamic modeling in the MATLAB environment. At each step of the load increase in the system, emergency events are modeled randomly. The disturbances included losses of generation and connection of a large consumer at specified nodes. A set of the obtained system states was used to calculate the values of global L-index, and on the basis of local indices L_j, the reactive power injections ΔQ were found for each load node. As result, we computed the attribute values and pre-classified based on the L-index the obtained states as "normal", "alarm", "emergency" and "collapse". These characteristics were applied as class marks for training and testing the models on the basis of the PDSRF algorithm. Global and local PDSRF models are implemented in C++.

In order to make simulation data more close to behaviour of real power systems we propose to add sharp change of power load on some buses (Fig. 5). These changes can be considered as a connection of big consumer. Thus the data will contain concept-drift which should be efficiently handled by classification algorithms (Fig. 6).

The simulation results based on computation of local L-index indicate that buses 103, 105, 106 and 110 are the critical buses (Fig. 7). At this time the bus 105 is the more critical bus for this system. This means that for IEEE 118 system the voltage stability margin is limited by the outage of line 105.

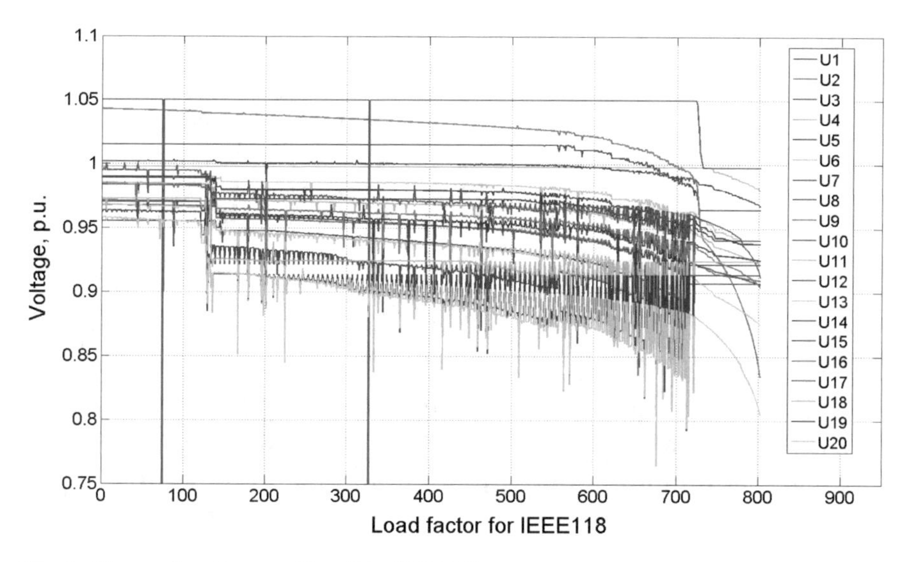

Fig. 5 Changes in voltage profiles of the IEEE 118 system

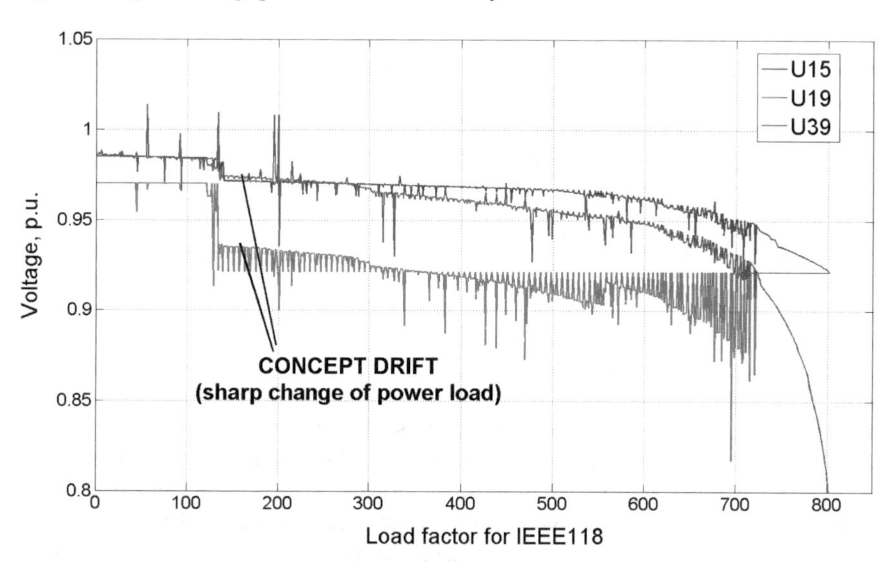

Fig. 6 Voltage profiles at load buses with a "concept-drift effect"

The testing results of the PDSRF algorithm compared to the other machine learning models are presented in Table 1.

As result, testing of the automatic intelligent system demonstrated that the developed on-line approach on the basis of PDSRF method makes it possible to highly accurately identify the system level of security in the test electric power system. For example, the accuracy of state identification for the scheme IEEE 118 made up

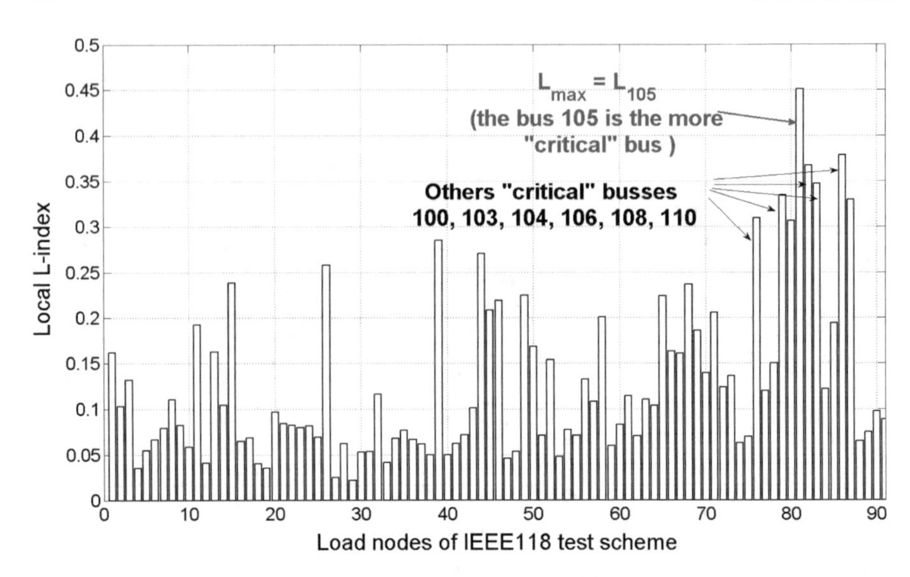

Fig. 7 The local L-index values for one of steady-state condition of IEEE 118

Table 1 Accuracy of various machine learning algorithms on repared dataset

Algorithm	Accuracy (%)	Kappa (%)
Global PDSRF model	97.24	95.30
Support Vector Machine	81.54	64.92
Random Forest	96.01	93.24
Gradient Boosting Trees	93.64	89.41
Extreme Learning Machine	80.80	65.08

97.24% (Table 1) which is approximately by 10% higher than the accuracy of other known intelligent approaches presented in some international peer-reviewed journals over the past years. These approaches include neural networks of Kohonen, supporting vector machines, hybrid neural network models, various algorithms of decision trees. This means that in all the models, except the PDSRF model, the accuracy declines at modeling of significant disturbances in the system, i.e. they could not adapt in real time and required additional training (updating).

Moreover, the calculations show that the proposed on-line approach on the basis of PDSRF provides lower errors (root-mean square error (RMSE) of order 13% for IEEE 118) and high speed of solving process (about centiseconds for each steady state of IEEE 118 compared to 30–40 min in the traditional approach)[1] when determining the additional reactive power injections (Table 2). This fact makes it possible

[1] All calculations were performed on the workstation c Intel processors (R) Core (TM) i7-4930K @ 3.40 GHz 3.30 GHz.

Table 2 The efficiency of various machine learning algorithms when determining the additional reactive power injections

Algorithm	RMSE	MAE	Train time (s)	Test time (s)
Local PDSRF model	0.1299	0.1116	4.812	0.00149
Support Vector Machine	0.1498	0.1254	3.441	0.00167
Random Forest	0.1502	0.1271	0.811	0.00153
Gradient Boosting Trees	0.1463	0.1234	4.671	0.00282
Extreme Learning Machine	0.1517	0.1282	0.021	0.00153

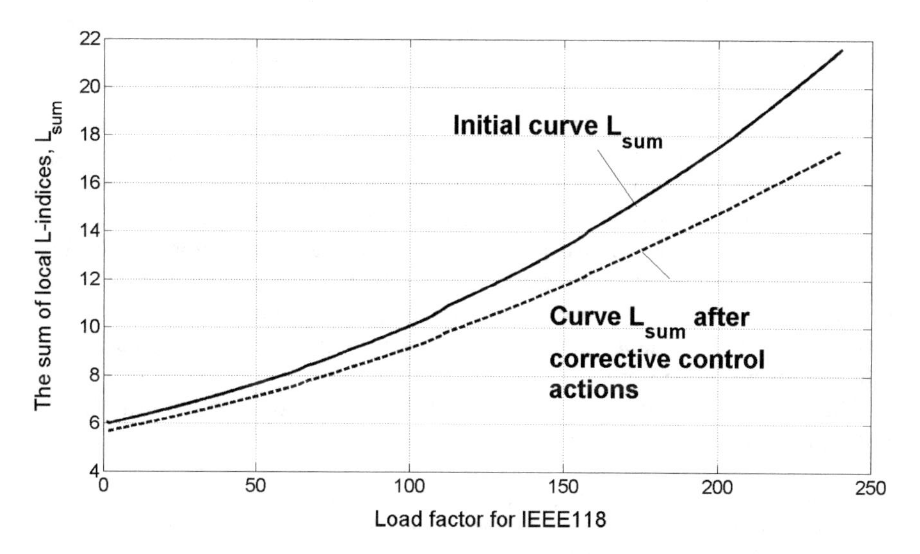

Fig. 8 The curves L_{sum} before and after self-healing control actions

to effectively apply the automatic intelligent system for monitoring to control security in power systems of large dimension in real time.

The obtained values of additional injections were used for reactive power compensation by using reactive power sources, which decreased L_{sum}, whose increase is indicative of even greater proximity of voltage collapse, first of all for the heavy load and dangerous conditions of IEEE 118 system (Fig. 8).

6 Conclusions

We devised an innovative on-line method for the assessment and control of voltage security of power system, using the technology of online decision trees, i.e. PDSRF, implemented in the language C++. The main qualitative distinction of this approach from the other modern approaches is the capability of PDSRF to independently and adaptively change in real time in case of serious changes in the received telemetry data without loss of accuracy while identifying the conditions of electric power system.

This paper presents a PDSRF model for voltage security assessment based on the L-index as security label. This voltage stability index employ fundamental Kirchoff-Laws and can reflect the weak point where to locate the vulnerable locations and can predict collapse point of the system. The bus with the highest L-index value will be the most vulnerable bus in the system and hence this method helps in identifying the weak areas in the system which need critical reactive power support.

Voltage stability L-index indicator have been employed for security control using the PDSRF algorithm in order to safely trace the concept drift in data stream and perform the security assessment of the whole system. Experiments with IEEE 118 system various failure scenarios have demonstrated the efficiency of proposed approach. The on-line PDSRF method can be to used for improving the voltage stability margin in real time using reactive power sources, generator excitation and on load tap changer transformers as controllers for different loading conditions. Proposed method can improve the voltage stability margins in real time using reactive power sources, generator excitation and on load tap changer transformers as controllers for different loading conditions.

The primary focus of our future work will involve voltage stability investigation of the power systems integrated with large-scale wind farms.

Acknowledgements This is the extended version of the manuscript of the First EAI International Conference on Computer Science and Engineering (COMPSE 2016), November 11–12, 2016, Penang, Malaysia. AZ, DS and DP are partly supported by the International science and technology cooperation program of China and Russia, project 2015DFR70850 and by the National Natural Science Foundation of China Grant No. 61673398.

References

1. Beiraghi, M., & Ranjbar, A. M. (2013). Online voltage security assessment based on wide-area measurements. *IEEE Transactions on Power Delivery*, *28*(1), 989–997. https://doi.org/10.1109/TPWRD.2013.2247426.
2. Diao, R., Sun, K., et al. (2009). Decision tree-based online voltage security assessment using PMU measurements. *IEEE Transactions on Power Systems*, *24*(2), 832–839. https://doi.org/10.1109/TPWRS.2009.2016528.
3. Sidorov, D. (2015). Integral dynamical models: Singularities, signals & control, T. 87. In L. O. Chua (Ed.), *World Scientific Series on Nonlinear Science Series A*. Singapore: World Scientific Publishing Co Pte Ltd. https://doi.org/10.1142/9789814619196_fmatter.

4. Pao, Y. H., & Sobajic, D. J. (1992). Combined use of unsupervised and supervised learning for dynamic security assessment. *IEEE Transactions on Power Systems*, *7*(3), 878–884. https://doi.org/10.1109/59.141799.
5. Rahimi, F. A., Lauby, M. G., Wrubel, J. N., et al. (1993). Evaluation of the transient energy function method for on-line dynamic security analysis. *IEEE Transactions on Power Systems*, *8*(2), 497–507. https://doi.org/10.1109/59.260834.
6. Tomin, N., Zhukov, A., Sidorov, D., et al. (2015). Random forest based model for preventing large-scale emergencies in power systems. *International Journal of Artificial Intelligence*, *13*(1), 211–228.
7. Wehenkel, L. (1995). Machine learning approaches to power system security assessment. PhD Dissertation, University of Liege.
8. Lachs, W. R. (2002). Controlling grid integrity after power system emergencies. *IEEE Transactions on Power Systems*, *17*(2), 445–450. https://doi.org/10.1109/TPWRS.2002.1007916.
9. Breiman, L. (2001). Random forests. *Machine Learning*, *45*(1), 5–32. https://doi.org/10.1023/A:1010933404324.
10. Wang, H., Fan, W., et al. (2003). Mining concept-drifting data streams using ensemble classifiers. In *Proceedings of the Ninth ACM SIGKDD International Conference on Knowledge Discovery and Data Mining* (pp. 226–235).
11. CIGRE Task Force. (2007). Coordinated voltage control in transmission network. C4.602, Technical Report.
12. Alimisis, V., & Taylor, P. C. (2015). Zoning evaluation for improved coordinated automatic voltage control. *IEEE Transactions on Power Systems*, *30*(5), 2736–2746. https://doi.org/10.1109/TPWRS.2014.2369428.
13. Corsi, S. (2000). The secondary voltage regulation in Italy. In *2000 Power Engineering Society Summer Meeting (Cat. No.00CH37134)*, Seattle, WA (Vol. 1, pp. 296-304). https://doi.org/10.1109/PESS.2000.867599.
14. Paul, J. P., Leost, J. Y., & Tesseron, J. M. (1987). Survey of secondary voltage control in France: Present realization and investigation. *IEEE Transactions on Power Systems*, *2*(2), 505–511. https://doi.org/10.1109/TPWRS.1987.4335155.
15. Piret, J. P., Antoine, J. P., Subbe, M., et al. (1992). The study of a centralized voltage control method applicable to the Belgian system. In *Proceedings of the CIGRE*, Paris, France (pp. 39–201).
16. Sancha, J. L., Fernandez, J. L., Cortes, A., & Abarca, J. T. (1996). Secondary voltage control: Analysis, solutions, simulation results for the Spanish transmission system. *IEEE Transactions on Power Systems*, *11*(2), 630–638. https://doi.org/10.1109/59.496132.
17. Corsi, S., Pozzi, M., et al. (2004). The coordinated automatic voltage control of the Italian transmission grid. Part I: Reasons of the choice and overview of the consolidated hierarchical system. *IEEE Transactions on Power Systems*, *19*(4), 1723–1732. https://doi.org/10.1109/TPWRS.2004.836185.
18. Taylor, C. W. (2006). Discussion of "The coordinated automatic voltage control of the Italian transmission grid"—Part I: Reasons of the choice and overview of the consolidated hierarchical system. *IEEE Transactions on Power Systems*, *21*(1), 444–450. https://doi.org/10.1109/TPWRS.2004.836262.
19. Breiman, L. (1996). Bagging predictors. *Machine Learning*, *24*(2), 123–140. https://doi.org/10.1007/BF00058655.
20. Ho, T. K. (1998). The random subspace method for constructing decision forests. *IEEE Transactions on Pattern Analysis and Machine Intelligence*, *20*(8), 832–844. https://doi.org/10.1109/34.709601.
21. Geurts, P., Ernst, D., & Wehenkel, L. (2006). Extremely randomized trees. *Machine Learning*, *63*(1), 3–42. https://doi.org/10.1007/s10994-006-6226-1.
22. Breiman, L., Friedman, J., Stone, C. J., & Olshen, R. A. (1994). *Classification and regression trees*. New York: Chapman and Hall/CRC.
23. Kuncheva, L. (2004). Classifier ensembles for changing environment. In F. Roli, J. Kittler, & T. Windeatt (Eds.) *5th International Workshop on Multiple Classifier Systems* (pp. 1–15).

24. Zhukov, A. V., Sidorov, D. N., & Foley, A. M. (2017). Random forest based approach for concept drift handling. *Communications in Computer and Information Science*, *661*, 69–77. https://doi.org/10.1007/978-3-319-52920-2_7.
25. Kessel, P., & Glavitsch, H. (1986). Estimating the voltage stability of a power system. *IEEE Transactions on Power Delivery*, *1*(3), 346–353. https://doi.org/10.1109/TPWRD.1986.4308013.
26. Kessel, P., & Glavitsch, H. (1986). Estimating the voltage stability of a power system. *IEEE Transactions on Power Delivery*, *1*(3), 346–354. https://doi.org/10.1109/TPWRD.1986.4308013.
27. Gong, X., Zhang, B., et al. (2014). Research on the method of calculating node injected reactive power based on L indicator. *Journal of Power and Energy Engineering*, *2*, 361–367. https://doi.org/10.4236/jpee.2014.24048.
28. Brzezinski, D. (2010). Mining data streams with concept drift. Dissertation, Poznan University of Technology.

Enhanced Security of Internet Banking Authentication with EXtended Honey Encryption (XHE) Scheme

Soo Fun Tan and Azman Samsudin

Abstract The rapid growth of security incidents and data breaches recently had risen concerns on Internet banking security issues. Existing Internet banking authentication mechanism that primarily relies on the conventional password-only authentication cannot efficiently resist to recent password guessing and password cracking attacks. To address this problem, this paper proposed an eXtended Honey Encryption (XHE) scheme by adding an additional protection mechanism on the existing user authentication mechanism. When the malicious user attempts to unauthorized access to online bank account by entering his guessed password, instead of rejecting the access, the XHE algorithm generates an indistinguishable bogus bank data, subsequently redirects attacker to fake user account, in which the attack could not determine whether the guessed password is working correctly or not. Therefore, increasing the complexity of password guessing and cracking attacks. This paper also provides an in-depth study of attack models on password-based authentication mechanism and their countermeasures. Subsequently, a preliminary study on Malaysian online banking authentication system is presented.

Keywords Internet banking security · Authentication
Password-based attack · Honey Encryption

1 Introduction

Internet banking has increasingly become a dominant delivery channel for financial services in Malaysia and worldwide. Malaysia Internet banking have seen a rapidly exponential growth with 2.6 million subscribers in 2005 up to 23.5 million subscribers at the end of February 2017, a 803.85% growth for the past 13 years [1].

S.F. Tan (✉) · A. Samsudin
School of Computer Sciences, Unversiti Sains Malaysia, Penang 11800, Malaysia
e-mail: soofun4818@yahoo.com

A. Samsudin
e-mail: azman.samsudin@usm.edu.my

I. Zelinka et al. (eds.), *Innovative Computing, Optimization and Its Applications*, Studies in Computational Intelligence 741,
https://doi.org/10.1007/978-3-319-66984-7_12

Although the convenience of Internet banking gained its current popularity, as the increasing numbers of security breaches and fast-changing nature of cybercrime incidents recently, public is shifting concerns on their online banking accounts' security and privacy issues. Generally, the security of Internet banking system consists of confidentiality, authentication, integrity and non-repudiation issues [2, 3]. It typically involves authenticating a user with pre-defined credentials, verifying the user's claimed identity is valid, granting access authorization to legitimate user, protecting the user account information privately and unmodifiable by third parties, as well as ensuring any transactions made by the user are traceable and verifiable. Secure Sockets Layer/Transport Layer Security (SSL/TLS) has become the de facto Internet banking standard to ensure its confidentiality and integrity, however, there is none single scheme has become pre-dominant yet for authentication [2].

A combination of username and password, or so called as Single Factor Authentication (SFA) or password-only authentication [4], however, is the most common method of authenticating user, and always serves as the only defensive line of providing security in controlling access to Internet banking account recently, due to its convenience and practicality. Traditionally, these hashed and salted passwords are considered as "computationally secure", if the best-known method of breaking the algorithms (e.g. MD5, SHA-1, SHA-2, PBKDF2, etc.) require an unreasonably large amount of computer processing time [5]. However, with exalted technological progress, parallelism techniques and distributed algorithms, existing online banking systems that relies on the conventional username and password as the means of authentication and access control mechanisms of theirs online user account pose extensive security risks of being challenged and broken [6, 7].

To address this problem, Juels and Ristenpart [8, 9] proposed a Honey Encryption (HE) scheme to increase the complexity of password guessing and cracking process. Considering the fact that the majority (63%) of confirmed data breaches incidents involved leveraging weak, default and stolen password [10], as well as these stolen passwords often do not set off any security alarms, this paper extends the HE scheme to enhance the security of user authentication and account access mechanisms. When the attacker attempts to access the online account with their guessing password, instead of rejecting their account access, the proposed eXtended Honey Encryption (XHE) scheme generates an indistinguishable bogus user account that closely resemblance the actual user account data. Subsequently, a security alert will be generated and typically reported to an administrator or computer security incident response team. This paper also reviews attack models on online password-based authentication systems and discusses its corresponding countermeasure. Subsequently, a comprehensive survey of user authentication mechanism that have been used in Malaysian online banking systems is further analyzed. The main contribution therefore are as follows. (i) An Extended Honey Encryption (XHE) scheme is proposed to enhance the security of password-based authentication mechanism in online banking system; (ii) An in-depth study of attack models on password-based authentication mechanism; (iii) A detailed discussion of recent countermeasure password-based attack models; (iv) A comprehensive survey of Malaysian online banking authentication system.

The rest of this paper is organized as follows. Section 2 overviews attack models of password-based authentication systems and its corresponding countermeasure. Subsequently, the survey of Malaysian online banking authentication systems and their countermeasure of password-based attacks is presented in Sect. 3. Section 4 describes preliminaries and algorithm of eXtended HE (XHE) scheme. Section 5 demonstrates the applicability of extended HE scheme to enhance the user authentication security of Internet banking system. Finally, this paper draws a conclusion in Sect. 6.

2 Attack Models on Password-Based Authentication System and Its Countermeasures

Most of the recent user authentication and account access control of the online banking systems are constructed based on the Single Factor Authentication—username and password. The security of password-based authentication, however, depends upon the strength of user selected password. The novel studies on password habits revealed that 86% of user selected passwords are extremely weak, in which consist either of dictionary words, digits, lowercase letters only or combination of these [11, 12]. And the threat is getting worse, in which 61% of the users are reusing the single password for multiple online accounts [13]. Several attack models on password-based authentication and its countermeasures are summarized in Table 1 and subsequently discussed in the following.

Table 1 Attack models on password-based authentication system and its recent countermeasures in banking sectors

Attack models	Countermeasures
Password guessing and cracking	
Brute force attack	Stronger Password Policy, Salted Hashing Password, Account Lockout Policy, Throttle Access Attempts, Ban IP Addresses, CAPTCHA, Two-Factor Authentication
Dictionary attack	
Rainbow table attack	
Password recovery/reset attacks	Security Email/SMS Alert
Social engineering	
Shoulder surfing attack	Two-Factor Authentication, Multi-Factor Authentication (MFA)
Phishing attack	SSL/TLS, Two-Factor Authentication, Security Verification Questions, A User Pre-Defined Personal Image Or Passphrases Before User Entering The Login Password, Spam Filters, Phishing Alert, Proper Email Client And Web Browser Configuration, Anti-Phishing Awareness Campaign, Phishing Simulation Tools
SMishing attack	
Vishing attack	
Spear phishing attack	
Pharming attack	
Malware	Two-Factor Authentication, Anti-malware software
Session hijacking	SSL/TLS, Two-Factor Authentication, OTP

i. **Password Guessing and Cracking Attacks**. Password guessing is a typical type of password attacks that trying all the possibilities to guess the combination of legitimate user's username and password, with the aim of compromising the authentication mechanisms and gaining access to legitimate users' resources. Meanwhile, password cracking is the process of recovering a password from a stored database or during data transmission. Both password guessing and cracking attacks can be conducted either locally or remotely, with a manual approach (e.g. shoulder surfing), or automated software programs (e.g. Hydra [14], RainbowCrack [15], etc.). These password guessing and cracking attack methods are further discussed in the following.

- **Brute Force Attack**. Also, known as exhaustive key search. It is a fundamental trial and error password guessing method that attacker trying every possible password until the legitimate users' password is identified. It is the most time-consuming password attack method, however, is considered as the most successful attack method [16]. For instance, the brute force attack on the eight-characters length password that consists of combination of any 95 characters (uppercase, lowercase letters, digits and symbols) only take 5.5 h for brute-forcing 95^8 possibilities, that is equivalent to the speed of 350 billion-guess-per-second on the processor [17]. Brute-force attack generally can be conducted online or offline mode. In the online mode of brute-force attack, the attacker attempts to verify whether a guessed password is correct by interacting with the Internet banking login server. This online brute-force attack, however, can be defeated effectively by throttling access attempts and banning IP addresses that attempt to log in many times, as well as implementing a lockout policy. For instance, three incorrect login attempts will lock the user's banking account for 24 h. Besides that, increasing the complexity of login procedure, such as CAPTCHA (Completely Automated Public Turing test to tell Computers and Human Apart), or using Two Step Verification (also known as Two-Factor Authentication), in which verification code (typically 4 or 6 digits) will be sent to user via SMS or a voice call, or alternatively can be retrieved from the Time-Based One Time Password apps, can be used to prevent the automated online brute-force attacks such as Hydra [14] and RainbowCrack [15]. On the other hand, much of the works on password attack have focused on offline attacks [18]. The aim of offline brute-force attack is fraudulently stealing the user's password file either by invading an insufficiently protected user's computer or banking server via malicious virus or Trojan horse, thus enabling unconstrained trials of guessing user's password. The main urge for brute-force attackers is always the notorious weakness of user-chosen passwords. Therefore, a stronger password is further needed to withstand both online and offline brute-force attacks, in which requires probably survive from 10^6 and 10^{14} guesses respectively [19].

- **Dictionary Attack**. It is a password guessing method that uses brute-force technique, however, instead of trying all possibilities, the guessed passwords are constructed based on words, sentences, numbers or dates that taken from a dictionary [20]. Thus, greatly reducing the computer processing time and storage requirement compared to conventional brute force attack [21]. The common countermeasures against dictionary attack includes increasing login complexity, delaying response and account locking [22].
- **Rainbow Table Attack**. It is a password cracking method that works like dictionary attack; however, focuses on the password that encrypted with hashing algorithm. Generally, the attacker steals the hashed password either from databases or sniffed from authentication traffic between a client and banking server. Subsequently, the attacker compares these stolen passwords with their pre-computed table—rainbow table—for reversing the hash functions and recovering the password in plaintext. A common countermeasure of this password cracking method is to attach a random string, called as salt, into the user's password before applying hashing algorithms. Applying the short-length salt or reusing the same salt for multiple times significantly reduce its strength to resists the rainbow table attack.
- **Password Recovery/Reset Attacks**. It is a password cracking method that attackers often find it much easier to be exploited compared to time-consuming other password guessing and cracking approach. To reset forgotten login password and recover access to targeted banking account, however, always require attacker to obtain targeted users' private information before answering security questions (such as mother maiden name, primary school, favorite pet's name, etc.). Also, it always triggers a security alert to be sent to legitimate user.

ii. **Social Engineering**. Euphemism for non-technical or low-technology methods, often involving psychological manipulation of divulging user's password or private information [20]. Several social engineering attacks techniques are further elaborated as follows.

- **Shoulder Surfing Attack**. It is an observation practice that looking over the targeted user's shoulder or spying on the user's computer or mobile devices to obtain their login password. Recently, the shoulder surfing attack can be prevented effectively with the Mobile Two-Factor Authentication (TWF) or Multi-Factor Authentication (MFA) schemes, in which the user authentication mechanism does not relies solely on the user's passwords, however, also on possession factors (e.g. 6-digit verification code sent to mobile phone) and inherence factors (e.g. biometric fingerprint, voice or retina recognition).
- **Phishing Attack**. It is a social engineering attack method in which attackers masquerading as a trustworthy banking or financial institutions, and sending a fake email that typically direct the user to visit fake or cloned

malicious sites. Subsequently, the user is asked to enter their password to login, update or reset their login password. Compared to other types of cybercrime, phishing attack is the most affordable attack method in terms of the investment and level of technical expertise required [23]. In 2016, Kaspersky Lab's anti-phishing technologies have detected a total of 154.96 million attempts to visit different kinds of phishing pages, and nearly half of them (47.48%) were targeted to banking and financial industries [23]. A combination of the Hypertext Transfer Protocol (HTTP) with the Secure Sockets Layer/Transport Layer Security (SSL/TLS), or called as Hypertext Transfer Protocol Secure (HTTPS), has become the de facto Internet banking standard to defense phishing attack. Other technical countermeasures of banking and financial institutions includes two-factor authentication, increasing the login complexity, such as answering personal verification questions or displaying a user pre-defined personal image or passphrases [24] to help the user distinguish a real banking site from a phishing site, before user entering the login password. On the other hand, technical countermeasures that focuses on users' side includes using spam filters, proper email client and web browser configuration. Meanwhile, non-technical countermeasures such as anti-phishing awareness campaign, and phishing simulation tools (e.g. PhishSim [25], SecuityIQ [26], AwareEd [27]), serve as a last defensive line of phishing attack, by training the users to better spot suspicious emails and identify phishing attempts via mock or simulated attacks.

- **SMishing Attack**. It is a variant of phishing attack, that using the Short Message Service (SMS) systems to lure the users and subsequently asked for login credentials or trigger to download a Trojan horse, playback, virus or other malware onto their mobile devices.
- **Vishing Attack**. Voice Phishing, or so called "Vishing" attack, is a variant of phishing attack that conducted via voice technology such as landline, cellular telephone, VoIP (voice over IP), etc. to bait users to divulge theirs banking information or further directing user to phishing site. Most of the time, caller identity spoofing will be employed to disguise legitimate source of bank or financial institutions.
- **Spear Phishing Attack**. It is a variant of phishing attack, however, instead of sending a high volume of generalized emails randomly, spear phishing attack sends a customized email to targeted victim, to increase success rate to have the link or attachment being clicked. The FireEye studies [28] further showed that 70% of victims' open spear phishing email and 50% of those who open the spear-phishing emails click on the links within the email, compared to 5% for phishing attack.
- **Pharming Attack**. It is a variant of phishing attack, however, focuses on inserting the malicious code onto user's computer, modifying host-files or hijacking the Domain Name System (DNS) server. Subsequently, the users are redirected to the attacker controlled sites; even after the users have entered the legitimate URL of the intended website.

iii. **Malware**. Short for Malicious software, is a malicious program or code that includes viruses, worms and Trojan horses, used to infiltrate on user computer, steals password and financial data without the user's consent. The malware attack has rapidly evolved and becoming more sophisticated hacking tools that able to intercept verification code sent to mobile device [29], thus thwarting Two-Factor Authentication (TWF) countermeasures. The Kaspersky Lab studies [23] further revealed the number of banking Trojans attacks was increased by 30.55% to reach 1.09 million in 2016, subsequently 2.87 million attempts to launch malware that capable of stealing money via online banking channel have been blocked. Among these banking malware, Zbot is still the most widespread banking malware family (44.08% of attacked users) but in 2016 it was actively challenged by the Gozi family (17.22%) [23]. A common countermeasure of malware attack is installing a real-time and up-to-date anti-malware software in both user and banking server.

iv. **Session Hijacking**. Also, known as Man-In-The-Middle or Main-In-The-Browser Attack. It is a form of active wiretapping on Transmission Control Protocol (TCP) and has the advantage of interfering in real time transaction, in which attacker seizes control of a previously established communication session between user and banking server [20]. Most of the time, session hijacking works together with phishing attack and malware injection. For instance, Zeus, Silent Banker, Cobalt Strike and SpyEye malware facilitate the hijacking of a victim's active online session by stealing session data or cookies, subsequently present them to the bank's server to gain access to a legitimate online banking session [30]. A common banking countermeasure is securing communication session with SSL/TLS, as well as enforcing two-step verification such as require user to provide a valid One-Time Password (OTP) upon request transaction to be made.

3 A Survey of Malaysian Online Banking Authentication Systems and Theirs Countermeasures

This section presents the preliminary survey of Malaysia personal Internet banking Authentication system and their countermeasure of password-based attacks, as summarized in Table 2. Note that all the personal Internet banking authentication systems were assessed and studied in March 2017. Ten major banking and financial institutions that provides personal Internet banking has been selected in terms of total assets [31].

Generally, the authentication system of Malaysia personal Internet banking is fundamentally constructed based on password-based approach, in which user is authenticated with username and password to access their banking resources. The username and password of all the banks satisfy the Bank Negara Malaysia (BNM) password minimum requirement of 6 alphanumeric characters. Most of the

Table 2 Overview of Malaysia online banking authentication system and their countermeasures

Malaysia personal internet banking	Security countermeasures of password-based attacks														
	User authentication mechanisms				Password recovery mechanisms							Communication channel		Security & awareness campaign	
	Username	Password	Account lockout policy	Others	Online					Offline		HHTP	HTTP with SSL/TLS	Security tips	Simulation tools
					ATM/debit card/credit	PIN	Identity card	CAPTCHA	TAC/OTP	ATM	Phone call				
Maybank	√	√	√	Security image and passphrase,	√	√	√						√	√	
CIMB	√	√	√	Secure word	√	√	√	√					√	√	
Public Bank	√	√	√	Person login phrase						√	√		√	√	
RHB Bank	√	√	√	Secretword	√	√							√	√	
Hong Leong Bank	√	√	√	Security image and passphrase	√	√	√	√	√				√	√	
Am Bank	√	√	√	Personal image			√		√				√	√	
United Overseas Bank	√	√	√	–			√						√	√	
Bank Rakyat	√	√	√	Security color and secure phrases			√						√	√	
OCBC Bank	√	√	√	OTP					√				√	√	
HSBC Bank	√	√	√	8-digit Secondary password or 6-digit OTP from security device					√				√	√	

Banks implement additional security features on user registration and authentication mechanisms, in which user is required to select pre-defined security feature during their registration and used it to distinguish a real banking site from a phishing site, before user entering the password during the login process, therefore, effectively prevent the disclosure of user password to unauthorized third-parties and attackers. These additional security features, includes security image (e.g. Maybank, Hong Leong Bank, Am Bank), passphrases (e.g. Maybank, CIMB, Public Bank, RHB Bank, Hong Leong Bank), security color (e.g. Bank Rakyat) and secondary password (e.g. HSBC Bank). To countermeasure password guessing and cracking attacks, either manually or with automated software (e.g. Hydra), most of them are enforcing account lockout policy, in which the user account will be locked after a specified number of invalid or failed login attempt (generally 3 attempts).

Meanwhile, the countermeasure of defensing password recovery attack is varying among these banks. Nine out of ten banks allow users to recover or reset their login password online, excludes Public Bank which only allows users to reset theirs password via ATM machine or operator assistance. Obviously, all the banks enforce Two-Factor Authentication on password recovery mechanism, in which possession factors (something only the user has) is subsequently used to confirming a user's claimed identity. Example of these possession factors includes identity card number (e.g. Hong Leong Bank, Am Bank, United Overseas Bank, Bank Rakyat, OCBC Bank), ATM or Debit Card or Credit Card Number and its PIN number (e.g. Maybank, CIMB, RHB Bank, Hong Leong Bank), as well as One-Time Password (OTP) generated with security device (e.g. HSBC Bank). Furthermore, some of the banks implements CAPTCHA (e.g. CIMB, Hong Leong Bank, OCBC Bank), Transaction Authorization Code (TAC) verification (CIMB, Hong Leong Bank) or security question verification (e.g. HSBC Bank) to countermeasure the automated password recovery attacks.

From the communication channel security aspect, all the banks secure their data transmission and communication with SSL/TLS to ensure their confidentiality and integrity, as well as defense session hijacking attacks. Also, a user is required to enter a valid 6-digit TAC or OTP upon a transaction request. Moreover, all the banks display security tips and alerts on their banking system to increase security awareness of users. However, none of them are providing a more sophisticated awareness campaign such as anti-phishing simulation tools.

4 Extended Honey Encryption (HE) Scheme

While existing Malaysia Personal Internet banking systems enforce two-factor authentication mechanism to countermeasure password-based attacks such as brute force attack, dictionary attack, rainbow table attack, etc., however, can be thwarted with recent malware attack which capable to intercept TAC/OTP sent to mobile device [29], as well as subjected to SSL stripping attack [32]. Furthermore, stolen passwords pose extensive security risks because attacks using stolen passwords often do not set off any alarms. To address this problem, this paper further extended the Honey Encryption (HE) scheme to enhance the security of user authentication and account access mechanisms. Section 4.1 describes preliminaries and some background works of HE scheme. Subsequently, Sect. 4.2 presents the algorithm of extended HE scheme.

4.1 Preliminaries

Honey Encryption (HE) scheme [8, 33] was firstly introduced by Juels and Ristenpart on 2014 to add the extra protection layer onto the password-based RSA encryption algorithm and credit card applications. Subsequently, extended by Tyagi et al. [34] and Huang et al. [35] to secure the basic text messaging and genomic data application respectively. Meanwhile, Joseph et al. [36] enhanced the security of HE scheme to resist the message recovery attacks. More recently, tan et al. [21, 37] further extended HE scheme (XHE) scheme to secure the cloud data storage. This section introduces some concepts and background which will be used in the construction of the extended HE scheme in Sect. 4.2.

Message Space $(\mathcal{M})$ [8, 33]. Since HE deceives the attackers by providing ambiguous looking messages, it requires a message space, $\mathcal{M}$, which contains all possible messages, M. The size of $\mathcal{M}$ must be customized for each scenario and dependent on the type of contents that need to be encrypted. The distribution over $\mathcal{M}$ is denoted as ψ_m, subsequently, sampling per this distribution is denoted as $M \leftarrow \psi_m \mathcal{M}$.

Seed Space (S) [8, 33]. Seed space, S, is the space of all n-bit binary strings for some predetermined n. Each message in $\mathcal{M}$ is mapped to a seed in S. The size of the seed is directly proportional to how likely a message is to appear. The size also must be large enough at such even the least likely messages have to be mapped to at least 1 seed. Similar to $\mathcal{M}$, S is predefined by developer which can be based on personal judgment, research or sampling results. The distribution on set S is denoted as a map p: $S \rightarrow [0,1]$ such that $\Sigma_{s \in S}\, p(s) = 1$. Subsequently, sampling according to such distribution is denoted as $s \leftarrow {}_p S$.

Distribution-Transforming Encoder (*DTE*) [8, 33]. A *DTE* consists of a pair of algorithms, such that $DTE = (encode, decode)$. The *encode* algorithm takes as input a message $M \in \mathcal{M}$ and outputs a set of seed value, S from seed space, S. The deterministic *decode* algorithm takes as input a message $S \in$ and outputs a message $M \in \mathcal{M}$. The correctness of DTE algorithm follows as for any $M \in \mathcal{M}$, $Pr[decode\ (encode\ (M)) = M] = 1$.

Inverse Sampling DTE (*IS-DTE*) [8, 33]. A *IS-DTE* consists of a pair of algorithms, such that IS-DTE = *(IS-encode, IS-decode).* The *IS-encode* algorithm runs the Cumulative Distribution Function (CDF), F_m such that with a pre-defined message distribution ψ_m and $\mathcal{M} = \{M_1, M_2, \ldots, M_{|\mathcal{M}|}\}$. Define $F_m(M_0) = 0$, subsequently generates M_i such that $F_m(M_{i-1}) \leq S < F_m(M_i)$, where $S\ \ _\$ [0,1)$. Lastly, encodes the input message M_i by selecting a uniformly random value from the range $[F_m(M_{i-1}), F_m(M_i]$. The *IS-decode* algorithm is the inverse of CDF, such that *IS-decode* $= F_m^{-1}(S)$.

DTE-then-Encrypt (*HE [DTE, SE]*) [8, 33]. A HE [DTE, SE] algorithm is a pair of algorithms (*HEnc, HDec*) that encrypts a message by using the *DTE* algorithm, subsequently re-encrypts the output of *DTE* algorithm with Symmetric Encryption scheme (*SE*) as follows.

***HEnc* (*K, M*)**. Given the symmetric key, *K* and a message *M*, let the *H* be the hashing algorithm and *n* is the number of random bits, select a uniformly random, $s \leftarrow_\$ encode(M)$ and $R \leftarrow_\$ \{0,1\}^n$, outputs the ciphertext, $C = H(R,K) \oplus s$. The process of HEnc **(*K, M*)** is illustrated in Fig. 1.

***HDec* (*K, R, C*)**. Given the *K*, *R* and *C*, computes $s = C \oplus H(R, K)$ and subsequently outputs the ciphertext, $M = decode(s)$. The seed, *s*, alone is insufficient to retrieve the message, *M*, unless it is a one-to-one mapping. In most cases, a *s* falls into a seed range, *S*. Therefore, Inverse sampling table comes into play for message lookup as illustrated in Fig. 2.

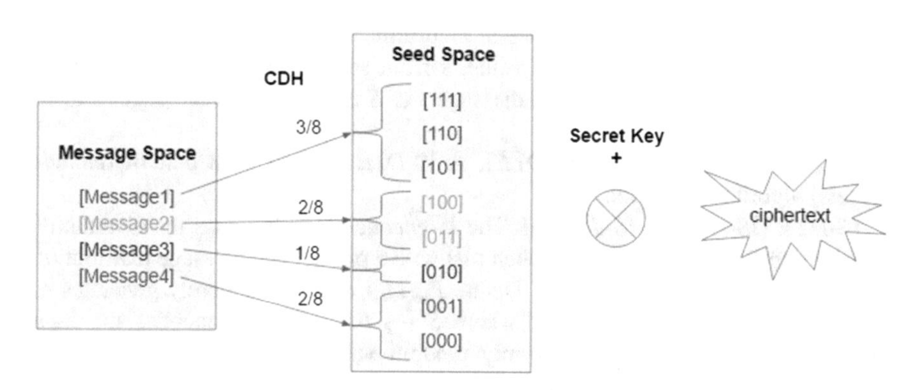

Fig. 1 The process of *HEnc* algorithm

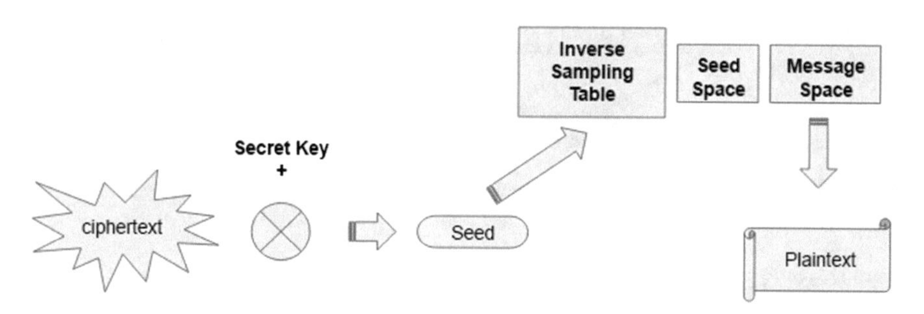

Fig. 2 The process of *HDec* algorithm

4.2 Extended Honey Encryption (XHE) Scheme

This section presents an extended version of the HE scheme [8, 33], so called as eXtended Honey Encryption (XHE) scheme. Similar to credit card applications, the length of the user password usually has a limit size. For instance, Maybank, CIMB and Public Bank user password has a maximum length of 12 characters, and most of time, these passwords do not exceed 30 characters (e.g. HSBC Bank). Next, the construction of the extended HE scheme is presented in the following.

Message Space (*$\mathcal{M}$*). Given the message, M is a user password that consists of 36 alphanumeric characters with the maximum length of 30 characters. With the total of 36^{30} possibilities, the distribution over $\mathcal{M}$ is denoted as ψ_m and the sampling according to such distribution is denoted as $M \leftarrow \psi_m\ \mathcal{M}$, as illustrated in Fig. 3.

Seed Space (*S_1, S_2*). Seed space, S, is the space of all n-bit binary strings. Each message in $\mathcal{M}$ is mapped to a seed in S such that $\Sigma_{s \in S}\ p(s) = 1$.

Distribution-Transforming Encoder (*DTE*). A *DTE* consists of a pair of *DTE (encode, decode)* algorithms as follows:

***DTE (encode, decode)*.** The *encode* algorithm takes a user password as input, $M \in \mathcal{M}$ and outputs a set of seed value, s from seed space, S. The deterministic *decode* algorithm takes as input a message $s \in S$ and outputs a message $M \in \mathcal{M}$ (Fig. 4).

Inverse Sampling DTE (*IS-DTE*). A IS-*DTE* consists of a pair of *(encode, decode)* algorithms as follows:

***IS-DTE (IS-encode, IS-decode)*.** The *IS-encode* algorithm runs the Cumulative Distribution Function (CDF), F_m, such that with a pre-defined message distribution, ψ_m and $\mathcal{M} = \{M_1, M_2, \ldots, M_{|\mathcal{M}|}\}$. Define $F_m(M_0) = 0$, subsequently generates M_i such that $F_m(M_1) \leq S < F_m(M_i)$, where $S \leftarrow_{\$} [0,1)$. Lastly, encodes the input message M_i by selecting a uniformly random value from the range $[F_m(M_{i-1}), F_m(M_i)]$. The *IS-decode* algorithm is the inverse of CDF, such that *IS-decode* $= F_m^{-1}(S)$.

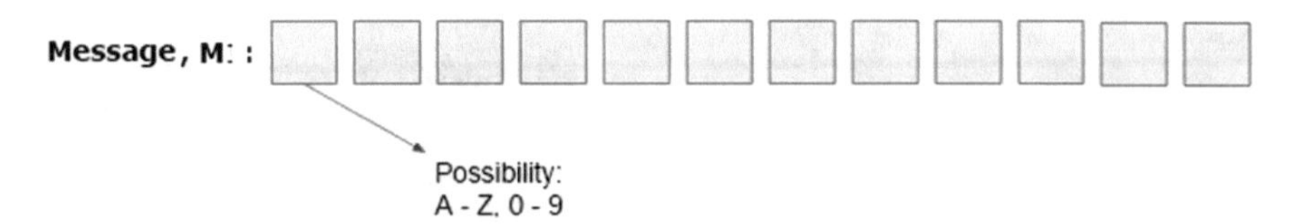

Fig. 3 The message space, M of the extended HE scheme

Fig. 4 The *DTE* algorithm of the extended HE scheme

DTE-then-Encrypt (*HE [DTE, SE]*). A HE [DTE, SE] algorithm is a pair of algorithms (*HEnc, HDec*) that encrypts a message by using the *DTE* algorithm, subsequently re-encrypts the output of *DTE* algorithm with a Symmetric Encryption scheme (*SE*) as follows.

HEnc (K, M). Given the symmetric key *K*, and a user password *M*, let the *H* be the hashing algorithm and *n* is the number of random bits, select a uniformly random, $s \leftarrow_{\$} encode(M)$, and $R \leftarrow_{\$} \{0,1\}^n$, outputs the ciphertext, $C_1 = H(R,K) \oplus s$.

HDec (K, R, C). Given the *K*, *R*, *C* computes $s = C \oplus H(R, K)$. Subsequently outputs the user password, $M = decode(s)$ with the lookup inverse sampling tables.

5 Extended Honey Encryption in Enhancing Security of Internet Banking Authentication System

This section describes the applicability of the XHE scheme to enhance the user authentication security of Internet banking system as illustrated in Fig. 5. When malicious user attempts to unauthorized access Internet banking account with the guessed password, instead of rejecting their account access, the proposed eXtended Honey Encryption (XHE) scheme generates an indistinguishable bogus user account that closely resemblance the actual user account data. Subsequently, a security alert will be triggered and reported to an administrator or computer security incident response team. The application scenario is further described in the following.

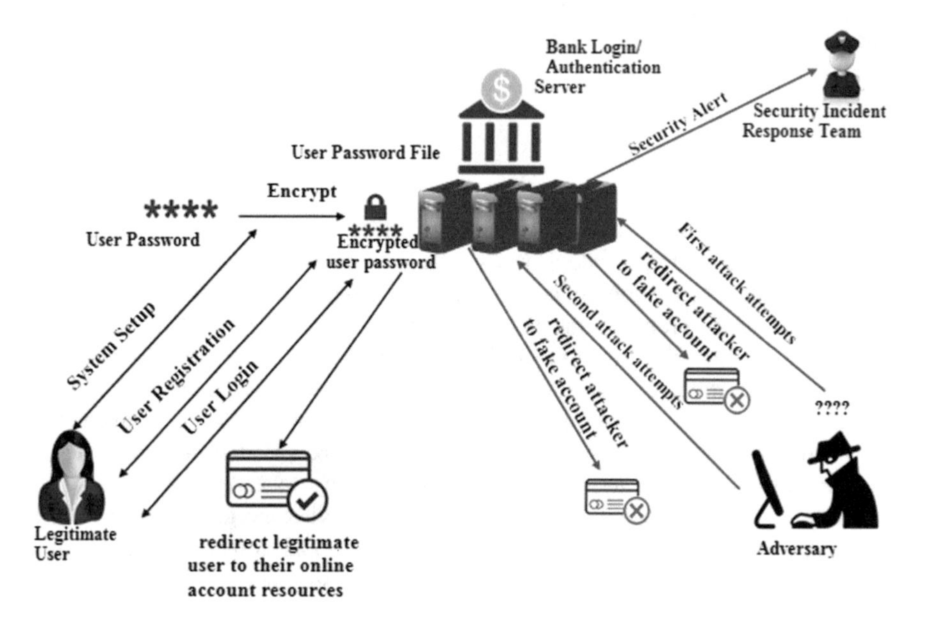

Fig. 5 Enhanced user authentication mechanism with the extended honey encryption

System Setup(λ). Given a security parameter, λ, define the distribution over the key space, ψ_k. Subsequently, takes the PIN number of users' bank card (e.g. ATM card/debit card/credit card) as an input, and outputs the shared secret key, *K*.

User Registration Mechanism with Encrypt (*K, F*) Algorithm. Given the secret key, *K* and a user password, *p* (e.g. word documents, images, etc.), takes a *p* as an input, as a message space *M*. Subsequently, define the distribution over the messages space ψ_m. Next, generates the seed *s* with *DTE_encode* algorithm. Then, runs the *IS-encode* algorithm to generate a series of fake user account, such that $F_i = \{F_m(M(_{i-1}), F_m(M_i)$. Lastly, outputs the encrypted user password, $p_e = C$ with *HEnc (K, M)* algorithm and stores in banking password file.

User Authentication and Granting Access Mechanisms with Decrypt (*K*, *F*$_e$) Algorithm. The user enters his password on login page. The login server runs *HDec* algorithm to verify user identity, if it is a valid user's password, then granting user an access control to his banking resource, otherwise, redirect to fake user account and triggers a security alert to an administrator or computer security incident response team.

Attack Scenario. Suppose that the adversary intercepts the encrypted password, p_e, either from the user's device (e.g. desktop, tablets, mobile devices, etc.), banking password file or during the data transmission, and subsequently attempts to decrypt it. With his guessed password, the *Decrypt* algorithm of login server redirects attacker to a fake account in response to every incorrect guess of the user's password or shared symmetric key, *K*. These fake account is generated with the *IS-encode* algorithm in which the fake user account distribution is closed related to actual user account distribution. Therefore, it is indistinguishable from the attacker perspective, thus increases the complexity of determining whether the attacker have guessed a password correctly or not.

6 Conclusion

While the existing Internet banking authentication systems relies on password-based approach, which is vulnerable to password-based attacks such as brute-force attack, dictionary attack, rainbow table attack, main-in-the middle attack, this paper proposed an eXtended Honey Encryption (XHE) scheme for enhancing the security of password-based authentication mechanism. When the malicious user attempts to unauthorized access the user banking account with his guessing password, instead of rejecting their account access as conventional password-based authentication mechanism, the extended HE algorithm generates an indistinguishable bogus user account that are closely related to the actual user account, in which the attack could not determine whether the guessed password is working correctly or not. It is noticeable that the additional storage space is required to keep the inverse sampling tables and fake account information. In future, whether the proposed XHE scheme can be further optimized to reduce its storage requirement is another interesting topic to be explored.

Acknowledgements This work was by a research grant from Universiti Sains Malaysia (USM) [1001/PKOMP/811334]. The authors also thank the COMPSE 2016, First EAI International Conference on Computer Science and Engineering, NOVEMBER 11–12, 2016, PENANG, MALAYSIA.

References

1. Department of Statistic, Bank Negara Malaysia. (2017). Malaysia's payment statistics: Internet banking and mobile banking subscribers. Available via BNM. Retrieved March 31, 2017, from http://www.bnm.gov.my/index.php?ch=34&pg=163&ac=4&bb=filed.
2. Hiltgen, A., Kramp, T., & Weigold, T. (2006). Secure internet banking authentication. *IEEE Security and Privacy, 4*(2), 21–29.
3. Hutchinson, D., & Warren, M. (2003). Security for internet banking: A framework. *Logistics Information Management, 16*(1), 64–73.
4. Boonkrong, S. (2017). Internet banking login with multi-factor authentication. *KSII Transactions on Internet and Information System, 11*(1), 511–535.
5. Diffie, W., & Hellman, M. E. (1976). New directions in crytography. *IEEE Transactions on Information Theory, 22*(6), 644–654. New Jersey: IEEE Press.
6. Wang, X., Feng, D., Lai, X., & Yu, H. (2004). Collisions for hash functions MD4, MD5, HAVAL-128 and RIPEMD, Cryptology ePrint Archive Report 2004/199, 16 Aug 2004, revised 17 Aug 2004. Retrieved March 30, 2017, from http://merlot.usc.edu/csac-f06/papers/Wang05a.pdf.
7. Stevens, M. (2013). New collision attacks on SHA-1 based on optimal joint local-collision analysis. EUROCRYPT 2013, Lecture Notes in Computer Science (Vol. 7881, pp. 245–261). Springer.
8. Juels, A., & Ristenpart, T. (2014) Honey Encryption: Encryption beyond the brute-force barrier. *IEEE Security and Privacy, 12*(4), 59–62. New York: IEEE Press.
9. Juels, A., & Ristenpart, T. (2014). Honey Encryption: Security beyond the brute-force bound. Advances in Cryptology—Eurocrypt, LNCS (Vol. 8841, 293–310). Heidelberg: Springer.
10. Verizon 2016 Data breach investigations report. Retrieved March 30, 2017, from http://www.verizonenterprise.com/verizon-insights-lab/dbir/2016/.
11. Bonneau, J. (2012) The science of guessing: Analyzing and anonymized corpus of 70 million passwords. In *Proceeding of the IEEE Symposium on Security and Privacy* (pp. 538–552), California, USA.
12. Florencio, D., & Herley, C. (2007). A large-scale study of web password habits. In *Proceeding of the 16th ACM International Conference on the World Wide Web* (pp. 657–666), Banff, Canada.
13. Consumer Survey: Password habits—A study of password habits among American consumers, CSID report 2012. Retrieved March 30, 2017, from https://www.csid.com/wpcontent/uploads/2012/09/CS_PasswordSurvey_FullReport_FINAL.pdf.
14. Hauser, V. Hydra THC. Retrieved April 30, 2017, from https://github.com/vanhauser-thc/thc-hydra.
15. Shuanglei, Z. Retrieved April 30, 2017, from http://project-rainbowcrack.com/.
16. Hole, K. J., Moen, V., & Tjostheim, T. (2006). Case study: Online banking security. *IEEE Security and Privacy, 4*(2), 14–20.
17. Gosney, J. M. (2012). Password cracking HPC. In *Pawword^12 Security Conferenc*e, Olso, Norway. Retrieved April 30, 2017, from http://passwords12.at.ifi.uio.no/Jeremi_Gosney_Password_Cracking_HPC_Passwords12.pdf.
18. Herley, C., & Florencio, D. (2008). Protecting financial institutions from brute-force attack. In *Proceedings of 23rd International Information Security conference* (pp. 682–685).

19. Herley, C., Florencio, D., & Oorschot, P. C. (2014). An administrator's guide to internet password research. *Journal of Usenix LISA*. https://www.microsoft.com/en-us/research/publication/an-administrators-guide-to-internet-password-research/; http://research.microsoft.com/apps/pubs/?id=227130.
20. RFC 4949—Internet security glossary, version 2.
21. Tan, S. F., & Samsudin, A. (2017). Enhanced security for public cloud storage with honey encryption. *Advanced Science Letters, 23*(5).
22. Pinkas, B., & Sander, T. Securing passwords against dictionary attacks.
23. Financial Cyberthreats in 2016. Kaspersky lab report, February 2017. Cited April 30, 2017, Retrieved 30 March 2017, from https://media.scmagazine.com/documents/287/kaspersky_lab_financial_cybert_71527.pdf.
24. Dhmija, R., & Tygar, J. D. (2005). The battle against phishing: Dynamic security skins. In *Proceedings of the ACM Symposium on Usable Security and Privacy, ACM International Conference Proceedings Series* (pp. 77–88). ACM Press.
25. PhshSim. InfoSec Institute. Retrieved April 30, 2017, from https://www.infosecinstitute.com/phishsim.
26. SecurityIQ. InfoSec Institute. Retrieved April 30, 2017, from https://securityiq.infosecinstitute.com/.
27. AwareEd. InfoSec Institute. Retrieved April 30, 2017, from https://www.infosecinstitute.com/aware-ed.
28. FireEye. Best defense again spear-phishing: Recognize and defend against the signs of an advanced cyber attack. FireEye resource archive. Retrieved March 30, 2017, from https://www.fireeye.com/current-threats/best-defense-against-spear-phishing-attacks.html.
29. Whigham, N. (2016, March). Sophisticated malware detected that steals online banking passwords, thwarts text authentication. Available via News.com.au. Retrieved 30 March 2017, from http://www.news.com.au/technology/online/security/sophisticated-malware-detected-that-steals-online-banking-passwords-thwarts-text-authentication/news-story/afa5cf65dfcd350acc069aaf41545e39.
30. Making sense of man-in-the-browser attacks: Threat analysis and mitigation for financial institutions. RSA report. Retrieved March 30, 2017, from https://www.rsa.com/content/dam/rsa/PDF/Making_Sense_of_Man_in_the_browser_attacks.pdf.
31. Top banks in Malaysia. Retrieved March 15, 2017, from http://www.relbanks.com/asia/malaysia.
32. Marlinspike, M. (2009). More tricks for defeating SSL in practice. Black Hat USA.
33. Juels, A., & Ristenpart, T. (2014). Honey Encryption: Security beyond the brute-force bound. Advances in Cryptology—Eurocrypt, LNCS (Vol. 8841, pp. 293–310). Heidelberg: Springer.
34. Tyagi, N., Wang, J., Wen, K., & Zuo, D. (2015). Honey Encryption applications. 6.857 Computer and Network Security. Massachusetts Institute of Technology. Available via MIT. Retrieved March 15, 2017, from http://www.mit.edu/~ntyagi/papers/honey-encryption-cc.pdf.
35. Huang, Z., Ayday, E., Fellay, J., Hubuax, J.-P., & Juels, A. (2015). GenoGuard: Protecting genomic data against brute-force attacks. In *IEEE Symposium on Security and Privacy* (pp. 447–462). California: IEEE Press.
36. Joseph, J., Ristenpart, T., & Tang, Q. (2016). Honey encryption beyond message recovery security. IACR Cryptology ePrint Archive (pp. 1–28).
37. Edwin, M., Samsudin, A., & Tan, S.-F. (2017). Implementing the honey encryption for securing public cloud data storage. In *Proceedings of First International Conference on Computer Science and Engineering*.

An Enhanced Possibilistic Programming Model with Fuzzy Random Confidence-Interval for Multi-objective Problem

Nureize Arbaiy, Noor Azah Samsudin, Aida Mustapa, Junzo Watada and Pei-Chun Lin

Abstract Mathematical models are established to represent real-world problems. Since the real-world faces various types of uncertainties, it makes mathematical model suffers with insufficient uncertainties modeling. The existing models lack of explanation in dealing uncertainties. In this paper, construction of mathematical model for decision making scenario with uncertainties is presented. Primarily, fuzzy random regression is applied to formulate a corresponding mathematical model from real application of a multi-objective problem. Then, a technique in possibilistic theory, known as modality optimization is used to solve the developed model. Consequently, the result shows that a well-defined multi-objective mathematical model is possible to be formulated for decision making problems with the uncertainty. Indeed, such problems with uncertainties can be solved efficiently with the presence of modality optimization.

Keywords Fuzzy random regression · Possibilistic programming Confidence-interval · Modality optimization

N. Arbaiy (✉) · N.A. Samsudin · A. Mustapa
Faculty of Computer Science and Information Technology, Soft Computing and Data Mining Center (SMC), Universiti Tun Hussein Onn, Parit Raja, 86400 Batu Pahat, Johor, Malaysia
e-mail: nureize@uthm.edu.my

J. Watada
Department of Computer & Information Sciences, Universiti Teknologi PETRONAS, 32610 Seri Iskandar, Perak, Malaysia

P.-C. Lin
Department of Information Engineering and Computer Science, Feng Chia University, No. 100, Wenhwa Rd., Seatwen, Taichung 40724, Taiwan, ROC

I. Zelinka et al. (eds.), *Innovative Computing, Optimization and Its Applications*, Studies in Computational Intelligence 741,
https://doi.org/10.1007/978-3-319-66984-7_13

1 Introduction

In standard application problems, a mathematical model is assumed to be provided. Therefore, many researches put much focus on problem solving based on pre-defined model. However, in real world actual practice, mathematical models are not given by some higher authority; hence, mathematical modelling requires skills in building mathematical models, determining parameters value, fitting model, and selecting among competing models. Mathematical programming is widely used to model real world problems by transforming the problem into a mathematical programming problem with variable's symbols and numerical values. Such numerical values in the model are commonly provided by the expert or generated by statistical tool. But, determining precise and rigid values is difficult [1–6] because in nature, problems in decision making process involve uncertainties such as human and machine errors, and are subject to ungrounded evaluations as well as lack of information [7].

Real world problem and decision making dealt with various uncertainties. It makes, modeling such problem needs appropriate approach. Decision variables with uncertainties usually approximately determined as crisp values. If the values are not appropriately determined as crisp, the developed model may yield an infeasible solution [8]. Existing solution usually handles crisp information or single uncertainty such fuzzy information only. In fact, multiple uncertainties may occur simultaneously in the information. For example, fuzzy and random happens simultaneously and are captured in the data used to model the problem. For that reason, a technique such as fuzzy random regression is significant to deal with a problem among simultaneous uncertainties ([9–12]).

In the other hand, the formulated mathematical model for real application problem may involve uncertain information. The coefficients and goals of the model are fuzzy or not known exactly in a multi-objective problem model due to imprecise judgment made by the decision maker. For instance, the manager of a manufacturing company decides to set their production profit for a year is more than 5 million dollars. The statement of '*more than*' shows unclear edge of the numerical values given which is '5 million dollars'. It makes the developed model contain fuzzy goal. The model should first treat the uncertainty before the mathematical models could be solved [12]. In this case, possibility theory has been widely used in building decision making models with uncertainty (see [8, 13–15, 26]).

Motivated by the above-mentioned situation, this paper emphasizes the building of problem model from real world problem which contain simultaneous uncertainties and provide a solution approach to solve the developed model. This paper follows the approach to first translate the real-world problem which contain fuzzy random uncertainties in its data into a problem model by estimating the coefficients by fuzzy random regression approach [9, 12]. Next, the research applied a modality optimization to solve multi-objective possibilistic problem (MOPP) model which is developed in prior based on fuzzy random based coefficient. The modality optimization uses necessity measure to evaluate the objectives and constraints in the

MOPP model with vague and ambiguous data. The necessity measure is useful to deal with ambiguous data and vague targets (goals) as well as to measure to what extent the decision makers aspiration/target can be achieved with certainty. The approach also uses a weighting scheme from the individual minimal and maximal solution. The performance of the model is demonstrated through a case study developed for palm oils production problem. Diverse solutions achieved by the proposed approach outperform deterministic solutions in terms of given performance measure.

The remainder of this paper is organized as follows. Section 2 reviews related works that bring to fuzzy random regression and possibilistic programming. Section 3 presents the solution of the possibilistic programming evaluation model using the necessity approach. Section 4 provides a case study to illustrate the proposed methodology. Sections 5 present the results and discussion, while Sect. 6 concludes with some indication for future works.

2 Preliminary Studies and Definitions

This section explains a Fuzzy Random Regression and Possibilistic Programming method as a ground studies for this work.

2.1 Fuzzy Random Regression

Statistical data that are observed from real world situation may contain imprecise lingual or vague value. Normal random variable is incapable to handle such data due to the presence of stochastic and fuzzy uncertainty [17–19]. Hence fuzzy random variable is integrated in the regression technique to cope with the fuzzy random data. Statistical analysis with fuzzy random data may require transformation of the vague parameter while making decision, whereby the fuzziness are transferred into coefficient value [18]. Confidence interval and expected value are used to describe fuzzy random in regression model. The interval is valuable to estimates a population parameter hence indicate the reliability of an estimate. The detail explanation of fuzzy random variable and fuzzy random regression are given elsewhere [1, 9, 12]. This method works efficiently with the data containing simultaneous fuzzy random uncertainties.

The data used for fuzzy random regression is formalized using LR-triangular fuzzy number (TFN). Fuzzy random data Y_j (dependent) and X_{jk} (independent) for all $j = 1, \ldots, N$ j = 1, …, N and $k = 1, \ldots, K$ are defined as $Y_j = \bigcup_{t=1}^{M_{Y_j}} \left\{ \left(Y_j^t, Y_j^{t,l}, Y_j^{t,r} \right)_\Delta, p_j^t \right\}$, and $X_{jk} = \bigcup_{t=1}^{M_{X_{jk}}} \left\{ \left(X_j^t, X_j^{t,l}, X_j^{t,r} \right)_\Delta, q_{jk}^t \right\}$, respectively. p_j^t and q_{jk}^t demonstrate the probability of the event happens in $j = 1, \ldots, N$, $k = 1, \ldots, K$ and $t = 1, \ldots, K$.

The fuzzy random regression model with $\sigma-$ confidence intervals [8] is described as follows:

$$\begin{aligned}
&\min_{A} \quad J(A) = \sum_{k=1}^{K} \left(A_k^r - A_k^l\right) \\
&A_k^r \geq A_k^l, \\
&\mathrm{Y}_j^* = \mathrm{A}_j^* I\left[e_{X_{j1}}, \sigma_{X_{j1}}\right] + \cdots + \mathrm{A}_K^* I\left[e_{X_{jK}}, \sigma_{X_{jK}}\right] \underset{\tilde{h}}{\supset} I\left[e_{Y_j}, \sigma_{Y_j}\right] \\
&j = 1, \ldots, N; k = 1, \ldots, K,
\end{aligned} \tag{1}$$

where $e_{X_{j1}}, \sigma_{X_{j1}}$ is the expected value and variance, respectively. The $\sigma-$ confidence interval is defined as $I[e_X, \sigma_X] \underline{\underline{\Delta}} \left[E(X) - \sqrt{\mathrm{var}(X)}, E(X) + \sqrt{\mathrm{var}(X)}\right]$. The confidence level shows us the frequency of an observed interval contains the parameter.

The fuzzy regression problem with $\sigma-$ confidence interval is written as follows:

$$\mathrm{Y}_j^* = \mathrm{A}_j^* I\left[e_{X_{j1}}, \sigma_{X_{j1}}\right] + \cdots + \mathrm{A}_K^* I\left[e_{X_{jK}}, \sigma_{X_{jK}}\right] \underset{\tilde{h}}{\supset} I\left[e_{Y_j}, \sigma_{Y_j}\right] \tag{2}$$

where $\left[e_{X_{jK}}, e\sigma_{X_{jK}}\right]$ are the one-sigma confidence interval.

2.2 Possibilistic Programming

Possibilistic programming concerns very much in expressions which are useful to formulate real-world problems with uncertainty. In this theory, a vague aspiration is denoted by a fuzzy goal G_i. A fuzzy goal is a fuzzy set whose membership function μG_i expresses a degree of satisfaction to a soft constraint such as '*considerably larger than* G_i' or '*considerably smaller than* G_i'. The membership function of linear fuzzy goal G_i is in the form of Eq. (1).

$$\begin{aligned}
&\mu G_i(r) = \max\left\{\min\left(1 - \frac{r - g_i}{e_i}, 1\right), 0\right\} : r \geq g_i \text{ or} \\
&\mu G_i(r) = \max\left\{\min\left(1 - \frac{g_i - r}{e_i}, 1\right), 0\right\} : r < g_i
\end{aligned} \tag{3}$$

Using the parameters of g_i and e_i, the linear fuzzy goals G_i defined by (3) are written as $G_i =]g_i, g_i + e_i)$ and $G_i = (g_i - e_i g_i[$, respectively.

Additionally, a possibility distribution $\prod_i$ presents ambiguous data, and this distribution is considered as a fuzzy restriction. Thus, a possibility distribution $\prod_i$

can be defined in terms of a fuzzy set A_i, representing the linguistic expression such as '*about* a_i' as $\prod_i = \mu A_i$, where μA_i is a membership function of '*about* a_i', A_i.

A symmetric triangular fuzzy number (TFN) $A_i = \langle a_i, d_i \rangle$ is used to define a possibility distribution $\prod_i$ with the membership function as shown in Eq. (4).

$$\mu A_i(r) = \max\left\{1 - \frac{r - a_i}{d_i}, 0\right\} \quad (4)$$

A possibilistic programming is written as in Eq. (5).

$$\begin{aligned} Y_i \triangleq \textstyle\sum_{j=1}^{n} A_{ij} x_j \widetilde{\geq} g_i, i = 1, \ldots, m \\ x_j \geq 0, j = 1, \ldots, n \end{aligned} \quad (5)$$

where A_{ij} is a possibilistic variable restricted by a possibility distribution that is defined by a triangular fuzzy number $A_i = \langle a_i, d_i \rangle$, with center a_{ij}, width d_{ij} and $\widetilde{\geq}$ is the fuzzy inequality. A fuzzy inequality demonstrates expression such as '*considerably larger than*'. Y_i is the dependent variable (output) for certain model. This will cater linguistic expressions such as 'considerably larger than g_i' that corresponds to a fuzzy goal G_i as defined by a fuzzy set with linear membership function $G_i = (g_i, g_i + d_i[$.

2.2.1 Possibility and Necessity Measures

The uncertainty adopted is written by using fuzzy sets and the concepts of possibility and necessity measures [18] to deal with the vagueness and ambiguity. Vagueness is a notion used in expressing the fuzziness of the degree to which extent an element of a set belongs to the set [18]. Meanwhile, ambiguity is related to the fuzziness of the value itself. The characterization of uncertainty of fuzzy sets enhances the ability to model real-world problems and gives a methodology for exploiting the tolerance for imprecision or uncertainties [14].

The interpretation of the possibility concept with possibility distribution $\prod_A(B)$ specified as $\prod_A(x) \triangleq \mu B(x)$ is given as follows.

Definition 1 Given a possibility distribution $\prod_A(x)$, the possibility measure of a fuzzy set B specified by $\mu B(x)$ is defined as Eq. (6).

$$\prod_A(B) = \sup_x \{\mu B(x) \wedge \prod_A(x) \quad (6)$$

The interpretation of the problem plays an essential role in formulating a model in mathematical programming. From the perspective of possibility theory, the interpretation of a model is based on the possibility measure and necessity measure.

Definition 2 Given a possibility distribution $\prod_A(x)$, the necessity measure of a fuzzy set B specified by $\mu B(x)$ is defined as Eq. (7).

$$N_A(B) = {}^{\inf \max}_{\ \ \ \ \ x}\{1 - \Pi_A(x)), \mu_B(x)\} \tag{7}$$

From Eqs. (6) and (7), the possibility measure $\prod_A(B)$ evaluates reasonable that it is possible for the possibility distribution $\prod_A$ to be under restriction, or the possibilistic variable α to be in the fuzzy set B. Likewise, $N_A(B)$ evaluates to what extent that it is certain for the possibility distribution $\prod_A$ under restriction contains the possibilistic variable α in the fuzzy set B.

Therefore, the relation as shown in Eq. (8) always holds:

$$\begin{aligned} N_A(B) &\leq \prod\nolimits_A(B) \\ N_A(B) &= 1 - \prod\nolimits_A(\overline{B}) \end{aligned} \tag{8}$$

where $\overline{B}$ is the complement of B.

In a case of applying such definitions in a multi-objective possibilistic programming problem, let α be a possibilistic variable. Let $B = (-\infty, g]$ be a non-fuzzy set of real numbers that is not greater than g. Under the possibility distribution $\prod_A(x)$, the possibility and necessity measures are defined as Eqs. (6) and (7), respectively.

$$\begin{aligned} \mathrm{Pass}(\alpha \leq g) &= \prod\nolimits_A((-\infty, g]) \\ &= \sup_r\{\mu A(r) | r \leq g\} \\ \mathrm{Nec}(\alpha \leq g) &= N_A((-\infty, g]) \\ &= 1 - \sup_r\{\mu A(r) | r \rangle g\} \end{aligned} \tag{9}$$

where $\mathrm{Pass}(\alpha \leq g)$ and $\mathrm{Nec}(\alpha \leq g)$ show the possibility and certainty degrees to what extent α is not greater than g.

2.2.2 Handling Uncertainties Through Necessity Measure

Following the previous works, this research represents uncertainties in the form of fuzzy sets [20] and characterizes them based on their membership functions. In general, the membership covers the range from having no membership (0.0) to having complete membership (1.0). Defining uncertainty based on fuzzy sets will enhance the ability of the model to cope with real-world problems, hence the methodology for exploiting the tolerance for imprecision or uncertainties [7].

The possibilistic programming problem with constraints is given in Eq. (10):

$$\begin{aligned} &\max \alpha(x_j)x \\ &\text{s.t.}: \bar{a} \times \widetilde{\leq} \bar{b}, \times \geq 0 \end{aligned} \tag{10}$$

where $f_j = \alpha(x_j)$ is an n – dimensional vector, $\widetilde{\leq}$ describes a fuzzy constraint, $\alpha, \tilde{\bar{A}}$ and $\tilde{\bar{b}}$ are possibilistic variable vectors. $f_j = \alpha(x_j)$ and $\tilde{\bar{A}} \times \widetilde{\leq} \tilde{\bar{b}}$ denotes objective function and constraints, respectively.

In solving the problem in (10), the constraints and objective functions are handled using the possibility and necessity measures. It is assumed that the decision maker specifies the possibility and necessity aspiration levels with respect to objective function values.

(A) Constraints

In order to treat the constraints, the constraints must first be translated according to the decision maker's desire as close as possible. The necessity measure allows incorporation of the certainty degree to which extent a decision maker achieves the constraint. Let $v^N \in [0, 1]^m$ be a necessity aspiration degree vector for m multi-objectives, each of which a decision maker is aspired or required to achieve with certainty. Using the necessity measure, the constraints $\tilde{\bar{A}} \times \widetilde{\leq} \tilde{\bar{b}}$ can be treated as shown in Eq. (11).

$$\mathrm{Nec}\left(\tilde{\bar{A}} \times \widetilde{\leq} \tilde{\bar{b}}\right) \geq v^N \tag{11}$$

The case in (11) is defined for the case that the decision maker assures that a certainty degree is not less than the symmetric fuzzy number and is written as $M = \langle \sum_{j=1}^{n} x_j a_j, \sum_{j=1}^{n} |x_j| c_j \rangle$.

Based on Eq. (11), assume that h is not less than v^N, hence Eq. (12).

$$h = \sum_{j=1}^{n} x_j a_j, \sum_{j=1}^{n} |x_j| c_j \tag{12}$$

Expressions in Eq. (12) are treated as a constraint, which is obtained from the necessity measure and considered as the certainty degree of decision maker's intention to satisfy the problem constraint.

(B) Objectives

In a fuzzy mathematical programming problem, each objective function is restricted by a possibility distribution $\prod(x)$. Therefore, the meaning of the objective and constraints should be properly interpreted during the development of the mathematical model. In solving the mathematical model in Eq. (8), the fuzzy goal is included in the objective function as a constraint by using a modality approach. A modality model corresponds to the minimum-risk approach to a stochastic programming problem [21], and is presented as a dual approach to the fractile optimization [22].

Consider the situation where the decision maker wants to maximize the certainty degree (necessity degree) such that some evaluation of the event is not smaller than v^N and written as max $\mathrm{Nec}(ax \geq v^N)$.

Using an additional variable, h, the decision maker's intention can be rewritten by using the expression given in Eq. (13).

$$\begin{aligned} &\max \mathrm{h}, \\ &\mathrm{s.t.: Nec}\left(\mathrm{a(x)} \geq \mathrm{v^N}\right) \geq \mathrm{h}. \end{aligned} \tag{13}$$

Next, taking Eq. (10) into consideration, the final model can be obtained from Eq. (14).

$$\sum_{j=1}^{n} x_j a_j + \sum_{j=1}^{n} |x_j| c_j \geq h \tag{14}$$

By setting h = 0, Eq. (15) shows the following relation.

$$\sum_{j=1}^{n} x_j a_j + \sum_{j=1}^{n} |x_j| c_j \geq 0 \tag{15}$$

Finally, the problem modeled in Eq. (13) is equivalent to the following model in Eq. (16).

$$\begin{aligned} &\max v^{N,} \\ &s.t: \frac{\sum_{i=1}^{n} x_i a_i}{\sum_{i=1}^{n} |x_i| d_i} \geq v^N \end{aligned} \tag{16}$$

Note that by extending the singular objective programming model (16) to p objectives is essentially a multi-objective programming. Figure 1 illustrates a triangular fuzzy number. The following section presents the proposed modality approach for solving the multi-objective possibilistic problem.

Given the above information, all prior knowledge for dealing with a possibilistic programming with confidence-interval of fuzzy random regression for multi-objective problem is presented.

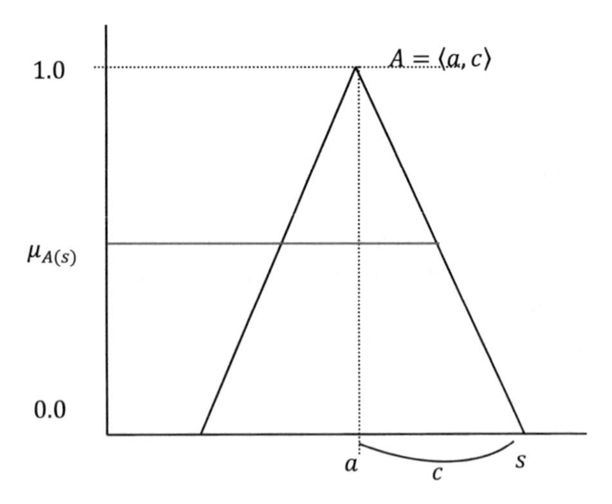

Fig. 1 Fuzzy number $A = \langle a, c \rangle$

3 Solution Approach: Modality Approach for Solving Multi-objective Possibilistic Problem with Fuzzy Random Confidence-Interval

This section is dedicated to explain solution approach of modality optimization for multi-objective possibilistic problem.

3.1 Terminology and Notation

Table 1 lists the terminologies and notations used to explain the modality approach in multi-objective possibilistic programming.

3.2 Solving Multi-objective Possibilistic Problem

The proposed multi-objective possibilistic model can be developed by using the following algorithm:

1. Describe the problem and build the model by using fuzzy random regression model. The readers may refer [9, 12] for estimating the coefficient and building the problem model using fuzzy random regression model.
2. Analyze the constraint as given in Eq. (11). Set the degree of certainty to v^N, and change the constraints into Eq. (12).

Table 1 Notations

Types of notation	
A	A fuzzy number denoted by $A = \langle a, d \rangle$ whose membership function $\mu_A(x)$ is defined as Eq. (17): $\mu_A = \max\left\{1 - \frac{\lvert x-a \rvert}{d}, 0\right\}$ (17) where a and d are a center value and a width, respectively (refer Fig. 1)
$\overline{A}$	Complement of fuzzy set A
$\propto$	Fuzzy coefficient $\langle a, d \rangle$ where a and d are a center value and a width, respectively
$\wedge$	Minimum operator
$\vee$	Maximum operator
$\prod(x)$	Possibility distribution of A
G	Fuzzy goal with $u_G = \langle g, e \rangle$ where g is a center value and e is a width
$\widetilde{\geq}$	Fuzzy inequality, approximately greater than or equal to
$N_A(B)$	Necessity measure of fuzzy set B under possibility distribution A
v^N	The threshold vector of necessity measure
x_j	Variable

3. Obtain the necessity aspiration level g^N of from the objective function. Convert the objective function into $\text{opt}\,\text{Nec}(ax \geq g^N)$.
4. Develop a multi-objective possibilistic programming model as follows:

$$\begin{aligned} & opt && \sum_{i=1}^{p} h_i^{\eta} \\ & \text{subject to:} && \sum_{i=1}^{p} \left(\frac{\sum_{j=1}^{n} x_j \alpha_j}{\sum_{j=1}^{n} |x_j| d_j} \right) \geq h \end{aligned} \tag{18}$$

The multi-objective possibilistic programming methodology is illustrated in Fig. 2.

A modality optimization approach is then used to solve the multi-objective possibilistic programming problem which results in fractional programming [23].

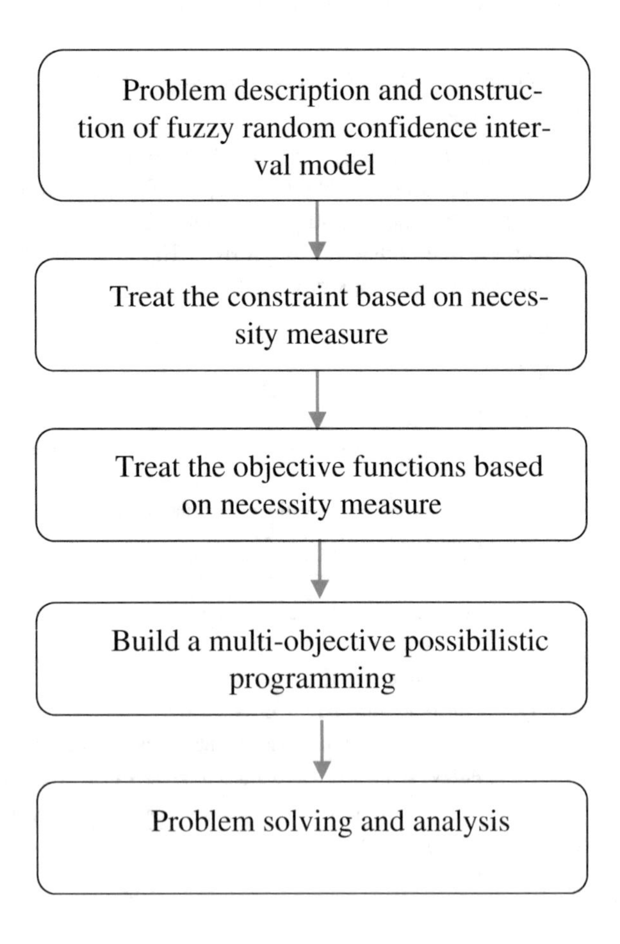

Fig. 2 Multi objective possibilistic programming methodology

The modality optimization takes the advantages of fractional programming in finding the problem's solution [24].

The multi-objective, fuzzy-random possibilistic programming problem (FR-PPP) model as shown in Eq. (18) is rewritten to Eq. (19) by using the treated constraints (12) and objectives (14) as follows:

$$\begin{array}{ll} opt & \sum_{i=1}^{p}\left(\frac{\sum_{j=1}^{n} x_j \alpha_j}{\sum_{j=1}^{n} |x_j| d_j}\right) \\ \text{subject to:} & \sum_{i=1}^{p}\sum_{j=1}^{n} x_j a_j + v^{\eta}\left(\sum_{j=1}^{n} |x_j| d_j\right), \\ & x_i \geq 0 \end{array} \tag{19}$$

Meanwhile, the mathematical model as shown in Eq. (18) is a linear fractional programming problem with multiple objectives. The general form of multi-objective linear fractional programming problem is shown in Eq. (20).

$$\begin{array}{ll} \text{opt} & Z(x) = \left[\frac{N_1 x}{D_1 x}, \cdots, \frac{N_p x}{D_p x}\right]^T, \\ \text{s.t.} & x \in X, X = \left\{x \in R^n : \tilde{A}x \leq \tilde{b}, x \geq 0\right\} \end{array} \tag{20}$$

In variable change technique [25], a linear fractional problem is modeled into a linear program that is used to solve problem modeled in Eq. (19).

The compatibility of a value of j of $P \leq \left(N_{ix}, N_i^0\right)$ is given as in [26] by the function in Eq. (21),

$$C_j^{N_i} = \begin{cases} 0 & \text{if } N_i x < p_i^j \\ \frac{N_i x - p_i^j}{N_i^0 x - p_i^j} & \text{if } p_i^j < N_i x < N_i^0, \\ 0 & \text{if } N_i x > N_i^0 \end{cases} \\ j = 1, 2, 3; i = 1, \ldots, p \tag{21}$$

where N_i, D_i and p_i^j are the minimum individual solution, maximum individual solution and threshold, respectively. Similarly, the compatibility of a value of j of $P \geq \left(D_i x, D_i^0\right)$ is given by Eq. (19).

$$C_j^{D_i} = \begin{cases} 0 & \text{if } D_i x < s_i^j \\ \frac{s_i^j - D_i x}{s_i^j - D_i^o} & \text{if } D_i^0 < D_i x < s_i^j, \\ 0 & \text{if } D_i x > D_i^0 \end{cases} \\ j = 1, 2, 3; i = 1, \ldots, p \tag{22}$$

Consider the relative importance w_i of objective i for $\mu_i^{N_i} = C_i^{N_j}$ and w_i' for $\mu_i^{D_i} = C_i^{D_j}$, such that $w_i > 0, w_i' > 0$ and $\sum_{i=0}^{n} w_i + w_i' = 1$. Thus, the simple additive weighting model could be obtain to solve the multi-objective linear fractional programming problem.

$$\begin{aligned} &\text{opt}V(\mu) = \textstyle\sum_{i=1}^{p}(w_i\mu_i^{N_i} + w_i' u_i^{N_i}) \\ &\text{such that: } \mu_i^{N_i} = C_i^{N_j}, \mu_i^{D_i} = C_i^{D_j}, \\ &Ax \le b, \boldsymbol{\mu}_i^{N_i} \le 1, \boldsymbol{\mu}_i^{D_i} \le 1, \\ &\boldsymbol{\mu}_i^{N_i} \ge 0, \boldsymbol{\mu}_i^{D_i} \ge 0, x \ge 0, i = 1, \ldots, p, \end{aligned} \tag{23}$$

where $V(u)$ is the achievement function.

3.3 Weighting Method for Multi-objective Linear Fractional Problem

The coefficient w_p of the pth objective function $z_p(x)$ is known as weight of the objective function. These weights can be obtained by various methods.

$$\text{opt } w_i z_i + \cdots + w_p z_p \tag{24}$$

The best maximal and minimal solution f_i^+ and f_i^- are defined as Eq. (25).

$$\begin{aligned} &f_i^+ = \max_{x \in X} f_i(C_i x), \\ &\text{s.t.: } A_x \ge b, x > 0, \\ &f_i^- = \min_{x \in X} f_i(C_i x), \\ &\text{s.t.: } A_x \le b, x \ge 0, \text{s.t.: } A_x \le b, x \ge 0, \end{aligned} \tag{25}$$

The weight w_i^* is calculated as $w_i^* = \frac{1}{f_i^+ - f_i^-} w_i^* = \frac{1}{f_i^+ - f_i^-}$ and the normalized weights $w_i^{*N_i}$ and $w_i^{*D_i}$ are defined as in Eq. (26),

$$w_i^{*N_i} = \frac{w_i^{N_i}}{\sum_{i=1}^{p} w_i^{N_i}} 0.5, w_i^{*D_i} = \frac{w_i^{D_i}}{\sum_{i=1}^{p} w_i^{D_i}} 0.5 \tag{26}$$

where $\Sigma_i \{ w_i^{*N_i} + w_i^{*D_i} \} = 1$.

w_i^* represents the weights for the membership functions of the decision vectors. The scheme expressed by Eq. (26) is used to obtain the normalized weight for the fuzzy objective of multi-objective linear fractional problem model. This normalization scheme provides the best normalization results as the objective functions are normalized by the true intervals of their variation over the Pareto optimal set.

4 Numerical Experimentation

To illustrate the feasibility of the FR-PP approach in solving multi-objective possibilistic programming problem, this section presents a case study on modelling palm oils production planning. The increasing demand in oil consumption has increased the request for more production of palm oil. The industry needs to improve its productivity and quality, and also targeted to increase its profits. Considering the targets and available resources, production planning to achieve this goal is constructed. In this study, the palm oil production planning problem is investigated with two decision variables and two functional objectives under four system constraints. The two decision variables are namely as crude palm oil, x_1 and crude palm kernel oil x_2.

The linear programming model was developed by means of fuzzy random regression method [9, 12]. The goal of the objective function and right hand side values of the constraints (b) are decided by experts. The triangular fuzzy number coefficients are estimated by a fuzzy random regression [9, 12] approach. The problem is then modeled as follows:

$$\begin{aligned}&\text{Maximize profit:}\\&\left.\begin{aligned}z_1 &= \langle 0.860, 0.100\rangle x_1 + \langle 1.100, 1.100\rangle x_2\\ z_1 &= \langle 1.126, 0.020\rangle x_1 + \langle 0.000, 0.000\rangle x_2\end{aligned}\right\}\text{Maximize production:}\end{aligned} \tag{27a}$$

$$\begin{aligned}&\text{Subject to:}\\&\text{Raw material:}\\&F_1 = \langle 3.75, 0.06\rangle x_1 + \langle 0.91, 0.08\rangle x_2 \le 87.75\\&\text{Labor:}\\&F_1 = \langle 0.65, 0.55\rangle x_1 + \langle 0.90, 0.09\rangle x_2 \le 4.42\\&\text{Mills:}\\&F_1 = \langle 17.35, 0.85\rangle x_1 + \langle 2.16, 0.27\rangle x_2 \le 95.20\\&\text{Capital:}\\&F_1 = \langle 0.87, 0.65\rangle x_1 + \langle 0.98, 0.68\rangle x_2 \le 20.15\end{aligned} \tag{27b}$$

In dealing with the constraints, assume that the decision maker decides that certainty degree not less than v^N is sufficiently high for the system constraint (b) in the Problem (27). To satisfy the decision maker aim, analyze the constraints under expression (11). Use expression (10) to transform the constraints based on decision maker aim as follows:

$$\begin{aligned} &\mathrm{Nec}(\langle 3.75, 0.06\rangle x_1 + \langle 0.91, 0.08\rangle x_2 \leq 87.75) \geq v^N, \\ &\mathrm{Nec}(\langle 0.65, 0.55\rangle x_1 + \langle 0.90, 0.09\rangle x_2 \leq 4.42) \geq v^N, \\ &\mathrm{Nec}(\langle 17.35, 0.85\rangle x_1 + \langle 2.16, 0.27\rangle x_2 \leq 95.20 \geq v^N, \\ &\mathrm{Nec}(\langle 0.87, 0.65\rangle x_1 + \langle 0.98, 0.68\rangle x_2 \leq 20.15) \geq v^N \end{aligned} \tag{28}$$

For the objective part (a), assume that the decision maker aims to maximize the certainty degree of profit is not smaller than 5.0 million dollars, and to maximize the certainty degree of production volume is not smaller than 5.2 million tones. According to (13), the decision maker aims are modeled as follows:

$$\begin{aligned} &\mathrm{Nec}(0.860x_1 + 1.10x_2 \geq 5.0) \\ &\mathrm{Nec}(1.126x_1 + 0.00x_2 \geq 5.2) \end{aligned} \tag{29}$$

Equations (28) and (29) are obtained by using the method explains in the Sect. 2.2.2 (treating the uncertainties using necessity measures). Finally, problem (27) can be rewritten as follows:

$$\mathrm{Max}\left\{\frac{0.86x_1 + 1.10x_2 - 5.0}{0.1x_1 + 0.1x_2}, \frac{1.126x_1 + 0x_2 - 5.2}{0.02x_1 + 0x_2}\right\}$$

Subject to:

$$\begin{aligned} &3.79x_1 + 1.1x_2 \leq 87.75, \\ &1.03x_1 + 0.96x_2 \leq 4.42, \\ &17.94x_1 + 2.34x_2 \leq 95.20, \\ &1.32x_1 + 1.43x_2 \leq 20.15, \\ &x_i \geq 0. \end{aligned} \tag{30}$$

Next, the equivalent linear programming problem is re-written as the following:

$$
\begin{aligned}
&\max v(\mu) = 0.258\mu_{N1} + 0.241\mu_{N2} + 0.072\mu_{D1} + 0.127\mu_{D2} \\
&\text{Subject to:} \\
&\quad 0.45\mu_{N1} - 0.860x_1 - 1.100x_2 = -5.3 \\
&\quad 0.37\mu_{N2} - 1.126x_1 - 0.000x_2 = -5.8 \\
&\quad 2.00\mu_{D1} + 0.100x_1 + 0.100x_2 = 2.0 \\
&\quad 3.00\mu_{D2} + 0.020x_1 + 0.000x_2 = 2.0 \\
&\quad \mu_{N1} \leq 1, \\
&\quad \mu_{D1} \leq 1, \\
&\quad \mu_{N1} \geq 1, \\
&\quad \mu_{D1} \geq 1, \\
&\quad x_i \geq 0
\end{aligned}
\tag{31}
$$

Problem (31) is solvable by using the equivalent ordinary linear programming model.

5 Discussion and Future Direction

As shown in model (27), the coefficient values were triangular fuzzy numbers that represented fuzzy judgments within the range of expected and the estimation target. A fuzzy random regression model [12] was used to statistically determine these coefficients due to the difficulty in estimating the coefficients in a multi-objective model. This model used the historical data to effectively determine the estimated coefficients, which in this case, the previous pattern of outcomes in predicting the future prediction or decision.

The mathematical programming problem as in model (27) was solved using the FR-PP. The constraints and objectives were treated using the necessity measures, which exemplified the decision maker's intention. The optimal solution of the problem model (30) was $(x_1, x_2) \approx (5.35, 1.03)$ whose objective value was $v(\mu) = 0.87$ with $\mu N_1 = 1.00$, $\mu N_2 = 0.62$, $\mu D_1 = 0.68$, and $\mu D_2 = 0.96$.

Taking the central value from the coefficient in model (27), the crisp multi-objective linear programming problem can be illustrated as follows:

$$
\begin{aligned}
&\max 0.86x_1 + 1.10x_2 \\
&\max 1.13x_1 + 0.00x_2 \\
&\text{subject to:} \\
&3.75x_1 + 0.91x_2 \leq 87.75 \\
&0.65x_1 - 0.90x_2 \leq 4.42 \\
&17.35x_1 + 2.16x_2 \leq 95.20 \\
&0.87x_1 + 0.98x_2 \leq 20.15
\end{aligned}
\tag{32}
$$

The problem was solved using a max-min operator approach, and achieved optimal solutions of $\lambda = 0.98$ and $D_1 = 0.68(x_1, x_2) = (5.37, 0.92)$. The following symmetric fuzzy number constraints in model (23) were used to generate the solution graph as shown in Fig. 3.

The possibility distributions corresponding to the solution of the proposed model (30) and the comparable model (32) are exemplified in Fig. 2. In the fuzzy mathematical programming system architecture [14], the obtained solution is evaluated by the achievement of the solution to the intentions of the decision maker. From the possibility distributions with respect to the solution from model (30), it is found that the certainty degree of the satisfaction of constraints on labor and mills is not high

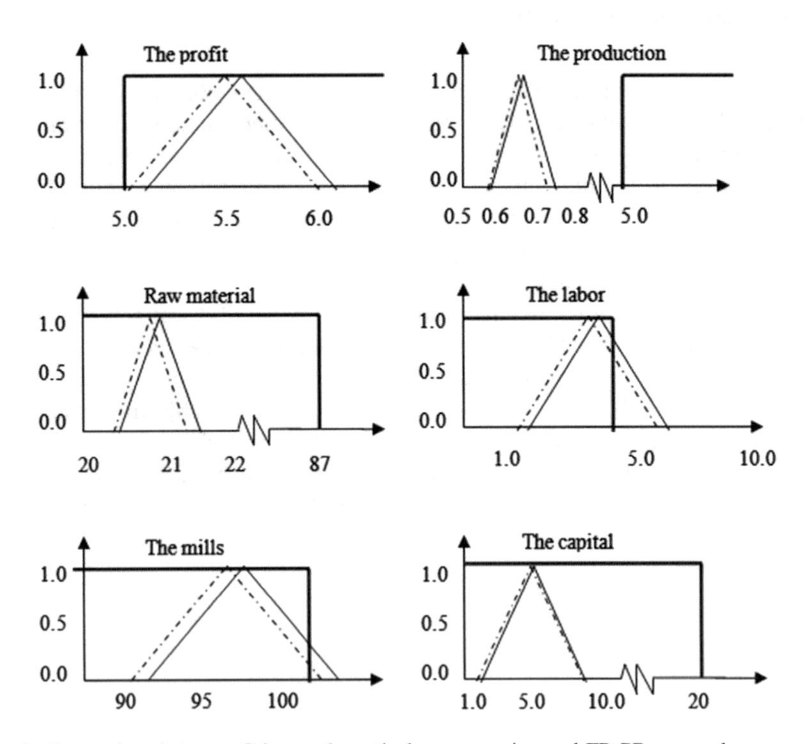

Fig. 3 Comparison between Crisp mathematical programming and FR-PP approaches

enough. This shows that the solution was ill-matched to the decision maker's intention. Even though the solution from the crisp mathematical problem in model (32) was slightly similar to the solution for the model (30), model (32) did not consider the decision maker's aspiration in its model evaluation.

Using the FR-PP method, the decision maker may be able to reconstruct the problem and redefine his/her new aspiration, for example by making the profit larger than 5.3 million. The figure proved that the FR-PP approach was able to produce various solutions on the basis of the decision maker's aim. The interactive system would be useful to verify the obtained solution against the decision maker's intention. Through such a system, the decision maker may understand to what extent the requirements by the decision makers could be satisfied.

In this study, the palm oil production planning problem is investigated with two decision variables and two functional objectives under four system constraints. The parameter values used in this study is shown in Table 2.

Taking the central value of the coefficient for problem (27), the result of the following crisp multi-objective linear programming problem is shown in the Table 3.

The results of this study indicate several interesting directions for future work. Although studies in decision making and evaluations have circulated steadily in the literature, there are numerous opportunities for further research beyond this study. From a theoretical perspective, it may be valuable to include risk analysis in the proposed methodology to ensure that the decision maker is aware of the consequences of the selected solution. In this case, more information can be presented to the decision maker to assist in selecting the best choice and while minimizing the risk of decision.

Table 2 Parameter values

Parameter	$P^0_{Z_1}$	$S^0_{Z_1}$	$N^0_{Z_1}$	$D^0_{Z_1}$
Z_1	0.3	2.0	0.75	0.00
Z_2	0.6	3.0	0.97	0.00

Table 3 Individual optimal solutions and normalized weights

Item	N_{Z_1}	N_{Z_2}	D_{Z_1}	D_{Z_2}
Minimal q_{ij}	−5.000	−5.200	0.000	0.000
Maximal μ_{ij}	0.753	0.978	0.639	0.109
Weight w	0.173	0.161	1.564	9.174
Normalized weight w^*	0.258	0.241	0.072	0.427

6 Conclusion

In this paper, the ambiguity inherent in the data and the vagueness of target goals were treated in the form of necessity measures, which in turn, expressed the decision maker's intention or aspiration. To solve the multi-objective programming problem, a modality approach was proposed based on a normalized weighting scheme that describes the relative importance of the membership functions. The proposed approach showed an effective solution under the presence of diverse uncertain situations by treating the inherent uncertainties into the mathematical programming model formulation. Various solutions can be obtained depending on decision maker's aims. The advantage of the proposed method is that the necessity measure was defined in the form of certainty degree in dealing with various uncertainties and imprecise information in order to ensure that the mathematical model represents the actual problem. The interactive system is useful where the decision maker can realize to what extent he can require membership grade for this problem to be satisfied in solving the problem. It is also useful to develop a method which can retain the uncertainty until an optimal solution is achieved.

Acknowledgements The authors express her appreciation to the University Tun Hussein Onn Malaysia (UTHM) and GATES IT Solution Sdn. Bhd. under its publication scheme. Also, thanks to the First EAI International Conference on Computer Science and Engineering (COMPSE 2016), NOVEMBER 11–12, 2016, PENANG, MALAYSIA.

References

1. Arbaiy, N., Watada, J., & Wang, S. (2014). Fuzzy random regression based multi-attribute evaluation and its application to oil palm fruit grading. *Annals of Operation Research, 219*(1), 299–315.
2. Zeleny, M. (1981). The pros and cons of goal programming. *Computers & Operations Research, 8*(4), 357–359.
3. Derghal, A., & Goléa, N. (2014). *Multi-objective generation scheduling using genetic-based fuzzy mathematical programming technique* (pp. 450–474). Handbook of research on novel soft computing intelligent algorithms: Theory and practical applications.
4. Vasant, P. (2013). *Hybrid linear search, genetic algorithms, and simulated annealing for fuzzy non-linear industrial production planning problems* (pp. 87–109). Meta-Heuristics optimization algorithms in engineering, business, economics, and finance.
5. Arbaiy, N., Watada, J., & Lin, P-C. (2016). *Fuzzy random regression-based modeling in uncertain environment* (pp. 127). Sustaining power resources through energy optimization and engineering.
6. Khan, L., Badar, R., Ali, S., & Farid, U. (2017). *Comparison of uncertainties in membership function of adaptive Lyapunov NeuroFuzzy-2 for damping power oscillations* (pp. 74). Fuzzy systems: Concepts, methodologies, tools, and applications: Concepts, methodologies, tools, and applications.
7. Jensen, H. A., & Maturana, S. (2002). A possibilistic decision support system for imprecise mathematical programming problems. *International Journal of Production Economics, 77*(2), 145–158.

8. Zadeh, L. A. (1978). Fuzzy sets as a basis for a theory of possibility. *Fuzzy Sets and Systems, 1*(1), 3–28.
9. Arbaiy, N., & Watada, J. (2010). Approximation of goal constraint coefficients in fuzzy goal programming. In *Proceeding of the Second International Conference on Computer Engineering and Applications* (Vol. 1, pp. 161–165).
10. Arbaiy, N., & Watada, J. (2012). Multi-objective top-down decision making through additive fuzzy goal programming. *SICE Journal of Control, Measurement, and System Integration, 5* (2), 63–69.
11. Akbari, M. G., & Khanjari, M. (2012). Sadegh estimators based on fuzzy random variables and their mathematical properties. *Iranian Journal Of Fuzzy Systems, 9*(1), 79–95.
12. Watada, J., Wang, S., & Pedrycz, W. (2009). Building confidence interval-based fuzzy random regression model. *IEEE Transactions on Fuzzy Systems, 11*(6), 1273–1283.
13. Julien, B. (1994). An extension to possibilistic linear programming. *Fuzzy Sets and Systems, 64*(2), 195–206.
14. Katagiri, H., Sakawa, M., Kato, K., & Nishizaki, I. (2008). Interactive multiobjective fuzzy random linear programming: Maximization of possibility and probability. *European Journal of Operational Research, 188*(2), 530–539.
15. Nematian, J. (2015). A fuzzy robust linear programming problem with hybrid variables. *International Journal of Industrial and Systems Engineering, 19*(4), 515–546.
16. Arenas, M., Bilbao, A., Perez, B., & Rodriguez, M. V. (2005). Solving a multiobjective possibilistic problem through compromise programming. *European Journal of Operational Research: Recent Advances in Scheduling in Computer and manufacturing Systems, 164*(3), 748–759.
17. Hasuike, T., & Ishii, H. (2009). Robust expectation optimization model using the possibility measure for the fuzzy random programming problem. In J. Mehnen et al. (Eds.), *Applications of soft computing: From theory to practice: Advances in intelligent and soft computing* (Vol. 58, pp. 285–294).
18. Inuiguchi, M., & Sakawa, M. (1996). Possible and necessary efficiency in possibilistic multiobjective linear programming problems and possible efficiency test. *Fuzzy Sets and Systems, 78*(2), 231–241.
19. Masahiro, I., & Ramık, J. (2000). Possibilistic linear programming: A brief review of fuzzy mathematical programming and a comparison with stochastic programming in portfolio selection problem. *Fuzzy Sets and Systems, 111*(1), 3–28.
20. Zadeh, L. A. (1965). Fuzzy sets. *Information Control, 8*(3), 338–353.
21. Stancu-Minasian, I. M. (1984). *Stochastic programming with multiple objective functions.* Dordrecht: D. Reidel Publishing Company.
22. Figueira, J., Greco, S., & Ehrgott, M. (2005). *Multiple criteria decision analysis: State of the art surveys.* Boston: Springer.
23. Torabi, S., Ali, & Elkafi, H. (2008). An interactive possibilistic programming approach for multiple objective supply chain master planning. *Fuzzy Sets and Systems, 159*(2), 193–214.
24. Chakraborty, M., & Gupta, S. (2002). Fuzzy mathematical programming for multi objective linear fractional programming problem. *Fuzzy Sets and Systems, 125*(3), 335–342.
25. Zhang, X., Huang, G. H., & Nie, X. (2009). Robust stochastic fuzzy possibilistic programming for environmental decision making under uncertainty. *Science of the Total Environment, 408*(2), 192–201.
26. Lotfi, M. M., & Ghaderi, S. F. (2012). Possibilistic programming approach for mid-term electric power planning in deregulated markets. *International Journal of Electrical Power & Energy Systems, 34*(1), 161–170.

A Crowdsourcing Approach for Volunteering System Using Delphi Method

Nurulhasanah Mazlan, Sharifah Sakinah Syed Ahmad and Massila Kamalrudin

Abstract Voluntary work is an important part of virtually every civilization and society. There are several versions of the volunteer management theory referenced in industry resources. Several organizations have developed organization-based systems designed to incorporate spontaneous volunteers. However, it can be difficult to find and recruit suitable candidates for volunteer organizations because the volunteers have many criteria to match with tasks. Also, we still have lacking information on the process of crowdsourcing in volunteering perspective. This paper, we conduct a review of volunteering management systems and crowdsourcing approach. Based on the insights derived from this analysis, we identify some issues for future research. To solve this problem, we designed a framework for the crowdsourcing approach in volunteering system to automate the process of selection volunteers and match with the criteria of volunteers and tasks. Crowdsourcing is one of the best approaches to get more information and faster from the crowd and to be more precise with the requirement from beneficiaries. We are using the Delphi method to criteria from volunteer organization. The implications of the findings for volunteering system are discussed, and future research directions suggested.

Keywords Volunteering system · Volunteering matching · Crowdsourcing Volunteering management · Volunteer selection · Delphi method

N. Mazlan (✉) · S. S. Syed Ahmad · M. Kamalrudin
Faculty of Information and Communication Technology, Universiti Teknikal Malaysia Melaka (UTeM), Melaka, Malaysia
e-mail: nurulhasanah@student.utem.edu.my; p031110013@student.utem.edu.my

S. S. Syed Ahmad
e-mail: sakinah@utem.edu.my

M. Kamalrudin
e-mail: massila@utem.edu.my

I. Zelinka et al. (eds.), *Innovative Computing, Optimization and Its Applications*, Studies in Computational Intelligence 741,
https://doi.org/10.1007/978-3-319-66984-7_14

1 Introduction

Volunteering is an altruistic activity where members of a community contribute time, resources, and services to fulfill a community need without being paid financially. There is the almost endless supply of volunteers; the unlimited amount of meaningful volunteer works [1]. Volunteering is making an important contribution to society, for instance, supporting social services, sporting events, religious activities, etc. Volunteering is often done in terms to collaboratively or corporately accomplish the task. A volunteer can help people regarding financial, physical, human and information. For example, the Projek Iqra' is one of volunteering organization that helps to strengthen the education and welfare of children and youth in welfare homes (orphanages, poor and beneficiaries) as well as needy families.

Every volunteer organization has their management. According to UN Worlds Volunteerism Report [47], information and communication technology (ICT) is the major enabler of volunteering process. Volunteer management system brings together potential volunteers with volunteering opportunities, allowing the scheduling, allocation, and execution of tasks, providing communication and coordination mechanisms for collaboration and cooperation and facilitating assessment and motivation strategies [43]. Recruitment and selection of volunteers are the most important process. However, the process to ensure the most appropriate match between volunteers and project are the complex task. Lacking the information for people who need help also the problem. The volunteer management has selection volunteers to suit with task. Every volunteer has their criteria such as profile, skills, interests, etc. The volunteer manager should assess the needs of the program and identify the types of volunteers available, then develop a description of the expectations and responsibilities of a specific volunteer position. The selection should also include a position description, including the title of the volunteer position, advisor, purpose of the position, benefits to the volunteer serving in this role, responsibilities of the volunteer, qualifications and skills needed, amount of time required and resources available. Once this information is prepared, volunteers can be recruited.

The collaboration with the crowd is the easiest way to get efficient and rapid results. Crowdsourcing is a technique that involves many volunteers to solve a problem that occurred. The crowdsourcing process is part of crowd management. Crowd management requires a collaboration of various areas of engineering and science, e.g. physics, computer science, civil engineering, management, etc. [44]. To ensure the crowdsourcing can impact on the volunteering, we need much information. However, we still lack information on the process of crowdsourcing in volunteering perspective.

This paper proposes to gain a better understanding of typical process on volunteering management system and crowdsourcing. We also focus on providing a crowdsourcing technique into volunteering system to find the best volunteers as the task solvers. The paper has been divided into four parts. Section 2 begins with the related works that definition of volunteer, volunteer management, the definition of crowdsourcing and identify the interaction among the volunteers is needed with their criteria. Section 3 describes the design of a framework for crowdsourcing in volunteering system. Section 4 conduct the criteria of volunteers using Delphi method. Section 5, we present the results obtained. The last part of this paper gives conclusions and discusses future research.

2 Related Work

This section describes some researchers about the definition of volunteer, volunteer management process and crowdsourcing process.

2.1 Volunteers

Volunteerism is defined as contributing one's time or talents for charitable, educational, social, political or other worthwhile purposes, usually in one's community, freely and without regard to compensation. Volunteering is the practice of people working on behalf of others without being motivated by financial or material gain [14, 50]. Volunteering may be done for other people or organizations, which can bring measurable benefits to the volunteer, individual beneficiaries, groups and organizations, communities, the environment and society as a whole. The variety of work done by these organizations and groups is enormous, and so the range of skills that potential volunteers can bring is almost limitless. Brett Williamson (Volunteering Australia Chief Executive Officer) [28] defines volunteering is time willingly given for the common good and without financial gain.

Like definitions found elsewhere, we can surmise that volunteering comprises activities that are unpaid and are entered into without compulsion. They freely offer up the time and service to help other people. Volunteerism fosters inclusiveness, empowerment, and sustainability evident in contributions to the physical, psychological, spiritual and economic well-being of the community.

2.2 Volunteer Management

Management can be described as making the most effective and efficient use of resources to achieve goals. A volunteer is a special kind of human resources, volunteer management subsystem contained in human resources management [8]. The volunteering management systems integrate and coordinate the core policies, processes, and activities that organization has in place for working with volunteers. It is made up of policies, procedures and work documents which control how the volunteer's program is implemented from day to day. When managing volunteers can bring stumbling blocks. The volunteers might feel burdened by multiple roles or the lack of funding or resources to fulfill their duties or frustrated because of the lack of communication and recognition, inappropriate placement, underused skills, talents and interests, and lack of training.

Zheng et al. [53] explain the volunteer management is to planning and implementation of planning carried out by a specific organization, by the need for volunteer activities. Event volunteers refer to those volunteers their time, energy, skills, and experience within the preparation and holding of the exhibition process, and try their best to the complete the task distributed to them. Several organizations have developed organization-based systems designed to rapidly incorporate spontaneous volunteers [21].

Figure 1 is the process of volunteering management in our perspective from the related study (*Yayasan Sukarelawan Siswa* (*YSS*)), [15] and interview session with several volunteer organization in Malaysia such as Mercy Malaysia, Spot Community, Projek Iqra', Hidayah Centre Foundation (HCF) and several universities.

The process in Fig. 1, the detailed procedures are started from volunteer organization makes planning and discussion to help beneficiary. Then organization makes announcement or publicity to attract volunteers to join the activities. They invite people using social media such as Facebook, website or phone. Volunteers whom interested need to register providing their profiles, including expertise, time available and location. Next, some organization needs interview session to filter the

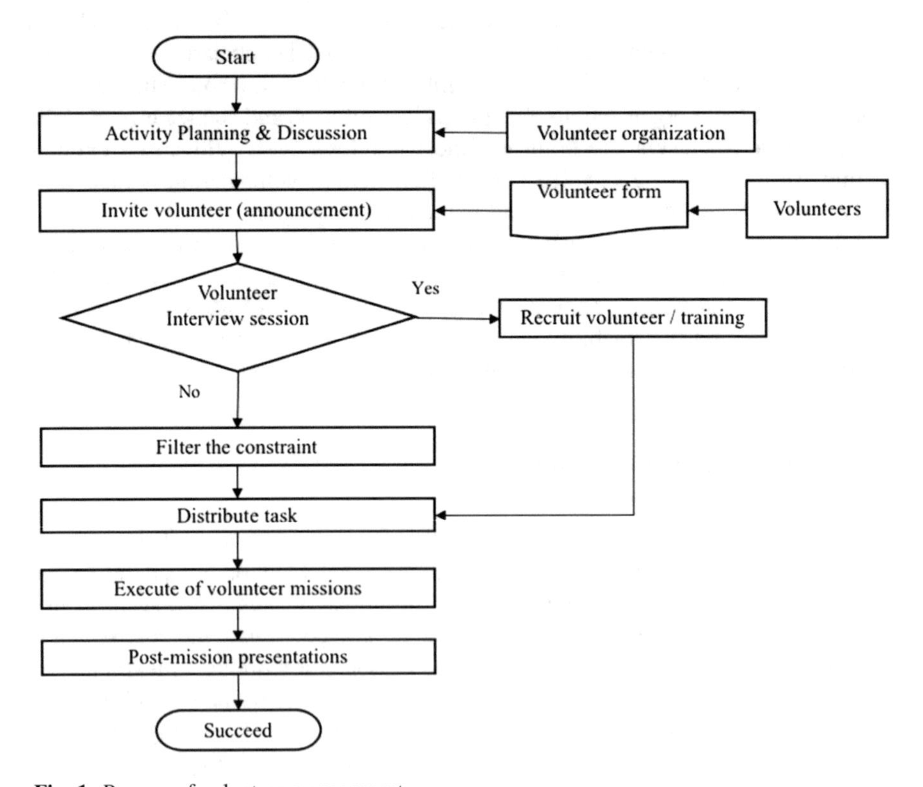

Fig. 1 Process of volunteer management

volunteer before proceed with the activity they do. The interview session provides not only an opportunity to talk to the potential volunteer about their background, skills but also to explore any doubts about the suitability of the candidate. If they have interviews, volunteers must join training sessions to qualify them to volunteer. Training is the process of helping volunteers develop the skills that will improve the quality of their work. Training helps volunteers develop basic skills, confidence and also provides support and opportunities for personal growth. If the organization does not have interview session, the volunteer filter based on constraints such as location, transportation, time availability, etc. The organization distributed their task when the volunteer was selected. Then the activities will be implemented. Finally, review the performance of volunteers and need to access the impact of the volunteer program. The volunteer gets acknowledge and reward from the volunteer organization. The recognition helps motivate volunteers to stay involved in the program. Recognition also provided through feedback on the job a volunteer is doing, challenging work assignments and opportunities to take on new responsibilities. Evaluation is the process of determining the results of volunteer performance by informal or formal methods and by giving feedback.

As shown in Table 1, we summarize the volunteer management model from numerous versions. The procedure is to ensure that volunteers are aware of what is required of them and management have a more coordinated approach. Without the support of appropriate procedures, there is the risk that policy will not be effective. The different situation has different volunteering process such as Olympic, health care, disaster management and also general process in general situation.

According to Fig. 1 and Table 1 in this paper indicated that recruitment and selection of volunteers are important to finding the right people to fill specific volunteer positions. Place the right volunteers in the right positions based on their interests, talents and schedules. Selecting volunteers for a job in which they are truly interested promotes success for the volunteer. This process should be open, fair and competitive. We analysis the content of this work from this point of view to

Table 1 Process of volunteer management based on review paper

Author	Process				
	Planning	Recruitment and selection	Induction and training	Supervision and evaluation	Recognition
Zheng et al. [53]	/	/	/	X	/
Ducharme [17]	/	/	/	/	/
Volunteer Glasgow Organization [50]	/	X	X	/	/
Howard [26]	/	/	/	/	/
Volunteer Australia [42]	/	/	/	/	/
Studer and von Schnurbein [46]	/	/	/	/	/
Dodd and Boleman [16]	X	/	/	/	/

make clear the focus because we want to automate the process of selection, which can do with crowdsourcing and match with the criteria of volunteers.

2.3 Recruitment and Selection for Volunteers

Recruitment is to identifying and attracting its potential volunteers. Selecting volunteers are to find a good match between the prospective volunteer and the opportunity on offer. It is an attempt to find a strong blend where the person has the right skills, experience, and enthusiasm and the opportunity satisfy the needs and interests of the volunteer. Cvetkoska et al. [12] present the process of recruitment and selection in an institution for selecting students as volunteers using Analytic Hierarchy Process Method (AHP) for decision-making methods on several criteria. They determined seven criteria, which is Curricula Vitae (CV), computer skills, languages, student's motivation to volunteer, skills, creativity, and initiative. Similarly as Howard [26] concerned about recruitment and selection process. Criteria they take are a profile of volunteers, reason to be volunteers, kind of voluntary work, skills, and availability. However, they focus on criteria that familiar used by volunteer organizations. Volunteers have many reasons for giving their time and resources, including the opportunity to help others, give back to the community, learn new skills and meet new people. Volunteers appreciate knowing that their time is well spent, that their work is meaningful and that their commitment is flexible. Based on this we know that is important on this part.

Besides that, matching technique is an important part of recruitment and selection of volunteer because to find the suitable volunteer before assigning the task. Different researchers have proposed matching algorithm based on various situation with volunteer such as OWL-S matcher, RF-based localization techniques, Web2.0 style, size-specified community creation method and classification method [8, 19, 25, 36, 39, 51]. Based on our review, we found that there is the various matching technique used in existing volunteering system. However, most of the techniques focus on to find the best volunteer with common characteristics and common using social network including Google calendar.

2.4 Volunteer Management Systems and Platforms

It has many volunteer management software products (*List of Volunteer Management Software*) [11]. Out of these, we selected five volunteer management systems, which have been empirically evaluated as most used systems. First of all, the system *Volgistics* (*Volgistics*) [4] allows non-profits can be customizing to fit their unique needs. For assignment matching compares the availability, characteristics, and preferences of each volunteer to selected assignment. *VolunteerLocal* (*VolunteerLocal*) [5] is a website used by volunteer coordinators to organize, recruit, schedule and communicate with their volunteers. *The Registration System*

(TRS) [3] has been specifically designed for events. *Voluntweeters* [45] system use social theory about self-organizing by features of coordination within a setting of interaction. The *GoVolunteer* [29] system can search, express interest, shortlist opportunities and find short term, on-going and even one-off opportunities to support a wide range of cause. *Travel2change* [32] was launched as an online crowdsourcing platform in 2011 for volunteer travelers can connect with locals.

Most of the volunteer management systems are commercial, each pursuing different goals and exhibiting different functionality to support the phases of a volunteering process. System engineering is an interdisciplinary approach that uses a structured development process from concept through production and operation. The techniques of systems engineering can be applied to volunteer management system that is comprehensive and adaptable.

2.5 *Crowdsourcing*

Since the prevalence of crowdsourcing in industry and academia, several surveys about crowdsourcing for the general purpose were published. There is a growing interest in 'engaging the crowd' to identify or develop innovative solutions to public problems. Crowdsourcing coined in 2006 by Howe [27] in Wired Magazine article which is crowdsourcing is not a single strategy, but "an umbrella term for a highly varied group of approaches." Crowdsourcing is human collective intelligence use the Internet to collect idea, solve complex cognitive problems and build high-quality repositories by self-organizing agents around data and knowledge [7]. According to Kittur et al. [31], crowdsourcing refers to the use of small amounts of time and effort from a large number of individuals to solve large problems, thus creating a network of such individuals collaborating together.

The advantage of crowdsourcing is that the solutions to members of the audience are directly involved in the ideation and proposed solutions. For examples of crowdsourcing platforms are Amazon Mechanical Turk, the reCaptcha, Wikipedia and the ESP game (for image labeling). This technique has long been used in various aspects [20, 52]. Understanding the factors that contribute to project success is necessary for crowdsourcing's continued adoption, efficient and effective implementation, and maximizing its potential. Website statistics shared by crowdsourcing project teams provide evidence that the potential of the crowd can be significant [40].

2.6 *The Process of Crowdsourcing*

The sequence of crowdsourcing process is independent of the order in which a seeker (an organization or individual) decides on the individual characteristics when it plans the process. According to Kucherbaev et al. [33] crowdsourcing is not only deploying a set of simple micro-tasks on a given platform but may comprise several different

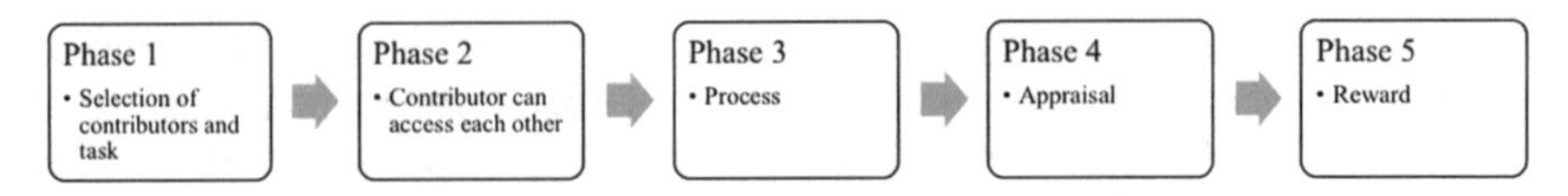

Fig. 2 Crowdsourcing process

tasks (writing, transcribing, classifying, aggregating, spell checking, and voting), actors (workers, and experts) and automated operations (data splitting, resolving redundancy of multiple delegations, making decisions, and synchronizing tasks).

Based on the literature [23, 34, 41], they present the process in crowdsourcing then we summarize the process of crowdsourcing in Fig. 2. The first phase is the process of choosing the right contributor for a specific task as outlined in a written job description. Besides that, some studies [30, 38, 51] involved a part of the process of crowdsourcing which is in the selection of contributors and task. Next step is contributor can access each other. In this process, contributors can be precise mechanisms to express their opinion on individual contributions. The third phase is how the crowd contribution within a crowdsourcing process is used by the crowdsourcing organization to achieve the desired outcome. The appraisal phase begins when the submitted ideas are clustered, rated, and best ideas will be rewarded. The final phase is a reward, determines how contributions are paid or otherwise compensated for their work.

Based on crowdsourcing process we can use this approach in volunteering management. This approach can get more information and faster from the crowd and be more precise with the requirement from beneficiaries.

3 Proposed Framework for Volunteering System

This section, we provide a general framework for volunteering system based on our findings. The design aims to solve the recruitment and selection suitable matches between volunteers, crowd, and organizations.

3.1 Framework

From the user's point of view, there are three kinds of end users: volunteer organization (those who manage the volunteers and projects), the crowd (those who need help or suggest the project) and volunteers (those who freely offer up their time and service).

The system framework is illustrated in Fig. 3, while Table 2 shows a procedure of the volunteering system by using crowdsourcing approach. We highlight that we involve crowdsourcing technique in volunteering system.

The basic idea of our approach is to use crowdsourcing approach in volunteering system. We used the crowd information to get the situation and project and volunteer also we get to solve the problem. We match the information from the crowd with

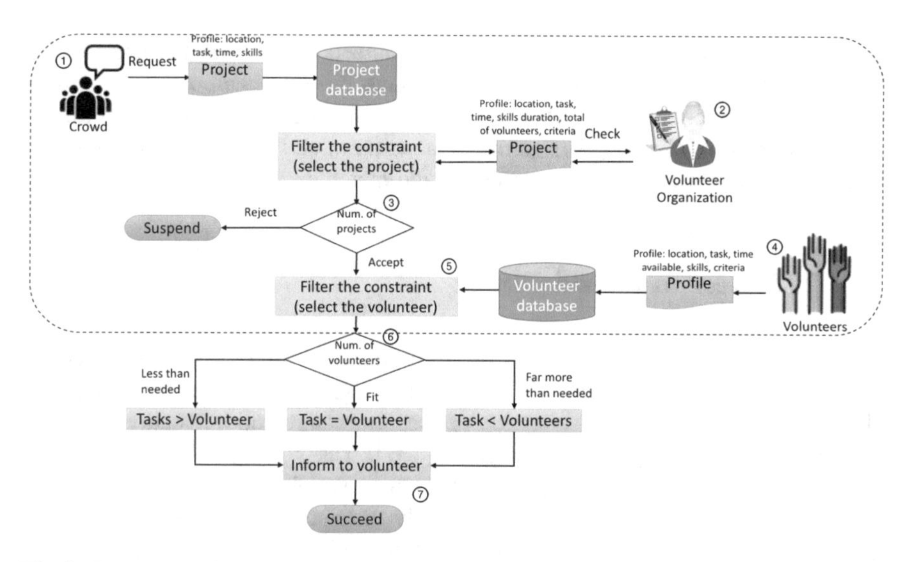

Fig. 3 Framework of volunteering system

Table 2 The procedure of volunteering system

Step 1:	The crowd will give their suggestion of their place who need help or their know other people need help to system. The crowd comes from local people, sometimes volunteers who had been involved in volunteer activities or people around that concerned. They send requests to the system. A request is transformed into a project and decomposed into many tasks. A project has constraints like expertise needed, task, time, criteria needed, etc
Step 2:	Volunteer organization filters and investigate the place that suggests from the crowd. The system filter based on the requirement for the beneficiary
Step 3:	If volunteer organization accepts the project, the organization continues the project mission. The organization announces to volunteers have freely time and service to help. If volunteer organization reject the project, the project has to suspend for more demand from the crowd or have its constraint changed and to be assigned again or give other volunteer organization that suitable with their objective
Step 4:	Volunteers whom interested need to register by providing their profiles, including expertise, available, location, criteria they have, etc. This form can help determine the potential volunteer
Step 5:	The system matches volunteers with the task regarding expertise, interests, talent, time available, location, transportation, and criteria
Step 6:	The project is accomplished through three situations: If the size of the candidate volunteers set fits with volunteers needed, then a task will give to a volunteer. Here "fit" means equal or a little larger for tolerance. If the number of volunteers is less than needed, then volunteers need to do more tasks in a project. If the number of volunteers is far more than needed, then a task does with many volunteers
Step 7:	After the system decides the volunteer task, the system informs to volunteers before execute the project mission

suitable volunteer criteria. The criteria of volunteers depend on the project. The crowd has potential to help the people and give the information for a volunteer organization to manage. The volunteers also come from the crowd itself. For finding a fit between several volunteers (profile and criteria) and the crowd (specification criteria needed) certain planning strategies and matching algorithm requirements.

3.2 Volunteer in Crowdsourcing

Volunteer management research has been an important topic in the social science and wanted to combine with ICT. There are many commercial crowdsourcing platforms but rarely with consideration of users' structure and interaction, which is the concern of socially aware computing [37]. Tasks are just broadcast to an unknown group of solvers in the form of an open call for solutions. Since crowdsourcing faces people who are geographically distributed and with diverse backgrounds, social unawareness will be subject to unpredictable performance because of the arbitrary composition of users who happen to accept an offered task. However, in reality, a group's suitability for performing a task varies from group to group, not only depending on the attributes of the individuals involved such as knowledge, skills, location, and available time but also the relationships among them.

The development of volunteers is an investment for the city, so it is important that resources be provided to develop capacity and enable volunteers to play a bigger role in their community. An organizer invites a targeted group of people to perform certain tasks to create value. With the new internet technologies, all the emergence of social networks and collaboration software can be organized by online platforms. With the rise of successful crowdsourcing, can act as the intermediary between volunteer and local communities. The empower both, volunteer and local communities to create a collective impact, as they can directly communicate and collaborate via the platform. Therefore, we use crowdsourcing techniques to get the information quickly and accurately to facilitate the management of volunteers. There are few studies on volunteers regarding motivation on volunteerism or use the latest technology to help people [14, 22, 49, 51] but they focus on the specific situation. In this paper, we focus on finding a specific group of volunteers that can do a project efficiently and relevant with their task and can be used in any situation.

4 Selecting Volunteer Criteria Using Delphi Method

The criteria in volunteer are important to determine the suitability of the volunteer with the task given by the volunteer organization. We conducted this needs assessment using Delphi method. The Delphi method is preferable in situations where a quick turnaround is required, or where directs the interaction may generate conflict. The focus group is preferable in situations where group interaction (e.g., "brainstorming") is more likely to be successful in breaking an impasse or generating an innovative solution.

4.1 Delphi Method

"Project Delphi" was the name given to an Air Force-sponsored Rand Corporation study focused on understanding the use of expert opinion [13]. The objective of the Delphi methodology was to "reduce the negative effects of group interactions" and to obtain the most reliable consensus of a group of experts [24]. It is a systematic and iterative process for obtaining opinions and if possible, a consensus of the experts taking part. Anonymity, controlled repetition and the statistical processing of the answers stand out among its main characteristics. It is methodology consists of selecting a group of experts who are asked their opinion about questions referring to future events. The experts' estimations are carried out in successive, anonymous rounds aiming to try and achieve a consensus whose final aim is stability in the panel of experts which enables predictions to be made [35]. On a practical level, the Delphi method is an alternative to formal meetings, interviews, or other face-to-face interactions. Unlike meetings where often not everyone can be present, the Delphi method allows all participants to have equal opportunity to be involved in the decision-making process.

4.2 Expert Panel

The major goal of the study was to determine the potential of Delphi study to assist an evaluator of volunteer depends on the project. The Delphi technique uses a panel of experts in a given field to develop consensus regarding the answer to a specific question or series of questions. Use of the Delphi method was examined and will be discussed in the context of goals clarification and developing evaluation questions.

The experts who met the profile were selected and they were contacted by email. The selection was initially comprised of seven experts, come from different areas. The experts consisted of assistant manager training, president, vice president, volunteer manager, and university faculty members. We have the interview with the expert from the volunteer organization. The experts come from Hidayah Centre Foundation (HCF), Spot Community, The National Department for Culture and Arts (JKKN), Projek Iqra', Malaysian Universities Volunteer Commission (MASKUM) and volunteer organization from universities (Universiti Teknologi Mara (UiTM) and Universiti Teknikal Malaysia Melaka (UTeM)).

4.3 Research Design

The empirical study is structured in three phases, Fig. 4: literature review, narrowing down and ranking.

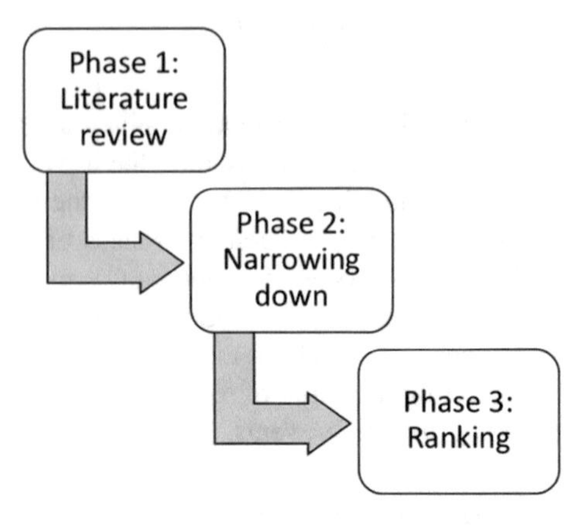

Fig. 4 Delphi method process

4.3.1 Phase 1: Literature Review

Preliminary content collected for the instrument using established quality filters, and criteria of the volunteer. This literature review indicated that existing models of volunteering system did not sufficiently overlap or interact with each other to provide a framework for the suitable volunteer.

4.3.2 Phase 2: Narrowing Down

In brief, the expert will narrow down criteria that reflect the suitable volunteer depends on their organization. The expert will get the questionnaire to randomly arrange to cancel out bias in the order of listing of the items.

4.3.3 Phase 3: Ranking

The goal of the final phase is to reach a consensus in the ranking of the relevant criteria within each expert. This phase of the procedure will involve each expert separately ranking the factors on each of their distinct pared-down lists.

5 Results

We have 69 criteria based on the related study [9, 10, 18, 48] (*Samaritan*) [2]. The results obtained from the preliminary analysis of criteria of volunteers that usually used are presented in Table 3.

According to the opinion of experts, that often use in volunteers' criteria is commitment. That important in volunteer because each volunteer must have this

Table 3 Criteria of volunteers

Criteria	Ranking
Leadership	6
Problem solver	2
Makes decision	1
Motivates other	4
Teamwork	5
Commitment	7
Skill	6
Communication	5
Organizes task	1
Knowledge	1
Age	3
Respectful	2
Quality patient care	1
Belief in the organization	2

Table 4 List of criteria and aspect

No.	Aspect	Criteria
1.	Leadership	Problem solver
		Makes decision
		Motivates other
		Organizes task
		Knowledge
2.	Teamwork	Spoken expression
		Facilitation of the discussion
3.	Commitment	Reliability
		Positively
		Initiative
4.	Skills	Quality patient care
		Years of working experience
		Ability to teach
5.	Communication	Written
		Oral
6.	Profile	Age
		Years of volunteer experience
		Criminal Offence
7.	Attribute	Respecful
		Belief in the organization
		Proactive
		Innovative

attitude to help people. The experts' opinion is that every one of criteria analyzed will important in the volunteerism. On the other hand, quality patient care, organize task, knowledge, makes decision is not every volunteer needed if they join voluntary work. Therefore, the experts predict that it will use the criteria if suit with the situation. After we do the interview, the organization have other criteria depending on their objective voluntary work. Based on Table 4, presents the results obtained from the expert and categorize criteria based on aspect. We can categorize into seven categories of aspect and have their criteria.

6 Discussion

In reviewing the literature, we can propose a framework and come out with the criteria to use in volunteering system. This work's main contribution is to identify, and analysis of the criteria was volunteering management system. We have based ourselves on some criteria to measure the impact of the voluntary work. The methodological background of the study is a synthesis of qualitative and quantitative research techniques that have been conducted through a collaborative process involving a panel of experts. As a result of this analysis, we find that the practices linked to commitment, leadership and skills will be the most relevant. This is due to their high level of expected and the positive assessment which emerges from the random situation of voluntary work.

Commitment is the important criteria which have the best result. The commitment is usually used to build up the volunteers. Secondly, leadership and skills attain a similar level of criteria after commitment. Thirdly, this study shows teamwork and communication are predicted. Fourthly, the motivates other is the criteria was medium important. Next criteria are age. Age is not important criteria based on voluntary work. There is some volunteer work with age limits, but there is volunteer work that does not have the age limit. They are free to do volunteer work no matter the child or the elderly. The second last of the ranking criteria are the problem solver, respectful and belief in the organization. Finally, the makes decision, organizes task, knowledge and quality patient care from a perspective expert in volunteering organization.

Each criterion is different depending on the situation voluntary work. Then we categorize the criteria into aspect, and we review it with the expert. We will collect many criteria of volunteers to get the best volunteer based on voluntary work.

7 Conclusion

In this paper, we extracted volunteering and crowdsourcing features from various definitions that found in the literature. To achieve this goal, different domain (healthcare, emergency, disaster, etc.) were considered. In this investigation, the aim of this research is to contribute crowdsourcing approach for volunteering system. To this intent, we first laid out an argument as to the opportunities of such approaches by drawing upon research on the process of crowdsourcing. We proceeded by conducting a review of the academic literature on recruitment and selecting method and matching technique in various volunteering contexts. We proposed a framework of volunteering system that includes the crowdsourcing approach. By searching through and reviewing the literature, we use Delphi method and supporting decision-making. Further research might investigate parameter for the volunteering system to construct the needed contents. We will complete the framework and develop a volunteering system likes our framework. We also would be evaluating our method in the real environment by user trials, especially the effectiveness of our approach.

Acknowledgements The authors would like to express their gratitude to the First EAI International Conference on Computer Science and Engineering (COMPSE) 2016 at Penang, Malaysia. The authors also would like to acknowledge Universiti Teknikal Malaysia Melaka (UTeM) and the Ministry of Higher Education Malaysia (MOHE) for the resources as well as the MyBrain15 scholarship.

References

1. Allen, K. (2006). From motivation to action through volunteer-friendly organizations. *The International Journal of Volunteer Administration, 24,* 41–44.
2. Anon. (2016). Samaritan. Retrieved January 23, 2016, from http://www.samaritans.org/.
3. Anon. TRS. www.theregistrationsystem.com.
4. Anon. Volgistics. https://www.volgistics.com/.
5. Anon. VolunteerLocal. www.volunteerlocal.com.
6. Anon. Yayasan Sukarelawan Siswa (YSS). https://prezi.com/_q0zqv0bja6g/yss-profile-english-ver2/.
7. Büecheler, T. A., et al. (2011). Modeling and simulating crowdsourcing as a complex biological system: Human crowds manifesting collective intelligence on the internet. In T. Lenaerts et al. (Eds), *The Eleventh European Conference on the Synthesis and Simulation of Living Systems. ECAL 2011* (pp. 109–116). Paris: MIT Press.
8. Chen, W., et al. (2011). Finding suitable candidates: The design of a mobile volunteering matching system. In *Human-computer interaction. Towards mobile and intelligent interaction environments* (pp. 21–29).
9. Cnaan, R. A., & Amrofell, L. (1994). Mapping volunteer activity. *Nonprofit and Voluntary Sector Quarterly, 23*(4), 335–351.
10. Cnaan, R. A., & Cascio, T. A. (1998). Performance and commitment. *Journal of Social Service Research, 24*(3), 1–37.
11. Cravens, J., & Jackson, R. (2016). List of volunteer management software. Retrieved January 1, 2016, from http://www.coyotecommunications.com/tech/volmanage.html.
12. Cvetkoska, V., Gaber, B. S., & Sekulovska, M. (2011). Recruitment and selection of student-volunteers: A multicriteria methodology. *Management (1820–0222)*, (61), 139–146.
13. Dalkey, N., & Helmer, O. (1963). An experimental application of the Delphi method to the use of experts. *Journal of the Institute of Management Science, 9*(3), 458–467.
14. Disability Equality (nw). (2013). *Good practice in supported volunteering*. www.disability-equality.org.uk.

15. Division of Industry and Community Network Universiti Sains Malaysia, 2013. *Volunteerism in Malaysia Fostering Civic Responsibility*, Penerbit USM.
16. Dodd, C., & Boleman, C. (2010). *Volunteer administration in the 21 st century: ISOTURE : A model for volunteer management*. http://agrilifecdn.tamu.edu/od/files/2010/06/Isoture-model-for-volunteer-management-E-457.pdf.
17. Ducharme, E. G. (2012). Our foundation—The basics of volunteer management. *Canadian Journal of Volunteer Resources Management, 20*(1), 2–4.
18. Endeavour Volunteer. (2009). Eligibility and assessment criteria (pp. 1–2). Retrieved January 23, 2016, from http://www.endeavourvolunteer.ca/volunteer/get-involved/consulting-opportunities/eligibility-and-assessment-critieria/.
19. Endo, D., & Sugita, K. (2010). A volunteer classification method for disaster recovery. *2010 International Conference on P2P, Parallel, Grid, Cloud and Internet Computing* (pp. 436–439). Fukuoka: IEEE.
20. Estellés-Arolas, E., & González-Ladrón-de-Guevara, F. (2012). Towards an integrated crowdsourcing definition. *Journal of Information Science, 38,* 1–22.
21. Fernandez, L. S. (2007). *Volunteer management system design and analysis for disaster response and recovery*. George Washington University.
22. Fuchs-Kittowski, F., & Faust, D. (2014). Architecture of mobile crowdsourcing systems. In N. Baloian et al. (Eds.), *Collaboration and technology: 20th international conference, CRIWG 2014, Santiago, Chile, September 7–10, 2014. Proceedings* (pp. 121–136). Cham: Springer International Publishing. http://dx.doi.org/10.1007/978-3-319-10166-8_12.
23. Geiger, D., et al. (2011). Managing the crowd: towards a taxonomy of crowdsourcing processes. In *Proceedings of the Seventeenth Americas Conference on Information Systems*, Detroit, Michigan.
24. Gupta, U. G., & Clarke, R. E. (1996). Theory and applications of the Delphi technique: A bibliography (1975–1994). *Technological Forecasting and Social Change, 53*(2), 185–211.
25. Hong, S., et al. (2007). Service matching in online community for mutual assisted living. *2007 Third International IEEE Conference on Signal-Image Technologies and Internet-Based System* (pp. 427–433). Shanghai: IEEE.
26. Howard, B. W. (1999). Managing volunteers. *Australian Journal of Emergency Management, 14*(3), 37–39.
27. Howe, J. (2006). The rise of crowdsourcing. *Wired Magazine, 14,* 1–5.
28. Hughes, K. (2015a). New definition of volunteering in Australia. In *Definition for Volunteering 2015*. www.volunteeringaustralia.org.
29. Hughes, K. (2015b). Opportunities in your hand—GoVolunteer goes mobile. *Volunteering Australia*. http://www.volunteeringaustralia.org/wp-content/uploads/041215-Media-Release-GoVol-App-launch_IVD.pdf.
30. Kittur, A., et al. (2013). The future of crowd work. In *Proceedings of the 2013 conference on Computer supported cooperative work (CSCW '13)* (pp. 1301–1317). New York, NY, USA.
31. Kittur, A., Chi, E. H., & Suh, B. (2008). Crowdsourcing user studies with Mechanical Turk. In *Proceeding of the Twenty-Sixth Annual CHI Conference on Human Factors in Computing Systems—CHI '08* (p. 453). New York, New York, USA: ACM Press.
32. Kohler, T., Stribl, A., & Stieger, D. (2016). Innovation for volunteer travel: Using crowdsourcing to create change. In R. Egger, I. Gula & D. Walcher (Eds.), *Open tourism: Open innovation, crowdsourcing and co-creation challenging the tourism industry* (pp. 435–445). Berlin, Heidelberg: Springer.
33. Kucherbaev, P., et al. (2016). Crowdsourcing processes: A survey of approaches and opportunities. *IEEE Internet Computing, 20*(2), 50–56.
34. Li, Z., & Hongjuan, Z. (2011). Research of crowdsourcing model based on case study. In *8th International Conference on Service Systems and Service Management (ICSSSM11)*. IEEE (pp. 1–5).
35. Linstone, H. A., & Turoff, M. (Eds.). (1975). *Delphi method: Techniques and applications*. Addison-Wesley Publishing.

36. Lo, C. C., et al. (2010). People help people: A pattern-matching localization with inputs from user community. In *ICS 2010—International Computer Symposium* (pp. 638–641).
37. Lukowicz, P., Pentland, S., & Ferscha, A. (2012). From context awareness to socially aware computing. *IEEE Pervasive Computing, 11*(1), 32–40.
38. Lykourentzou, I., et al. (2013). Guided crowdsourcing for collective work coordination in corporate environments. In C. Badica, N. T. Nguyen & M. Brezovan (Eds.), *Computational Collective Intelligence. Technologies and Applications: 5th International Conference, ICCCI 2013, Proceedings*, Craiova, Romania, September 11–13, 2013 (pp. 90–99). Berlin, Heidelberg: Springer.
39. McCann, R., Shen, W., & Doan, A. (2008). Matching schemas in online communities: A Web 2.0 Approach. In *2008 IEEE 24th International Conference on Data Engineering* (pp. 110–119). IEEE.
40. Mckinley, D. (2013). *How effectively are crowdsourcing websites supporting volunteer participation and quality contribution?*, New Zealand.
41. Muhdi, L., et al. (2011). Crowdsourcing: an alternative idea generation approach in the early innovation process phase of innovation. *International Journal of Entrepreneurship and Innovation Management, 14*(4), 315–332.
42. NHMRC. (2003). *Working with volunteers and managing volunteer programs in health care settings.*
43. Schönböck, J., et al. (2016). A survey on volunteer management systems. In *49th Hawaii International Conference on System Sciences (HICSS)* (pp. 767–776).
44. Sharma, D. (2016). A review on technological advancements in crowd management. *Journal of Ambient Intelligence and Humanized Computing.*
45. Starbird, K., & Palen, L. (2011). Voluntweeters. In *Proceedings of the 2011 Annual Conference on Human Factors in Computing Systems—CHI '11* (p. 1071).
46. Studer, S., & von Schnurbein, G. (2013). Organizational factors affecting volunteers: A literature review on volunteer coordination. *VOLUNTAS: International Journal of Voluntary and Nonprofit Organizations, 24*(2), 403–440.
47. UN Volunteers. (2015). State of the world's volunteerism report. http://www.volunteeractioncounts.org/.
48. Voluntary Sector National Training Organisation (VSNTO). (2004). *Management of volunteers national occupational standards*,
49. Volunteer Centre Dorset. (2010). *The good practice guide to volunteer management.* http://www.volunteeringdorset.org.uk.
50. Volunteer Glasgow. (2010). *Glasgow's strategic volunteering framework.* http://www.volunteerglasgow.org/partners/svf/.
51. Yu, Z., et al. (2012). Selecting the best solvers: Toward community based crowdsourcing for disaster management. *2012 IEEE Asia-Pacific Services Computing Conference* (pp. 271–277). IEEE: Guilin.
52. Yuen, M., & Chen, L. (2009). A survey of human computation systems. In *CSE '09: Proceedings of IEEE International Conference on Computational Science and Engineering* (pp. 723–728). IEEE Computer Society.
53. Zheng, Y. (Lydia), Deng, L., & Li, M. (2009). Study on the event volunteer management based on the service blueprint. In *2009 International Conference on Information Management, Innovation Management and Industrial Engineering* (pp. 408–411). IEEE.

One Dimensional Vehicle Tracking Analysis in Vehicular Ad hoc Networks

Ranjeet Singh Tomar and Mayank Satya Prakash Sharma

Abstract VANET system is a trending research area now days. Accident avoidance and congestion control are very big task in VANET System. In this paper we have performed initially prediction forecasting and measurement is done by GPS and the accurate estimation through Kalman filter. We have observed and predicted the position and velocity of vehicle using Kalman filter and also calculated its different parameter with the help of Kalman filter. Different vehicle velocities (slow, medium and high) have been considered as an important parameter affecting the accuracy of estimation. The application of Kalman filter in estimating the vehicle parameters in highways has been successfully demonstrated.

Keywords Vehicular ad hoc network (VANET)
Mobile ad hoc network (MANET) · Real time tracking system
Global positioning system (GPS) · Kalman filter

1 Introduction

In a society which is becoming travel oriented, collisions and traffic congestion are a serious problem. Annual traffic congestion costs are estimated at over 1000 billions of dollar per year in the world [16]. The highway system touches the life of every person [17]. The roads, which are the economic lifelines of our nation, are prone to congestion. This is negatively impacting the comfort of travelers and the economic competitiveness of businesses. The impact of crashes or other incidents on traffic-carrying capacity is also very significant. The highway capacity manual defines the theoretical number of vehicles that can use a section of highway. This is called the capacity of the road. The theoretical maximum capacity for a lane of a

R.S. Tomar (✉) · M.S.P. Sharma
ITM University Gwalior, Gwalior, India
e-mail: er.ranjeetsingh@gmail.com; ranjeetsingh@itmuniversity.ac.in

M.S.P. Sharma
e-mail: mayanksintal@gmail.com

I. Zelinka et al. (eds.), *Innovative Computing, Optimization and Its Applications*, Studies in Computational Intelligence 741,
https://doi.org/10.1007/978-3-319-66984-7_15

freeway under ideal conditions is 2400 vehicles per hour. Based on data from field studies, the Highway Capacity Manual defines how much the capacity of a road may be reduced because of crashes or lane closures due to incidents [18].

There has been a growing recognition that constructing new road capacity and intelligent transportation system at a rate fast enough to counteract increasing congestion is cost intensive and almost impossible. The immediate attention is, therefore, on maximizing the operational efficiency of existing facilities and intelligent transportation system. Agencies have been investing in operational improvements such as providing real-time traveler information, incident management programs, and improved traffic signal operations. Having timely and accurate data on traffic conditions is a basic requirement to develop and implement intelligent transportation system; so traffic management programs is a must [15].

There are several safety applications of Intelligent Transportation System (ITS). As a part of ITS, Vehicular Ad hoc Network (VANET) is becoming a growing development to satisfy the varied demands in real time application. In this system, vehicles are interconnect with one another and probably through roadside unit to produce a large list of applications varied from transportation protection to drive help and web access issue of concern to Mobile Ad hoc Networks (MANETs). Instead of running arbitrarily, vehicles tend to shift in an organized trend on road. The communications through roadside unit can likewise be characterized purely accurate. Many vehicles are block in their range of motion. Such a model may propose for safety issue as an example; one cannot security type such as email during drive. Navigation and GPS model might profit, as they could be joined through traffic reports to giving the quick route to work. Currently have a real time application based vision for prediction and measurement of vehicle in VANET System. There are many vehicle tracking algorithms but we proposed a new method based on Kalman filter to predict the vehicle during moving position [14]. In this paper, an approach of prediction and measurement of vehicle from a moving platform is presented, with an aim being to provide situational awareness to autonomous vehicle.

2 Background

Traffic congestion is a major problem in roadways at present transportation scenarios [3]. Vehicles move in single or multi lanes on the road with different speeds and accelerations. They also turn in different directions at junctions. Every vehicle starts moving from definite segment, travels on the road and can be changing the lanes and passing the other vehicles on the road. Events such as vehicular crashes and breakdowns in travel lanes are the most common form of traffic incidents are responsible for congestion. Construction activities on the roadway result in physical changes to the highway environment are also potential cause of congestion on the road. These changes may include a reduction in the number or width of travel lanes, lane shifts, lane diversions and even temporary roadway closures. All these conditions will increases the congestion on the road or sometime collisions on the road. In the transportation realm, congestion usually relates to an excess of vehicles on a portion of roadway at a particular time resulting in speeds that are slower, sometimes much slower than normal speeds. Congestion often means stopped or stop and go traffic on the road. Road accidents, construction work on roads, weather condition, increasing volume of vehicles and poor traffic signal timing are some reasons for traffic congestion. One of the main reasons for traffic congestion is the increase in traffic volume as per the road capacity. Congestion adversely affects mobility, safety and air quality. These factors directly impose the losses due to delays and accidents, and indirectly due to environmental impact. Mostly, the capacity of the existing highways cannot be increased by additional lanes due to space, resource, or environmental constraints. Most risky maneuvers that a driver has to perform in a conventional highway system are to merging the traffic on road, perform a lane changing, and overtaking maneuver. Lane changing, lane merging, overtaking and collisions are responsible for traffic delays often resulting in congestion on the road. Traffic delays and congestion will increase travel time and have a negative economic impact in the surrounding environments. The inter vehicle spacing or headway also affects the congestion and collisions on the road. For collision free vehicle following, the spacing should be large enough in order to no collisions during performing of all possible vehicle maneuvers on the road. Traffic management in the congested roadways or highways is also very complex, because every driver employs different technique to travel while interacting with other drivers. These factors are encourages the congestion, accidents or various types collision like front end/ rear end collisions and turning collisions on the road. Furthermore, they affect the economic and environmental conditions of the human society [11, 12]. For estimation, prediction and forecasting purpose, Kalman filter based estimation techniques will be applied on vehicles at various speed.

3 Estimated Filter

The Kalman filter [9] is a more than one input and output digital filter that can optimally evaluate, in real time environment, the states of a system depend on its unwanted signal outputs [6]. Kalman filter is an optimal estimator [2, 6]. The reason of a Kalman filter is to evaluate the Condition of a Framework from measurements which contain random errors [6]. An illustration is evaluating the location and speed of a satellite from radar Information. There are 3 Segments of location and 3 of speed so there are at least 6 variables to Evaluate [2, 6]. These variables are called state variables [14]. With 6 state variables the subsequent Kalman filter is called a 6-dimensional Kalman filter [6]. To Give present estimates of the framework variables—for example location coordinates—the filter utilizes statistical system to appropriately weight every new measurement relative to previous data [2, 6]. In Theory the Kalman Filter is an good estimator for Which is allude the linear-quadratic problem, which is the issue of evaluating the immediate “state” of

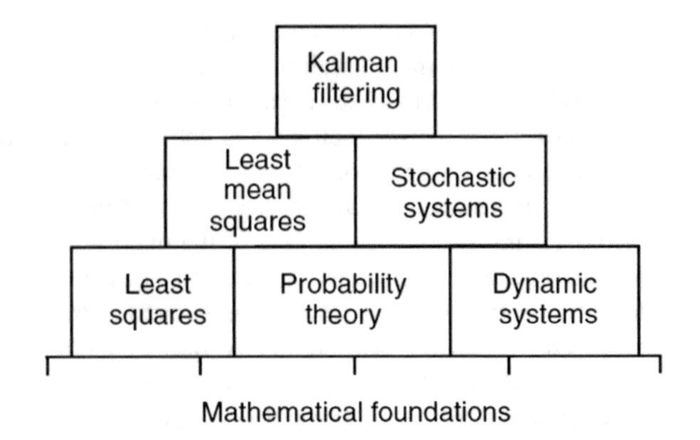

Fig. 1 Mathematical foundations of Kalman filter

linear dynamic device perturbed by way of white noise [2, 6] by way of use of the measurement linearly related with the state but distorted by white noise [2, 6]. Today Kalman filter is an well-known technique mostly applied in navigation [1, 6], missile and vehicle tracking [2, 6]. Kalman Filter depend on linear mathematical application [2, 6]. The resulting estimator is statically optimal with respect to any quadratic function of the estimator error. Kalman filter is usually represent using vector algebra as a smallest mean squared estimator [2, 6]. It is predictor, filter and smoothing technique and estimate's parameter values using present past measurement [6]. The Kalman filter is a recursive predictive filter that is depends on use of the state space algorithm and recursive technique. It estimates the state of a dynamic model can be distorted by some noise [6], usually describe as white noise to improve the estimate state the Kalman filter uses measurement that are related to the state but distorted as well [6]. It is only a device, it does not solve any difficulty all by itself although it can make is easier for You to do it. It is not a physical device, but an algebraically one.

- A Kalman filter is an standard recursive information set of rule
- Kalman filter carries all data information that can be distributed to it. It method all present measurements, irrespective of their precision, to estimate the present value of the variables of interest
- Computationally accurate due to its recursive algorithm
- Suppose that all variables being estimated are time based (Fig. 1) [1, 4].

4 Kalman Filter on VANET System

The Kalman filter system [4, 10, 13] accept that the condition of a framework at a time t evolved from the starting state at time t − 1 as per the condition

$$s_t = V_t s_{t-1} + B_t g_t + n_t \tag{1}$$

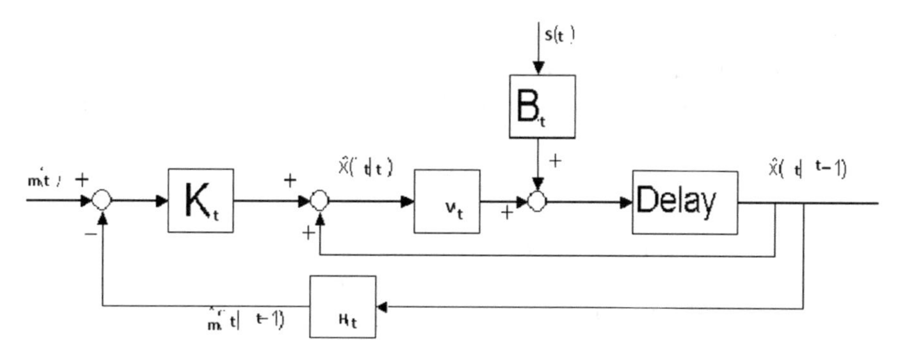

Fig. 2 Block diagram of Kalman filter

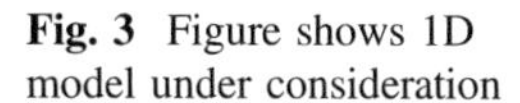
Fig. 3 Figure shows 1D model under consideration

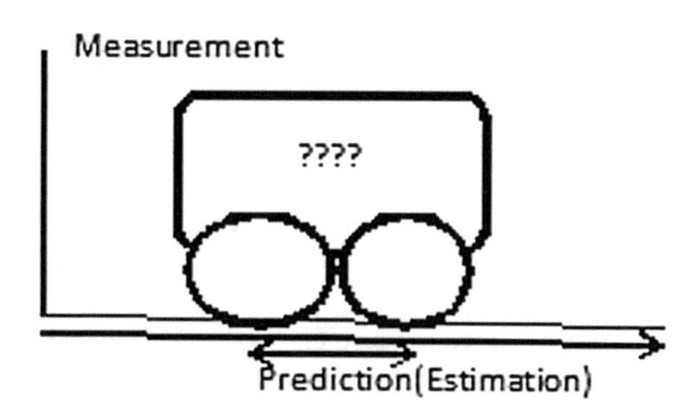

Measurement of the model can also be implemented, based to the system

$$m_t = H_t s_t + O_t \tag{2}$$

In paper, we will contain a generally one-dimensional tracking issue; let's take that vehicle as a car running along a road line (Figs. 2 and 3) [2].

5 Mathematical Overview of Kalman Filter

As we have taken example of tracking of a vehicle on road [7, 8]. We can include few example vector and matrices in this tracking issue. State vector S_t included the displacement and the speed of the vehicle

$$s_t = \begin{bmatrix} s_t \\ \dot{s}_t \end{bmatrix} \tag{3}$$

The vehicle human driver may acting braking or accelerating input to the model, In this paper we will focus here as a function of an the mass of the vehicle car m and acting force f_t. This control data information is recorded in control vector g_t

$$g_t = \frac{f_t}{m} \tag{4}$$

The connection between the throttle during the short period Δt or the force acting by the brake and the displacement and the speed of the vehicle is denoted by the following mathematical equation of motions.

$$s_t = s_{t-1} + (\dot{s}_{t-1} \times \Delta t) + \frac{s_t(\Delta t)^2}{2m} \tag{5}$$

$$\dot{s}_t = \dot{s}_{t-1} + \frac{f_t \Delta t}{m} \tag{6}$$

These straight equations can be written in the matrix form as given below

$$\begin{bmatrix} s_t \\ \dot{s}_t \end{bmatrix} = \begin{bmatrix} 1 & \Delta t \\ 0 & 1 \end{bmatrix} \begin{bmatrix} s_{t-1} \\ \dot{s}_{t-1} \end{bmatrix} + \begin{bmatrix} \frac{(\Delta t)^2}{2} \\ \Delta t \end{bmatrix} \frac{f_t}{m} \tag{7}$$

And now if we compare the above equation and Eq. (1) we can see that

$$V_t = \begin{bmatrix} 1 & \Delta t \\ 0 & 1 \end{bmatrix} \text{and } B_t = \begin{bmatrix} \frac{(\Delta t)^2}{2} \\ \Delta t \end{bmatrix} \tag{8}$$

The real state of the model s_t cannot linearly predict so Kalman estimator giving a technique to identify an estimate $\hat{s}_t$ by mix models of the system and unwanted signal measurement of different parameters. Parameter estimator is a Important task in this algorithm and estimation of the parameters in the state vector are giving by probability density function (PDF), rather than discrete values. The Kalman filter is depending on Gaussian probability Density function. But to fully represent the Gaussian function, we need to know their variance and covariance, and these are recorded in the covariance matrix P_t [5].

Kalman filter technique contain two steps: measurement and prediction upgrade. Kalman filter equations for the prediction steps are given below:

$$\hat{s}_{t|t-1} = V_t \hat{s}_{t-1|t-1} + B_t g_t \tag{9}$$

$$P_{t|t-1} = V_t P_{t-1|t-1} V_t^T + C_t \tag{10}$$

C_t Is the noise Covariance matrix included Noisy Control Input. Variance included with the prediction $\hat{s}_{t|t-1}$ Of an Unknown real value s_t is denoted by

$$P_{t/t-1} = E\left[\left(s_t - \hat{s}_{t|t-1}\right)\left(s_t - \hat{s}_{t|t-1}\right)^T\right] \tag{11}$$

Measurement and update equation are given below:

$$\hat{s}_{t|t} = \hat{s}_{t|t-1} + K_t\left(m_t - H_t\hat{s}_{t|t-1}\right) \tag{12}$$

$$P_{t|t-1} = P_{t|t-1} - K_t H_t P_{t|t-1} \tag{13}$$

Kalman gain: The Kalman gain is denoted by:

$$K_t = P_{t|t-1} H_t \left(H_t P_{t|t-1} H_t^T + R_t\right)^{-1} \tag{14}$$

Then T = 1, we additionally become an measurement of the Position of the Vehicle utilizing the radio positioning model, and it is Described by the Second Gaussian probability Function(Measurement Graph) in Fig. 4. The accurate estimation we obtain position of the vehicle is giving by Mix our information from the measurement and the prediction. This is accomplished by convolve the two Gaussian probability density Function together. This is Describe by the Third Probability density Function (Highest peak graph) in Fig. 4. Gaussian function important Property is exploited at the position [5] convolves of two Gaussian function result is different Gaussian function [5]. It is basic as it allows an endlessness number of Gaussian probability Density Function to be convolve over the time period, and resulting model doesn't increment in number of terms [3] or Complexity; after every time Duration the new Probability Density Function is completely describe by a Gaussian function [3]. It is the important recursive properties of the Kalman filter [5]. The steps Represent in Fig. 4 are currently considered again algebraically to derive the Kalman filter [5] Measurement upgrade equation [3]. Steps to describe in the figures are currently considered again

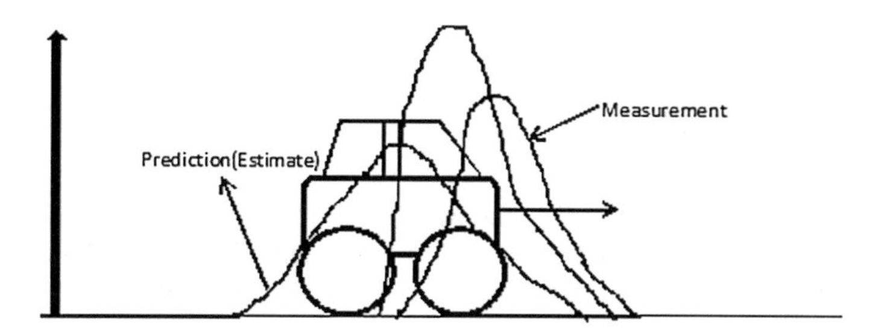

Fig. 4 Shows the new probability density function (Highest peak) developed by mix the probability density function included with the measurement and prediction of the vehicle position at time $t = 1$. This new probability density function gave the accurate estimate of the position of the vehicle by merging information from the measurement and the prediction of vehicle

algebraically to expand the Kalman filter measurement upgrade equation [5]. Prediction probability density Function depict by the prediction graph in Fig. 4 is given by the condition

$$y_1(r, \mu_1, \sigma_1) = \frac{1}{\sqrt{2\pi\sigma_1^2}} e^{\frac{(r-\mu_1)^2}{2\sigma_1^2}} \tag{15}$$

$$(r, \mu_2, \sigma_2) = \frac{1}{\sqrt{2\pi\sigma_2^2}} e^{\frac{(r-\mu_2)^2}{2\sigma 2^2}} \tag{16}$$

6 Vehicle Tracking Algorithm

Following steps are considered for vehicle tracking in vehicular environment:

Step 1—Make the mathematical model of the framework under tracking using differential equations.
Step 2—Define the state vectors utilizing that model, for next state and pervious state of the framework under following.
Step 3—Using the past state values; discover the next predicted value of the state, using prediction Eq. (1).
Step 4—Calculate the variance associated with the prediction of an unknown accurate value what we have to find out.
Step 5—finds the measurement vector using Eq. (2).
Step 6—Find the Kalman Gain using equation and with the support of variance matrix in step 4, and covariance of measurement noise.
Step 7—Now utilizing measurement update equation find estimated value of the state.

7 Simulation and Results

The simulation is performed using the MATLAB software with different speed of vehicles (low, medium and high speed of vehicles). For simulation work the acceleration of vehicles are considered as given in the table (Table 1).

Table 1 Acceleration of different speed vehicles

Vehicle	Speed	Acceleration
1	Low	U = 3 m/s^2
2	Medium	U = 4 m/s^2
3	High	U = 5 m/s^2

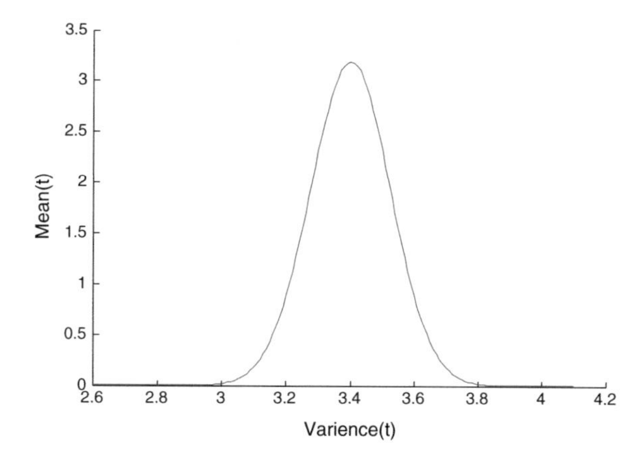

Fig. 5 Measurement through GPS

In Fig. 5 Measurement at different time with respect to mean and variance and optimal estimate of position is: $\hat{s}_{t|t} = \hat{s}_{t|t-1} = m_t$.

Variance of error in estimate: $\sigma_x^2(t_1) = \sigma_{m1}^2$.
Train in same position at time t_2—*Predicted* position is m_1.

7.1 *When New Estimation not Calculated (Initial Condition)*

In Fig. 6 blue graph is show initial prediction and red graph is show initial measurement through GPS.

In Fig. 6, we have the prediction $\hat{s}_{t|t-1}$.

GPS Measurement at t: Mean = m and Variance = σ_m.

7.1.1 Accurate Estimation (Medium Speed Vehicle)

Kalman filter helps you merge measurement and prediction on the basis of how much you trust each.

In this Figs. 7, 8 and 9 blue curve shows initially prediction of vehicle parameters, red shows the probability density Function (PDF) of GPS Measurement output, Green shows the accurate estimation of Kalman filter obtained after analyzing both the other curves. New variance is smaller than either of the previous two variances. Above graph Show that Conditional density function of Prediction and measurement for medium speed vehicle.

Corrected mean is the new optimal estimate of position (basically you've 'updated' the predicted position by using GPS.

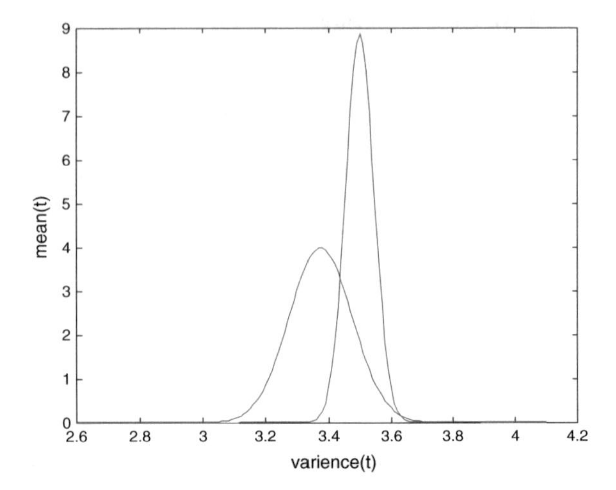

Fig. 6 Initial prediction and measurement

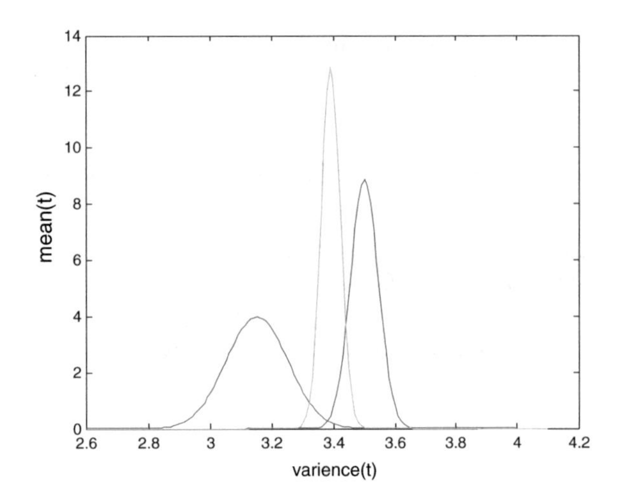

Fig. 7 Prediction and measurement probability density function for medium speed vehicle

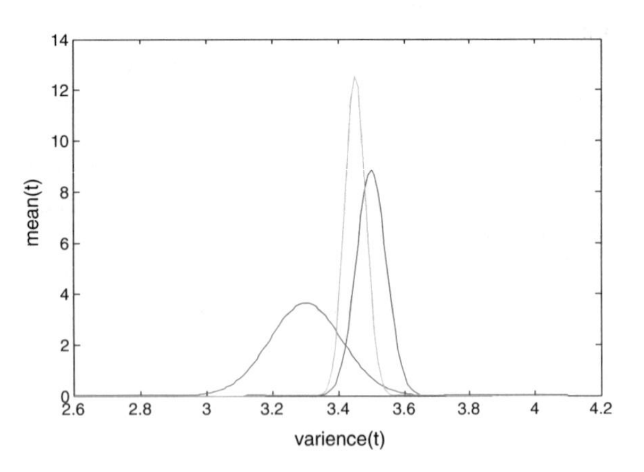

Fig. 8 Prediction and measurement probability density function for medium speed vehicle

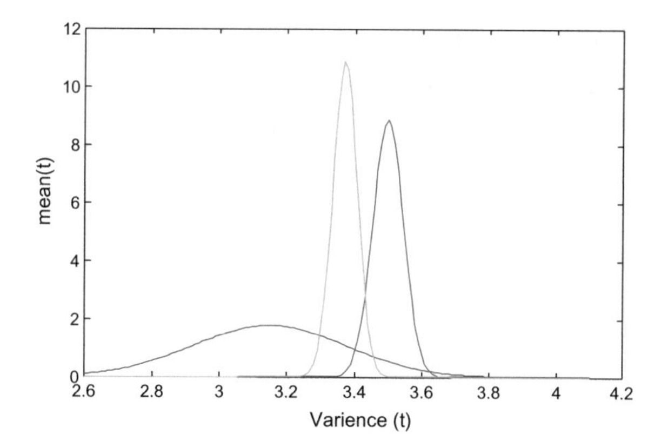

Fig. 9 Prediction and measurement probability density function for medium speed vehicle

7.1.2 Less Accurate Estimation (Lower Speed Vehicle)

Above Figs. 10 and 11 shows the mean versus variance plot of the vehicle in lower Speed. The curves have the main meaning as stated before. The curves are plotted in the case of lower velocity of the vehicle. The accuracy is lesser that the case of medium velocity vehicle.

7.1.3 Less Accurate Estimation (High Speed Vehicle)

The Figs. 12 and 13 shows the mean versus variance plot for vehicle velocity larger that the first two cases. Hence the accuracy is less from medium velocity Vehicle. We can see prediction of vehicle show in blue graph probability shifted in the right direction. The above graph is plotted for high velocity of vehicle. Here the curve for the GPS vanishes, so the accuracy is very low for high speed vehicles.

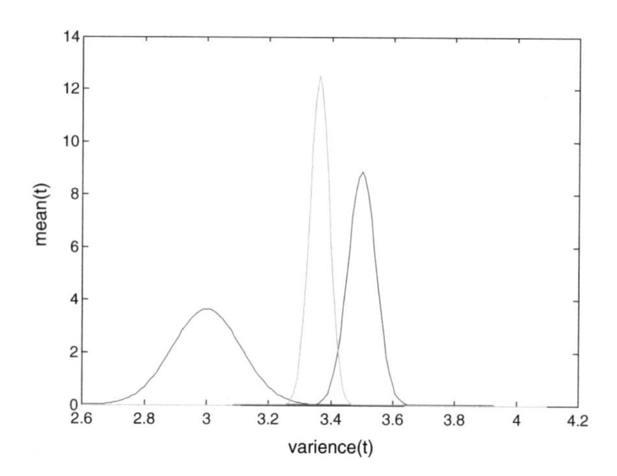

Fig. 10 Prediction and measurement Probability density function for lower speed vehicle

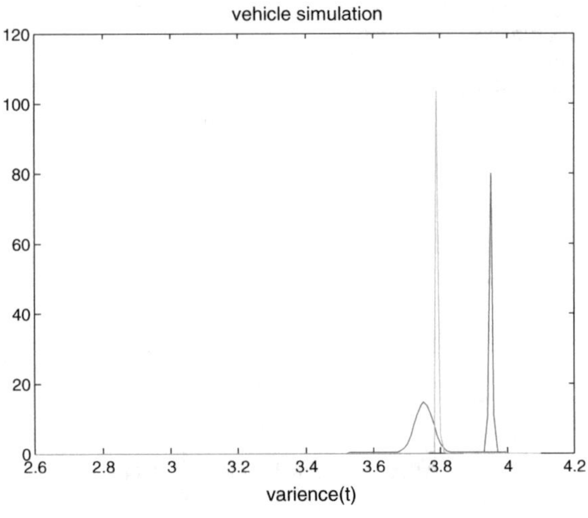

Fig. 11 Prediction and measurement Probability density function for lower speed vehicle

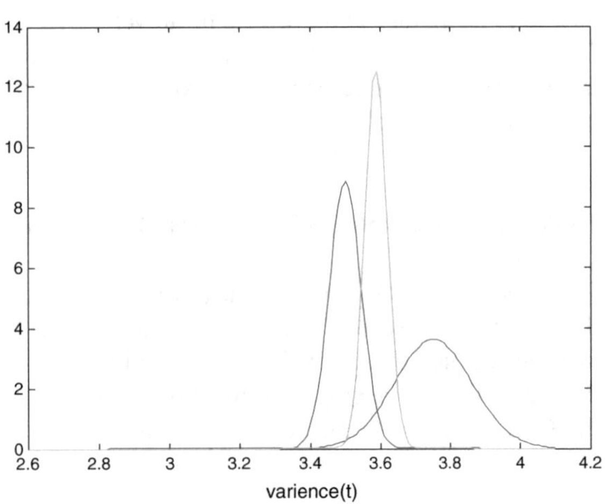

Fig. 12 Prediction and measurement Probability density function for high speed vehicle

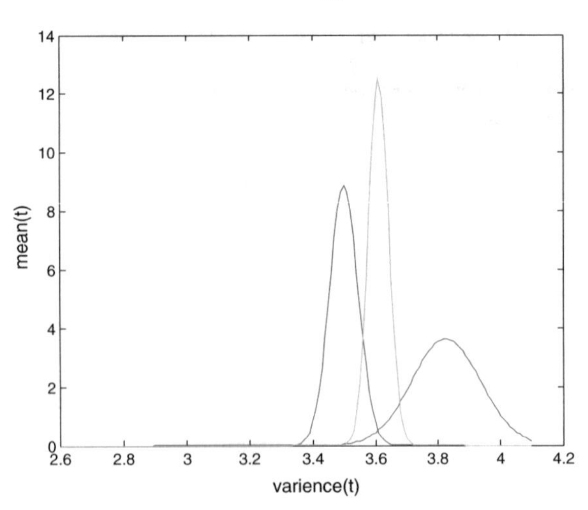

Fig. 13 Prediction and measurement at high speed vehicle

7.1.4 Case of Moving Velocity Vehicle

The Figs. 14 and 15 shows the mean versus variance plot for vehicle velocity larger that the first three cases. Hence the accuracy is less. In this case, prediction shows by green graph and probability shift in the right direction. Above Graph is shown that conditional density function of prediction and measurement for high speed vehicle

- At time t, Vehicle running with velocity ds/dt = $\hat{s}_t$
- Naïve approach: Shift probability to the right to predict
- This would work if we Identify the velocity perfectly (perfect model).

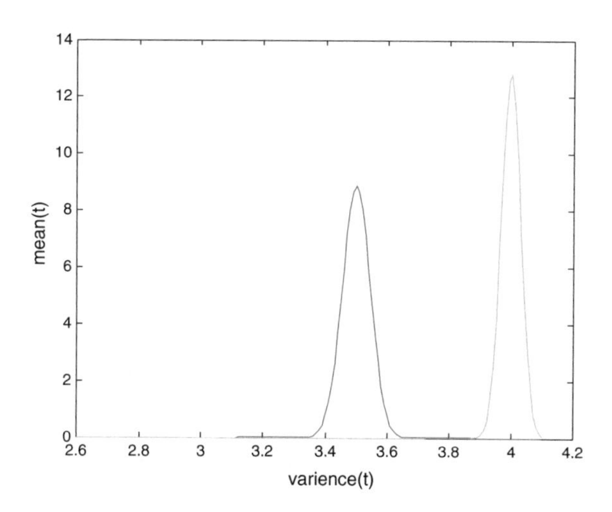

Fig. 14 Prediction and measurement Probabilty density function for moving vehicle

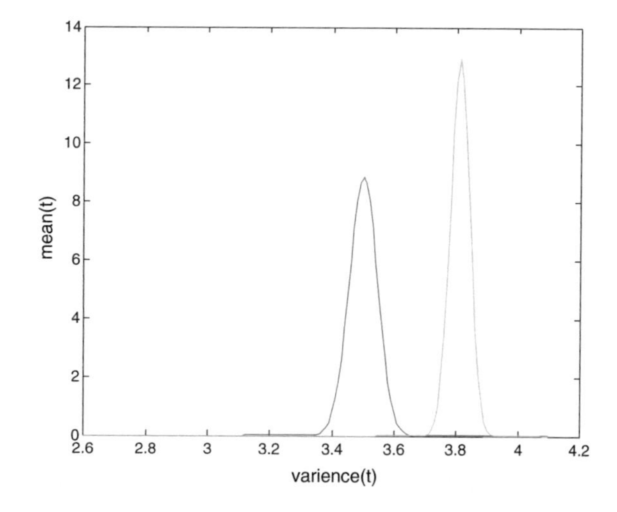

Fig. 15 Prediction and measurement probability density function for moving vehicle

7.1.5 Case of Noisy Prediction (Don't Sure About the Exact Velocity)

Better to suppose imperfect model by adding Gaussian noise. Distribution for prediction runs and spreads out.

In Figs. 16 and 17 is shows that noisy prediction of vehicle. This graph is plotted for medium-low velocity vehicle. Above graph is shown that conditional density function of prediction and measurement for medium-low speed vehicle. Accuracy is less medium speed vehicle due to Gaussian noise.

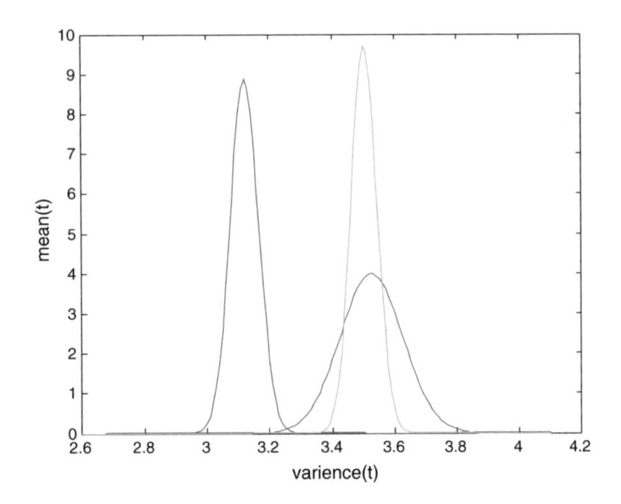

Fig. 16 Prediction and noisy measurement of vehicle movement

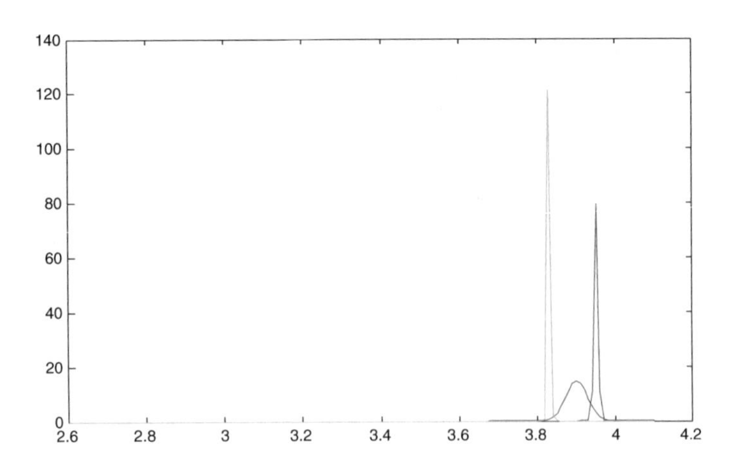

Fig. 17 Prediction and noisy measurement of vehicle movement

8 Conclusion

In this paper, we have predicted the position and velocity of the vehicle through GPS and measured the same with the help of GPS technology. Then, the best estimation of the vehicle parameters is done through Kalman filter by merging of the two probability density function of the estimation and measurement data of vehicles. The probability density function best estimation of Kalman filter so obtained is also Gaussian in nature. This paper demonstrates the application of Kalman filter in the tracking of vehicle moving in one dimension. In this paper, we have presented the accurate estimation of vehicle through Kalman filter. In other words, how can be Kalman filter applies on VANET System. We have simulated prediction and measurement graph on moving vehicles. Prediction and measurement is based on GPS device. Measured all parameter and best estimation is based on Kalman filter calculation.

Acknowledgements This manuscript is dedicated to COMPSE 2016 International Conference on Computer Science and Engineering, Gold Sands Resort, Nov 2016, Penang, Malaysia.

References

1. Anderson, B. D. O., & Moore, J. B. (1979). *Optimal filtering*. Prentice-Hall, New Jersey: Eaglewood Cliffs.
2. Anderson, B. D. O., & Moore, J. B. (2005). Optimal filtering. Dover.
3. Andelin, J., Carson, N., Page, E. B., et al. (1989). Advanced vehicle/highway systems and urban traffic problems. *Science, Education and Transportation Program* (pp. 1–32). Washington, DC: Congress of United States.
4. Bibby, J., & Toutenburg, H. (1977). *Prediction and improved estimation in linear models*. New York: Wiley.
5. Faragher, R. (2012). Understanding the basis of the Kalman filter via a simple and intuitive derivation. *IEEE Signal Processing Magazine, 29*(5), 128–132.
6. Grewal, M. S. (2011). *Kalman filtering*. Berlin, Heidelberg: Springer.
7. Groves, P. D. (2008). *Principles of GNSS, inertial, and multi-sensor integrated navigation systems*. Norwood, MA: Artech House.
8. Julier, S. J., & Uhlmann, J. K. (2004). Unscented filtering and nonlinear estimation. *IEEE Proceedings, 92*(3), 401–422.
9. Kalman, R. E. (1960). A new approach to linear filtering and prediction problems. *Journal of Basic Engineering, 82*(1), 35–45.
10. Kalman, R. E. (1960). A new approach to linear filtering and prediction problems. *Transaction of the ASME, Journal of Basic Engineering*, 35–45.
11. Lee, W. P., Osman, M. A., & Talib, A. Z., et al. (2008). Dynamic traffic simulation for traffic congestion problem using an enhanced algorithm. *World Academy of Science, Engineering and Technology*, 271–278.
12. Ran, B., Huang, W., & Leight, S. (1996). Solving the bottleneck problem at automated highway exits. *3rd ITS World Congress* (pp. 1–8). Medison, USA: University of Wisconsin.
13. Ribeiro, & Isabel, M. (2004). Kalman and extended Kalman filters: Concept, derivation and properties. *Institute for Systems and Robotics, 43*.

14. Sujitha, T., & Punitha Devi, S. (2014). Intelligent transportation system for vehicular ad-hoc networks. *International Journal of Emerging Technology and Advanced Engineering*, 2250–2459.
15. The CAMP Vehicle Safety Communications Consortium. (2005). Vehicle safety communications project task 3 final report-identify intelligent vehicle safety applications enabled by DSRC.
16. Traffic Road Safety: A Public Health Issue. (2004). http://www.who.int/features/2004/road_safety/en.
17. National Highway Authority of India. (1988). http://www.nhai.org/act1988.html.
18. International Road Traffic and Accident Database (IRTAD). (2017). http://internationaltransportforum.org/irtadpublic/index.html.

Parallel Coordinates Visualization Tool on the Air Pollution Data for Northern Malaysia

J. Joshua Thomas, Raaj Lokanathan and Justtina Anantha Jothi

Abstract The paper explains the contents of particles on the air pollution data through parallel coordinate visualization. This approach involves graph-plotting algorithms with parallel coordinates that explore the raw data with interactive filtering that facilitates the insight of the materials that mixed and harm the population in northern Malaysia. By presenting, the parallel coordinates method to visualize the parameter space that influence and visually identify the hazardous, moderate, unhealthy gaseous content in the air. The visual representation presents the large amount of data into single visualization. The paper discussed the performance of the chosen visualization method and tested with northern region datasets.

Keywords Parallel coordinates · Visual representation · Clustering data

1 Introduction

Air quality in Malaysia has much attention especially in the North of Malaysia, which are Penang, Perlis, and Kedah has more attention during last year haze [9]. These states affected with heavy haze whereby many victims from young and old admitted to hospitals. Most of these days, people used to wear facemask before and during the normal working time. Especially the tourism industry of Malaysia has badly affected because of the haze [15]. The neighbouring cities has covered with haze, it affected the visibility of ferries travelling from Penang Island to Mainland Butterworth as it also affect the port everyday transportation of goods in and outside

J.J. Thomas (✉) · R. Lokanathan · J.A. Jothi
Department of Computing, School of Engineering, Computing and Built Environment, KDU Penang University College, Penang, Malaysia
e-mail: joshopever@yahoo.com

R. Lokanathan
e-mail: raaj5671.r1@gmail.com

J.A. Jothi
e-mail: justtina@kdupg.edu.my

I. Zelinka et al. (eds.), *Innovative Computing, Optimization and Its Applications*, Studies in Computational Intelligence 741,
https://doi.org/10.1007/978-3-319-66984-7_16

Fig. 1 Penang Island to Butterworth visibility

Malaysia (Fig. 1). In 2016, the trend expected to continue rising with the increasing number of predicted weather conditions, where the Malaysian Ministry of Meteorological had expected the number cases to spike between the months of September and October.

In order to study the air pollution and investigate the bad air qualities we have gathered the information from the various weather monitoring stations from Malaysia. However, the work has focused on the data from the northern regions that is the economic hub of Malaysia. The weather data, in Malaysia recorded automatically from the data monitoring stations, which are located in various places around each states. There are total of 52 data monitoring stations in Malaysia, which recorded the weather data during each time interval for at least 6 times a day [8]. The information stored in the form of multivariate dimensions that is up to 10 dimensions. The data collection centers are have the facilities to store more than multidimensional data that includes wind speed, and wind directions [13].

In order to study the air quality problems mention above, the main aim of the work is to contribute to the community of researchers to develop a data visualization tool. To manipulate the huge multivariate data, into readable visual patterns the visual tool will explain the level of hazards presented in the atmosphere in terms of various pollutions. The visualization algorithm are used to plot the multivariate datasets. The implementation work of the visual tool has involved with parallel coordinates plots to illustrate significant pattern of particles, gaseous contents are measured from the existing data and it is classified based on its properties. In this work, we have used a research database to visual the data with the parallel coordinate's visualization tool to classify with colour coded, filtered represented of the data.

The outline of the rest of the content is as follows: Sect. 2 Literature Review—describes related work done by the researchers in parallel coordinates as an information visualization technique. Section 3 Methodology—describes the overview of the prototype, process flow of the system. Section 4 Visualization techniques—describes the implementation techniques of the components of the plotting of parallel coordinates; Sect. 5 experimental results—Explains the end products of the system followed by Conclusion and future work discussion in Sect. 6.

2 Literature Review

2.1 Related Work

(i) **Air Pollution** For the past years, many researchers are developing a stand-alone prototype that used to visualize the air pollution data. There are different data visualization methods have used to develop the prototype to visualize the air quality. We have published parallel coordinates visualization for examination timetabling data [16] and it is more suitable for the categorical data to be visualized before execute the computational algorithms.

Based on the research done from University of Seoul, Korea, they have developed a prototype, which uses the cloud computing technology to visualize the urban air quality data [19]. This prototype is more towards the predicting the air pollution using cloud computing technology. Moreover, there is another research done by student of Central South University, China collaborated with the Arizona State University, where they have developed a visual analytics system is a web based. The web-based system uses the parallel coordinate's method to visualize the air quality data. However, a GIS view used for spatial analysis, which includes a scatter plot view [6]. This research paper has taken as an inspiration to develop a visualization prototype. On the other hand, the Microsoft Research students have developed a mobile app, which forecasts the air quality based on the big data [3].

(ii) **Data Visualization**

Data visualization research are being widely carried out by researchers all around the world. During the research phase, we have studied the research paper [16]. The work has visualized the student and subject data from the university and created a parallel coordinates visualization framework. With the assistance of the work to get more idea on the implementation of the Air pollution visualization tool (APVT) software. After reading their research paper, we came to our conclusion that visual cues play a major role in data visualization software. We have applied some of the visual cues on our APVT software. This makes the APVT software more realistic. Besides that, Nambiar [7] also have carried out multiple researches on weather data visualization. However, his research was more towards the scientific visualization. APVT software was more towards information visualization. We have modified to develop the APVT software. Whenever it comes to data visualization, data mining plays an important role. Previously, proposed a data mining system, which analyzes the air quality data. However, it is known that using data mining techniques has its own weakness. Class visualization of high-dimensional data with applications [4] to analyze the data it is hard to understand the data because the analyzing process is almost same to the "black-box" testing.

Luo et al. [6] have extended the existing methods to handle the spatial distribution data includes the weather data. Most of the approaches are used are more towards visualization rather than analysis. Parallel coordinates is a well-established visualization technique first proposed by Inselberg [11]. It is a scalable framework in the sense that the increase of the dimensions corresponds to the addition of extra axes. The visual clutter caused by an excessive overlaying of polylines limits the effectiveness of parallel coordinates in visualizing a dense data set. However parallel coordinates visualization method has adopted to visualize the timetabling input matrices and output matrices in its multi-variable [17]. Visual clustering, filtering, and axis reordering are the common methods to reduce clutters in parallel coordinates. Some filtering techniques remove part of data in a pre-processing. Axis reordering computes the best axis order with minimal clutters, but it is very likely that the best axis order still leads to an unsatisfactory and cluttered display.

(iii) **Data Clustering and Optimization**

Various modern optimization algorithms has used for modularization of software systems. This paper contributes a fully functional visualization tool with minimal optimization in plotting the data. Bio-inspired algorithms are one of the familiar and recent for software modularization is the process of allocating modules to subsystems or clusters. Software modularization is particularly useful during the maintenance phase of the software development life cycle [12]. In another angle of working with pollutants and global warming. Global warming and fossil fuel depletion are two of the most important issues of this century. The considerations of energy security and climate change force increased societal interest in technologies that enable a reduction in the use of fossil fuels. It has well recognized that an effective solution to these issues is to develop non-carbon-dioxide-emitting and inexhaustible energy resources and energy technologies.

Recently, discovering of nuclear power have provide both to diminish our dependence on fossil fuel resources, and to provide electricity without any emissions of harmful air pollutants [2]. In data clustering from real-time is to maintain the reliability of sophisticated systems to a higher level, the systems' optimum structural design or highly reliable components of these systems are required, rather both of them may be sought simultaneously. Implementation of these methods to improve the system availability or reliability will normally consume resources such as cost, weight, volume etc. So the system reliability cannot be further improved effectively by considering these constraints [5]. Optimization is the core to many real world problems. Numerous techniques have proposed to solve optimization problems viz. linear programming, quadratic programming, integer programming, nonlinear programming, dynamic programming, simulated annealing, evolutionary algorithm etc. [18]. Clustering deals with unlabeled data i.e., no class labels is provided. It does not involve any training. On the contrary, classification is a two-step process training and testing. During training, it deals with label data while it deals with unlabeled data for testing [18].

With the widespread use of network, online clustering and outlier detection as the main data mining tools have drew attention from many practical applications, especially in areas where detecting abnormal behaviors is critical, such as online fraud detection, network instruction detection, and customer behavior analysis. These applications often generate a huge amount of data at a rather rapid rate.

Manual screening or checking of this massive data collection is time consuming and impractical. Because of this, online clustering and outlier detection is a promising approach for such applications. Specifically, data mining tools are used to group online activities or transactions into clusters and to detect the most suspicious entries. The clusters are used for marketing and management analysis. The most suspicious ones are investigated further to determine whether they are truly outlier. Numerous clustering and outlier detection algorithms have been developed but the majority of them are intended for continuous data [20].

Class Visualization of High Dimensional Data [13] is a visualization tool for clustering and analysis of high dimensional data sets. CViz utilizes a k-means clustering algorithm to find interesting concepts in the data. It then drawstwo-dimensional scatter plots by selecting pairs or triples of concepts and relating these concepts to the data examples and the other concepts. By using an animation technique called, touring, CViz allows the analyst to quickly cycle through all the different pairsor triples of concepts and see how the data changes from one perspective to another. This can provide new insight into underlying structure of the data. We found this idea very good, unfortunately we did not implemented it in our framework yet, but this is certainly something to think about for future work.

3 Methodology

3.1 APVT Parameters Organization and Display

Our interface provides a customizable overview of the parameter space designed as axial information. We use parallel vertical axes to display visualization parameters in terms of various material presented in the atmosphere. Each point within the parameter space (i.e. a set of parameters used to produce an image) is denoted by a polyline connecting the axes. We divide the parameters that displayed into two major categories.

- Visual parameters
  - Position
  - Orientation
  - Zoom and filter
  - Interactive rendering (Colour coded)
- Data parameters
  - Mapping the data between axis value and colour
  - Highlight the cluster of data for better insight
  - Dataset of Northern region of Penang.

Figure 2 further explored in the implementation of this paper; we believe it is important for dealing with multivariate data. In such data sets, one will have different data-specific parameters for each variable; however, the view-specific parameters must be identical. This will explored in our future research. In addition, this division creates a conceptual framework at this level that allows extensions to implement further. E.g. other data manipulation tools, such as segmentation tools, need to data specific parameters, for example to view the hazards components by regions. Users create or load preset nodes (CSV files) for each parameter and then connect a set of nodes to "generate lines". In this way, we use parallel coordinates not only as a method to display data items (parameter combinations), but also as an interface to create data items. Further, one can imagine wanting to split a parameter axis. It is possible to create additional axis. For example, the colour transfer function could be split into axes for different colour components. Similarly, a user

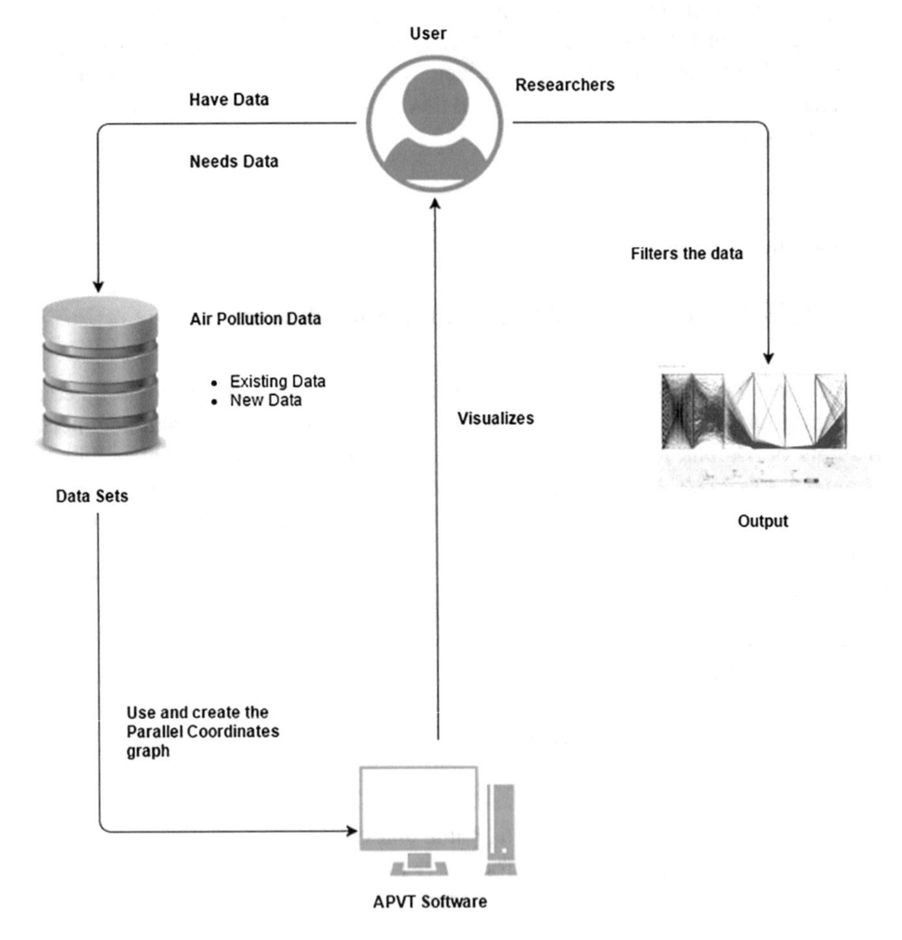

Fig. 2 Process flow of APVT

could simplify the display by grouping several parameter axes into one super-axis. Our interface allows such split with the base parameters of colour transfer function, opacity transfer function, dataset, renderer, zoom and translation, and orientation.

3.2 Design Goals

We designed our interface to support the visualization tasks. Note that we apply this visualization seeking mantra to parameter representation and data representation: Our design organizes all visualization parameters together in one space so that users can quickly gain an overview of the possibilities. Filter: Visualization parameters can be easily accessed a changed to zoom and filter data sets. Selecting a polyline highlights the polyline in blue so users can quickly identify the parameter settings that produced a given plot. Users can simplify the parameter space or fine-tune combined parameters. Effects of changing parameter settings can studied by making changes with a simple mouse action and viewing consecutive results. Our tool explicitly represents all parameters that influence rendering. It shows current and previous settings of all visualization parameters simultaneously. Parameters explored immediately within the context of previous settings. To the best of our knowledge, this never tried before, and we are convinced that it offers substantial benefits to the air-pollution data exploration.

3.3 Data Collection

The Department of Environment (DOE) have installed the monitoring stations all around Malaysia it retrieves the weather data real time. These weather data is then calculated and shown in the APIMS website is part of the DOE. The APIMS website does shows only the Air Pollution Index (API) of each monitoring stations only. These details are available to public, whereas the other dimensions of the weather data are not available for the public. An official letter needed to written to the DOE before accessing the real time weather data. Table 1 shows the current 10 dimensions of data retrieved by the data monitoring station around Malaysia.

The APVT prototype is mainly focused on visualizing the air pollution data only. Below are the list of the features the APVT has: Visualizing the gaseous content by different ranges of colors the prototype has 5 different types of colors. Table 2 each colors has its own meaning the list of colours and the descriptions of it. Visualizing the air pollution data from different monitoring stations around Malaysia. Filtering the air pollution data by date.

Table 1 List of weather data collected at the monitoring stations in Malaysia

Name	Units
Wind speed	m/s
Wind direction	Bearing
Year	y
Month	m
Day	d
Carbon monoxide (CO)	ppb
Nitrogen Oxide (NO)	ppb
Ozone (O_3)	ppb
Particulate matter 10 (PM_{10})	$\mu g/m^3$
Air pollution index (API)	Scale 100

Table 2 Gas ranges colors with its descriptions

Color	Descriptions
Black	Default
Blue	Good
Green	Moderate
Pink	Unhealthy
Orange	Very unhealthy
Red	Hazardous

3.4 Relationship with Parallel Coordinates

Although our interface is based on the concept of parallel coordinates, it has substantial differences from ordinary parallel coordinates displays. Many of the axes do not display one-dimensional variables. For example, a plot function (see in Sect. 4.1) is a complex concept consisting of variables such as data intensities, opacities, and sometimes intensity derivatives. Furthermore, axes can merged to produce nodes that are even more complex. Sorting and interpolating nodes therefore requires slightly more complex techniques than ordinary parallel coordinates. In our current implementation, we did focus on such high-level trends; instead, we display multiple lines that users can easily distinguish line colours and relate to their parameters.

4 Visualization Techniques

4.1 Parallel Coordinates

Parallel coordinate is one of the most popular tool which is used for visualizing and analyzing data [21]. Most of the researchers are using this visualization method to carry out their research analysis. Based on the article from Wikipedia, parallel coordinate was used before 1885 by Henry Gannetts. He used parallel coordinate to show the rank of states by ratios. Later after 79 years, which is during the 1959, a researcher developed the parallel coordinate as a system [11]. After that, many applications were developed by developers to do data mining or to analyze the data. Examples of application created was a collision avoidance algorithm for air traffic control.

In order for the Parallel Coordinate to work data is needed to be feed to it. Without data the parallel coordinate could not plot the coordinates. Given a small sets of data, the user can displayed it using a table form. Unless the size of the data is huge, which means it has more than million rows, the user cannot view it. A data sets with multiple attributes are named as multivariate data sets [14]. First of all, the user needs to add in the raw data into the visualization tool. After that, the preparation process will be started. This is where the process of filtering, extracting and calculating will be carried out. Filtering is a process where the amount of data will be reduced. Finally, the plotted coordinates can be viewed. However, it might be hard to find out the data which has been left out [14]. This is because raw data has hundreds of rows.

The user can interact with the data using some of the interaction tool which is available for the parallel coordinate data visualization method. The reason interaction tool is created is because, users can view or analyze the information at their first sight. There are many interaction tool such as brushing, and axis. These interaction are one of the important interaction. Brushing is a process where the user is able to select a sample of subset which originally referred to an axis aligned rectangle for selection [1]. The other interaction tool is axis. The user is allowed to move the axis of the parallel coordinate by flipping it or reordering it. The value of the axes will be changed according to the way the user flips the axis [10]. However, there are some limitations for the parallel coordinate data visualization method. First of all, if the number of data is huge, then there is a chance of over plotting. This can cause the user hard to see. Besides that, if the data is in a categorical data format then the user could not view the plotted coordinates. This is because parallel coordinates only works well with numerical values. Thus, interaction needed to be used in order to view the parallel coordinate.

4.2 *Implementation*

Drawing the axis and plotting the CSV data takes alot of time. The code block below shows the partial code on how the parallel coordinates are plotted. Drawing the axis and plotting the CSV data takes higher load time. The code block below shows the partial code on how the parallel coordinates plots.

```
public void paintComponent(Graphics g) {
        super.paintComponent(g);
        Graphics2D g2d = (Graphics2D) g;

g2d.setRenderingHint(RenderingHints.KEY_ANTIALIASING,
RenderingHints.VALUE_ANTIALIAS_ON);
        g2d.setColor(Color.WHITE);
        g2d.fillRect(0, 0, getWidth(), getHeight());
        g2d.setFont(font);
        line2ds = new HashMap();
// draw the data as parallel coordinates from data
        if (csvData != null) {
            if (plot == TWO_AXIS_PLOT) {
                plotTwoAxis(g2d);
            } else if (plot == PARALLEL_AXIS_PLOT) {
                if (selectedAttributes.size() == 1) {
         selectedAttr =  selectedAttributes.get(0);
                    plotAttribute(g2d);
                } else {
                    plotAllParallelAxis(g2d);
                }
            } else if (plot == SINGLE_ATTRIBUTE_PLOT) {
                plotAttribute(g2d);} } }
```

For our prototype, we have added a tooltip interaction for the parallel coordinates. The researchers can hover on the data line to know more details about the selected gaseous contents. Moreover, the way the APVT prototype plots the parallel coordinates is similar to the Alfred Inselberg plots the data. The each rows of the CSV data are stored in an arraylist. The stored data are plotted using the Java Line2D method. We have implemented the HashTable data structure for our prototype. The way we draw the axis is at first, we read the CSV file to get the columns name and then we draw the axis using a oval shape. After that, we plot the lines from the arraylist. Further plotting the parallel coordinates works the same way we plot the raw data.

5 Experimental Results

The whole prototype is developed using Java programming language. A computer with JDK1.8 installed can be used to run the APVT prototype. Minimum of 2 GB of memory is needed for the computer. This is because, parallel coordinates involves a lot of background processing. By installing the .jar file the APVT prototype can be initiated successfully. During the testing phase we have tested our prototype with 2 different data monitoring station data loaded into the prototype. From here we did some comparison on the selected gaseous content. First of all, we load the benchmark data and then we further plot the parallel coordinates. Figure 3 shows the result of the parallel coordinates data plotted with the overall pollutants (CO, NO, PM_{10}, and API) plotted and visually highlighted with colour coded. These colour exemplifies the level of safe air in the atmosphere at the day and time.

Suppose the problem shown in Fig. 4 is the target regions from the dataset. These dataset classified and visually represented from (a) to (f). In Fig. 4a the observation of data is between 12 months plotted in first axis named as month, day. The visual filter check boxes activated or enabled to plot the API index, Nitrogen Oxide (NO) and indicated in the colour-coded lines. Visually interpret as the level of particles present in the atmosphere and shows the safety levels. The blue lines indicates the safe mode of data observed). In Fig. 4b the observation of PM10 data is of months plotted in third axis named as PM10. The visual filter check boxes activated or enabled to plot the PM10 and indicated in the colour-coded lines. The red lines indicates the hazard level of data observed. In Fig. 4c the observation of data is between day, months. We plotted multi-dimension of data at one time by extending the axis and visually analyse, which month has the highest level of affected API. As you can see from the plot Fig. 4c the colour coded lines indicated O_3, PM10, are nearer to the hazard level however the overall API is safe plotted with blue lines and green lines. In Fig. 4d–f indicated the various observation from the two dataset as chosen to fine the pollutant within a year. The parallel coordinates plot has performed with the clustering algorithm and visually analyse that PM_{10} is the most hazards that cause the haze ticker. If the air content with the PM_{10}

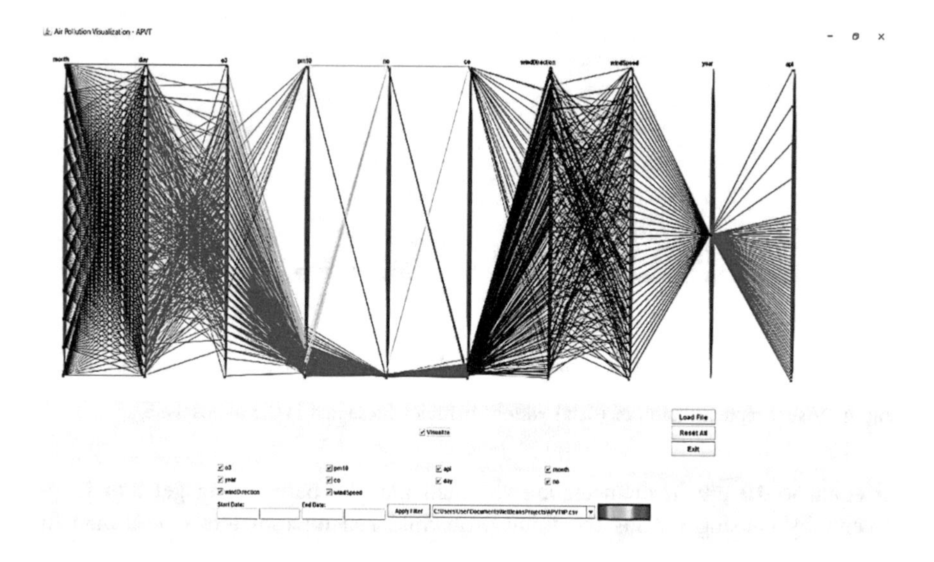

Fig. 3 Visualization of the 10-pollutants information plotted with parallel axis

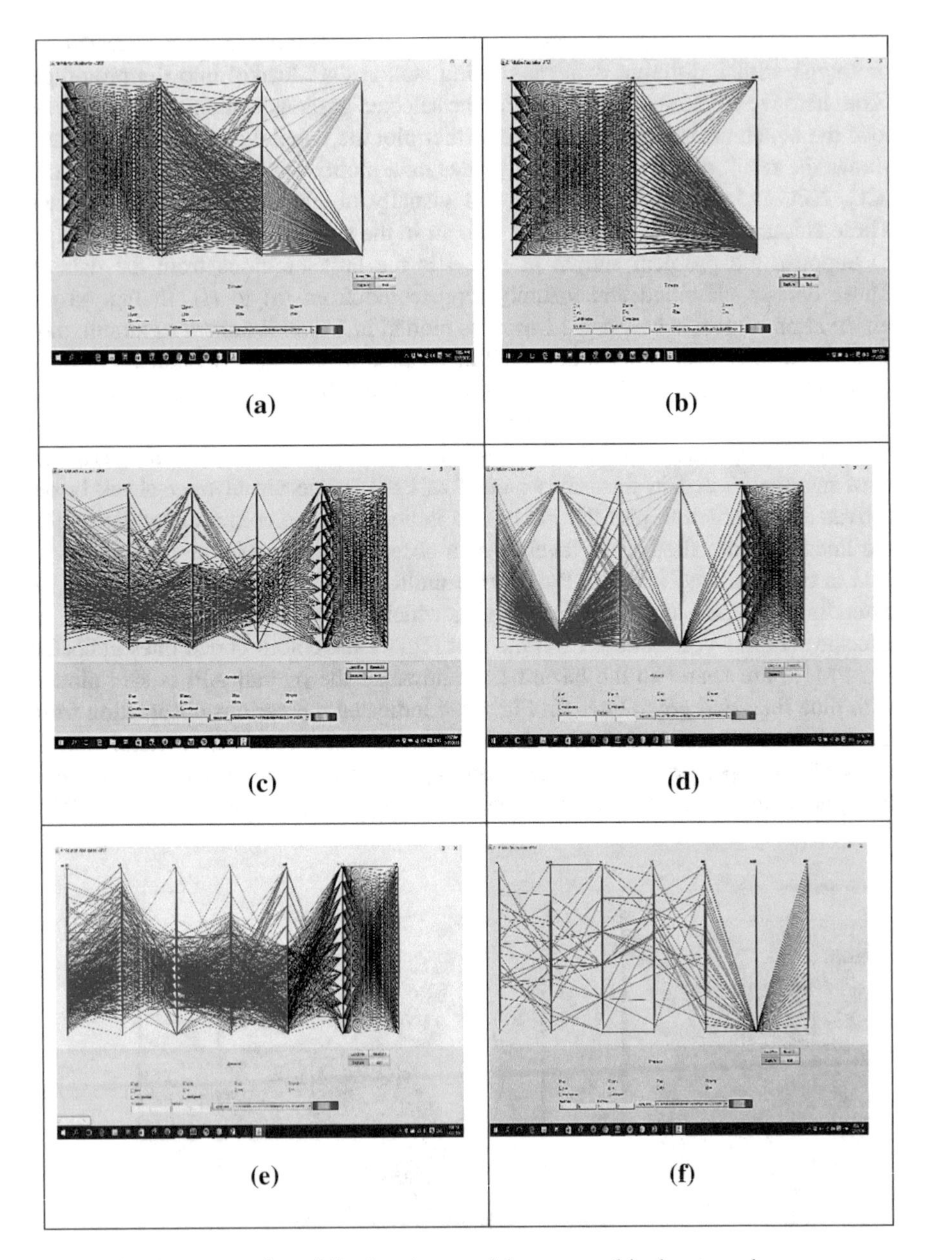

Fig. 4 Visual representation of (data) various particles presented in the atmosphere

or equal to 10 μm in diameter are so small that the particle can get into lungs, potentially causing serious breathing problems. Ten micrometers is less than the width of a single human hair in diameter. These materials forms when gases emitted from motor vehicles and industry undergo chemical reactions in the atmosphere.

The twelve months data in a particular year has confirmed that visual analyse and have the capacity to view the data in one visual-interface and able to manipulate each of the months data even drill down to days and also the time-period. We did not plot the time measure due to the clutter of data that cause the complex readability.

6 Conclusion and Future Work

In this paper we have shown a complete prototype which visualizes the air pollution in Malaysia. For this first implementation we have implemented only the parallel coordinates techniques with offline datastore. For the future recommendation, we have planned to implement the prototype into real time data retrieval with cloud services. This means by collecting the weather data in real time we can even start to predict the air pollution in Malaysia for the next year. The main required method is regression needed to be used. From here we can analyze the trend of the air pollution data and then predict the API value. Moreover, we also have planned to integrate more data visualization techniques into our prototype. Once all this features are implemented, we can make our prototype into a web based system, where publics can start to use visualize the air pollution data. Besides that, we also planned to developed our own method to visualize the air pollution data. Currently, we are using the existing approach to visualize the gaseous contents. Developing our own method makes the APVT more reliable.

References

1. Becker, R., & Cleveland, W. (1987). Brushing scatter plots. *Technometrics, 29*(2), 127. https://doi.org/10.2307/1269768.
2. Cebi, S., Kahraman, C., & Kaya, I. (2011). Soft computing and computational intelligent techniques in the evaluation of emerging energy technologies. *Innovation in Power, Control, and Optimization: Emerging Energy Technologies: Emerging Energy Technologies*, 164.
3. Chang, E., Li, T., Yi, X., Li, M., Li, R., Zheng, Y., & Shan, Z. (2015). Forecasting fine-grained air quality based on big data. In *KDD'15*. China: ACM—Association for Computing Machinery.
4. Dhillon, I. S., Modha, D. S., & Spangler, W. S. (2002). Class visualization of high-dimensional data with applications. *Computational Statistics & Data Analysis, 41*(1), 59–90.
5. Garg, H., Rani, M., & Sharma, S. P. (2013). Predicting uncertain behavior and performance analysis of the pulping system in a paper industry using PSO and Fuzzy methodology. *Handbook of Research on Novel Soft Computing Intelligent Algorithms: Theory and Practical Applications* (pp. 414–449).
6. Liao, Z., Peng, Y., Li, Y., Liang, X., & Zhao, Y. (2014). A web-based visual analytics system for air quality monitoring data. In *2014 22nd international conference on geoinformatics* (pp. 1–6). Changsha: IEEE.

7. Nambiar, P. (2015). Penang air quality at unhealthy levels, NST Online. Retrieved May 30, 2016, from http://www.nst.com.my/news/2015/10/penang-air-quality-unhealthy-levels.
8. Malaysia Air Pollutant Index (n.d.). Apims.doe.gov.my. Retrieved May 30, 2016, from http://apims.doe.gov.my/v2/faq.html.
9. Haze: Six areas in Perlis, Kedah and Penang record very unhealthy air quality (2015). Themalaymailonline.com. Retrieved May 30, 2016, from http://www.themalaymailonline.com/malaysia/article/haze-six-areas-in-perlis-kedah-and-penang-record-very-unhealthy-air-quality.
10. Hauser, H., Ledermann, F., & Doleisch, H. (2002). *Angular brushing of extended parallel coordinates* (p. 127). IEEE Computer Society.
11. Inselberg, A. (1985). The plane with parallel coordinates. *The Visual Computer, 1*(2), 69–91.
12. Jeet, K., & Dhir, R. (2016). Software module clustering using bio-inspired algorithms. *Handbook of Research on Modern Optimization Algorithms and Applications in Engineering and Economics, 445*.
13. Qu, H., Chan, W., Xu, A., Chung, K., Lau, K., & Guo, P. (2007). Visual analysis of the Air pollution problem in Hong Kong. *IEEE T*ransactions *on Visualization and Computer Graphics, 13*(6), 1408–1415. https://doi.org/10.1109/tvcg.2007.70523.
14. Steinparz, S., Aßmair, R., Bauer, A., & Feiner, J. (n.d.). *InfoVis—Parallel coordinates*. Graz University of Technology.
15. Tourist arrivals to Malaysia affected by haze, says deputy Tourism minister (2015). Themalaymailonline.com. Retrieved May 30, 2016, from http://www.themalaymailonline.com/malaysia/article/tourist-arrivals-to-malaysia-affected-by-haze-says-deputy-tourism-minister.
16. Thomas, J. J., Khader, A. T., & Belaton, B. (2011). A parallel coordinates visualization for the uncapaciated examination timetabling problem. In *Visual informatics: sustaining research and innovations* (pp. 87–98). Heidelberg: Springer.
17. Thomas, J., Khader, A., Belaton, B., & Ken, C. (2012). *Integrated problem solving steering framework on clash reconciliation strategies for university examination timetabling problem* (Vol. 7666, pp. 297–304). Berlin: Springer. Retrieved September 3, 2016, from http://link.springer.com/chapter/10.1007%2F978-3-642-34478-7_37#page-1.
18. Vora, M., & Mirnalinee, T. T. (2017). From optimization to clustering: A swarm intelligence approach. In *Nature-Inspired computing: concepts, methodologies, tools, and applications* (pp. 1519–1544). IGI Global.
19. Won Park, J., Ho Yun, C., Sun Jung, H., & Woo LEE, Y. (2011). Visualization of Urban Air pollution with cloud computing. In 2011 *IEEE world congress on services* (pp. 578–583). https://doi.org/10.1109/SERVICES.2011.111.
20. Wang, B., & Dong, A. (2012). Online clustering and outlier detection. *Data Mining: Concepts, Methodologies, Tools, and Applications: Concepts, Methodologies, Tools, and Applications, 1,* 142.
21. Zhang, J., Wang, W., Huang, M., Lu, L., & Meng, Z. (2014). Big Data density analytics using parallel coordinate visualization. *International conference on computational science and engineering*, p. 1115.

Application of Artificial Bee Colony Algorithm for Model Parameter Identification

Olympia Roeva

Abstract In this chapter, the Artificial bee colony (ABC) algorithm, based on the foraging behaviour of honey bees, is introduced for a numerical optimization problem. The ABC algorithm is one of the efficient population-based biological-inspired algorithms. To demonstrate the usefulness of the presented approach, the ABC algorithm is applied to parameter identification of an *E. coli* MC4110 fed-batch cultivation process model. The mathematical model of *E. coli* MC4110 cultivation process is considered as a system of three ordinary differential equations, describing the two main process variables, namely biomass and substrate dynamics, as well as the volume variation. This case study has not been solved previously in the literature by application of ABC algorithm. To obtain a better performance of the ABC algorithm, i.e. high accuracy of the solution within reasonable time, the influence of the algorithm parameters has been investigated. Eight ABC algorithms are applied to parameter identification of the *E. coli* cultivation process model. The results are compared, based on obtained estimates of model parameters, objective function value, computation time and some statistical measures. As a result, two algorithms are chosen—ABC1 and ABC8, respectively, with 60×500 number and 20×400 (population $\times$ maximum cycle number), such as algorithms with the best performance. Further, the best ABC algorithms are compared with four population-based biological-inspired algorithms, namely Genetic algorithm, Ant colony optimization, Firefly algorithm and Cuckoo search algorithm. The results from literature of metaheuristics applied for the considered here parameter identification problem are used. The results clearly show that the ABC algorithm outperforms the biological-inspired algorithms under consideration, taking into account the overall search ability and computational efficiency.

O. Roeva (✉)
Bioinformatics and Mathematical Modelling Department, Institute of Biophysics and Biomedical Engineering Bulgarian Academy of Sciences,
105 Acad. G. Bonchev Str, 1113 Sofia, Bulgaria
e-mail: olympia@biomed.bas.bg

I. Zelinka et al. (eds.), *Innovative Computing, Optimization and Its Applications*, Studies in Computational Intelligence 741,
https://doi.org/10.1007/978-3-319-66984-7_17

1 Introduction

Modelling approaches are central in system biology and provide new ways towards the analysis and understanding of cells and organisms. A common approach to model cellular dynamics employs sets of non-linear differential equations. Real parameter optimization of cellular dynamics models has become a research field of particularly great interest, as these problems are widely applicable. The parameter identification of a non-linear dynamic model is more difficult than that of a linear one, as no general analytic results exist. Due to the non-linearity and constrained nature of the considered systems, these problems are very often multimodal. Thus, gradient-based techniques in many cases fail to find the good solution. Although many different global optimization methods have been developed, the efficiency of the optimization method is always determined by the specificity of the particular problem addressed [65].

In the process of searching for new, more efficient metaphors and modeling techniques, nature-inspired metaheuristic methods receive great attention [3, 5, 26, 68, 71, 74–76, 88]. Among population-based nature-inspired metaheuristics are genetic algorithms (GAs), ant colony optimization (ACO), firefly algorithm (FA), cuckoo search (CS), etc., all of which use multiple agents or "particles".

Holland's book [33], published in 1975, is generally acknowledged as the beginning of the research of GAs. The GA is a model of machine learning which derives its behavior from a metaphor of the natural processes of evolution [28]. GA can cope with a great diversity of problems from different fields. It can quickly scan a large solution set. Owing to its inductive nature, the GA does not have to know any rules of the problem—it works by its own internal rules [22]. GAs are highly relevant for industrial applications, because they can handle problems with non-linear constraints, multiple objectives, and dynamic components—properties that frequently occur in real-life problems [28, 40]. Since their introduction and subsequent popularization, GAs have been frequently used as an alternative optimization tool to the conventional methods [28], and have been successfully applied to a variety of areas, and still find increasing acceptance [1, 4, 10, 12, 16, 25, 55, 56, 73, 75]. The effectiveness and robustness of GAs have already been demonstrated in fed-batch cultivation processes [60, 62, 64].

Another population-based metaheuristic that is a rapidly gaining attention is ACO. It can be used to find approximate solutions for a broad range of difficult optimization problems [13, 17, 20, 21, 50]. ACO has been successfully applied to complex biological problems and dynamic applications (as it adapts to changes such as new distances, etc.) [23]. ACO is on par with other metaheuristic techniques such as genetic algorithms and simulated annealing. ACO algorithms have been inspired by the ants' behavior in nature: ants usually wander randomly and when they discover food, they return to their colony, while leaving pheromone trails. If other ants find such a path, they follow the trail instead, returning and reinforcing it if they eventually encounter food. However, pheromone trails tend to evaporate over time. A shorter path, in comparison will be visited by more ants and thus the pheromone density remains high for a longer time. ACO is implemented as a team of intelligent agents, which simulate the ants behavior, traversing the graph representing the problem to be solved using mechanisms of cooperation and adaptation [2].

Metaheuristic algorithm CS was developed in 2009 by Xin-She Yang of Cambridge University and Suash Deb of C. V. Raman College of Engineering [80]. There are already several applications of CS for different optimization problems [7, 29, 31, 32, 34, 46, 48, 49, 53, 58, 79, 81, 82]. According to [Mohamad], the major category considered for the CS algorithm, is engineering, followed by object-oriented software (software testing), pattern recognition, networking, data fusion in wireless sensor networks, and job scheduling. Based on bibliography results, it is evident that the CS is a powerful novel population-based method for solving different engineering and management problems [58].

A relatively new metaheuristics is the Firefly algorithm (FA). This algorithm was proposed by Xin-She Yang [87]. Although the FA has many similarities with other swarm intelligence based algorithms, it is indeed much simpler both in concept and implementation [84–86]. There are already several applications of FA for different optimization problems [8, 14, 24, 41, 51, 63, 66, 89]. According to the published results, the population-based method FA is successfully applied to solving various optimization problems [70].

A more efficient population-based biological-inspired algorithm is Artificial bee colony (ABC) optimization. In the literature, there are three continuous optimization algorithms based on intelligent behavior of honeybee swarm [39, 57]. Yang developed a Virtual bee algorithm [83] to optimize only two-dimensional numeric functions. Pham et al. [57] introduced the Bees algorithm in 2005, which employs several control parameters. For optimizing multi-variable and multi-modal numerical functions, [39] described an ABC algorithm. The known results indicate that ABC can efficiently be used for many optimization problems [9, 27, 30, 31, 37, 38, 42–45, 54, 72, 78, 90].

This chapter presents the use of ABC algorithm to solve a model parameter identification problem. As a case study, *E. coli* MC4110 fed-batch cultivation process model is considered. The bacterium *E. coli* is crucial in modern biotechnology. It has a long history of use in the biotechnology industry and is still the microorganism of choice for most gene cloning experiments [15]. Mathematical modeling has the potential to provide a coherent and quantitative description of the interplay between gene expression, metabolite concentrations, and metabolic fluxes. It can provide tools for analyzing and predicting of cultivation process behavior [18, 19]. Mathematical modelling has been key in the understanding of such systems, providing tools for analysis and prediction of cultivation process behavior, although the models used are often rather large and complex.

The chapter is organized as follows. In Sect. 2, the background of ABC algorithm is given and the considered model parameter identification problem is commented. The numerical results are presented in Sect. 3 and discussed in Sect. 4. Conclusion remarks are made in Sect. 5.

2 Background

2.1 *Artificial Bee Colony Optimization*

Karaboga has described the Artificial bee colony algorithm for numerical optimization problems on the basis of foraging behaviour of honey bees [39].

In ABC algorithm, each cycle of the search consists of three steps [36]:

- sending the employed bees onto their food sources and evaluating their nectar amounts; after sharing the nectar information of food sources;
- selection of food source regions by the onlooker bees and evaluating the nectar amount of the food sources;
- determining the scout bees and then sending them randomly onto possible new food sources.

These three steps are repeated through a predetermined number of cycles called maximum cycle number (*MCN*) or until a termination criterion is satisfied. An artificial onlooker bee chooses a food source depending on the probability value associated with that food source, p_i, calculated by the following expression [1, 36]:

$$p_i = \frac{f_i}{\sum_{n=1}^{SN} f_n} \tag{1}$$

where f_i is the fitness value of the solution i, which is proportional to the nectar amount of the food source in the position i, and SN is the number of food sources, which is equal to the number of employed bees or onlooker bees. The SN equals the half of the colony size (employed bees + onlooker bees) NP.

In case of a minimization problem, the fitness value of solution (f_i) is calculated as:

$$f_i = \begin{cases} \frac{1}{1+f_i}, & f_i \geq 0 \\ 1 + \text{abs}(f_i), & f_i < 0. \end{cases} \tag{2}$$

The probability of an onlooker bee selecting a food source is based on the degree of fitness. So, for relatively low adaptation value, the probability of this food source to be selected also becomes low. In order to increase the probability of the lower fitness individuals to be selected, in [44] the following modification of the probability equation is proposed:

$$p_i = \frac{0.9 f_i}{\max(f_i)} + 0.1 \tag{3}$$

In order to produce a candidate food position from the old one in memory, the ABC uses the following Eq. (4) Karaboga and Akay [36]:

$$v_{i,j} = x_{i,j} + \phi_{i,j}\left(x_{i,j} - x_{k,j}\right) \tag{4}$$

where k = 1, 2, …, SN and j = 1, 2, …, D are randomly chosen indexes, D is the number of parameters of the problem to be optimized. Although k is determined randomly, it has to be different from i. $\phi_{i,j}$ is a random number between [–1, 1]. It controls the production of neighbour food sources around $x_{i,j}$ and visually represents the comparison of two food positions by a bee. As it can be seen from Eq. (4), as the difference between the parameters of the $x_{i,j}$ and $x_{k,j}$ decreases, the perturbation on the position $x_{i,j}$ decreases, as well. Thus, as the search approaches the optimum solution in the search space, the step length is adaptively reduced. If a parameter value produced by this operation exceeds its predetermined limit, the parameter can be set to an acceptable value. In this work, the value of the parameter exceeding its limit is set to its limit value. The food source which nectar is abandoned by the bees is replaced with a new food source by the scouts. In ABC, this is simulated by producing a position randomly and replacing it with the abandoned one. In ABC, if a position cannot be improved further through a predetermined number of cycles, then that food source is assumed to be abandoned. The value of predetermined number of cycles is an important control parameter of the ABC algorithm, which is called "limit" for abandonment.

Assume that the abandoned source is x_i and j = 1, 2, …, D, then the scout discovers a new food source to be replaced with x_i. This operation can be defined as in Eq. (5)

$$x_i^j = x_{\min}^j + rand\left(x_{\max}^j - x_{\min}^j\right) \tag{5}$$

After each candidate source position $v_{i,j}$ is produced and then evaluated by the artificial bee, its performance is compared with that of its old one. If the new food source has equal or better nectar than the old source, it is replaced with the old one in the memory. Otherwise, the old one is retained in the memory. In other words, a greedy selection mechanism is employed as the selection operation between the old and the candidate one.

Four different selection processes are employed in ABC algorithm [36]:

- a global probabilistic selection process, in which the probability value is calculated by Eq. (1) used by the onlooker bees for discovering promising regions;
- a local probabilistic selection process carried out in a region by the employed bees and the onlookers depending on visual information such as the color, shape and fragrance of the sources for determining a food source around the source in the memory in the way described by Eq. (4);
- a local selection process called greedy selection carried out by onlooker and employed bees in that if the nectar amount of the candidate source is better than

that of the present one, the bee forgets the present one and memorizes the candidate source produced by Eq. (4). Otherwise, the bee keeps the present one in the memory;

- a random selection process carried out by scouts as defined in Eq. (5).

The flowchart of the ABC algorithm is presented in Fig. 1.

Major advances and disadvantages of the ABC algorithm can be summarized as follows [11, 36]:

Advantages:

- algorithm flexibility— allows adjustments with few control parameters;
- algorithm simplicity—ease of implementation;
- algorithm ability to handle the objective cost with stochastic nature
- both exploration and exploitation;
- broad applicability—even in complex functions with continuous, discrete or mixed variables.

Disadvantages:

- high number of objective function evaluations;
- increasing the computational cost with an increase in the number of population;
- search space limited by initial solutions;
- algorithm requires new fitness tests on the new algorithm parameters to improve performance.

2.2 Problem Formulation

Application of the general state space dynamical model to the fed-batch cultivation process of bacteria *E. coli* leads to the following nonlinear differential equation system [67]:

$$\frac{dX}{dt} = \mu_{\max} \frac{S}{k_S + S} X - \frac{F_{in}}{V} X \tag{6}$$

$$\frac{dS}{dt} = -\frac{1}{Y_{S/X}} \mu_{\max} \frac{S}{k_S + S} X + \frac{F_{in}}{V} (S_{in} - S) \tag{7}$$

$$\frac{dV}{dt} = F_{in} \tag{8}$$

where X is biomass concentration, [g/l]; S is substrate concentration, [g/l]; F_{in} is feeding rate, [l/h]; V is bioreactor volume, [l]; S_{in} is substrate concentration in the feeding solution, [g/l]; $\mu_{\max}$ is the maximum value of the specific growth rate, [1/h]; k_S is saturation constant, [g/l]; $Y_{S/X}$ is yield coefficient, [-].

For the parameter estimation problem, real experimental data of an *E. coli* MC4110 fed-batch cultivation process are used. Measurements of biomass and glucose concentration are used in the identification procedure.

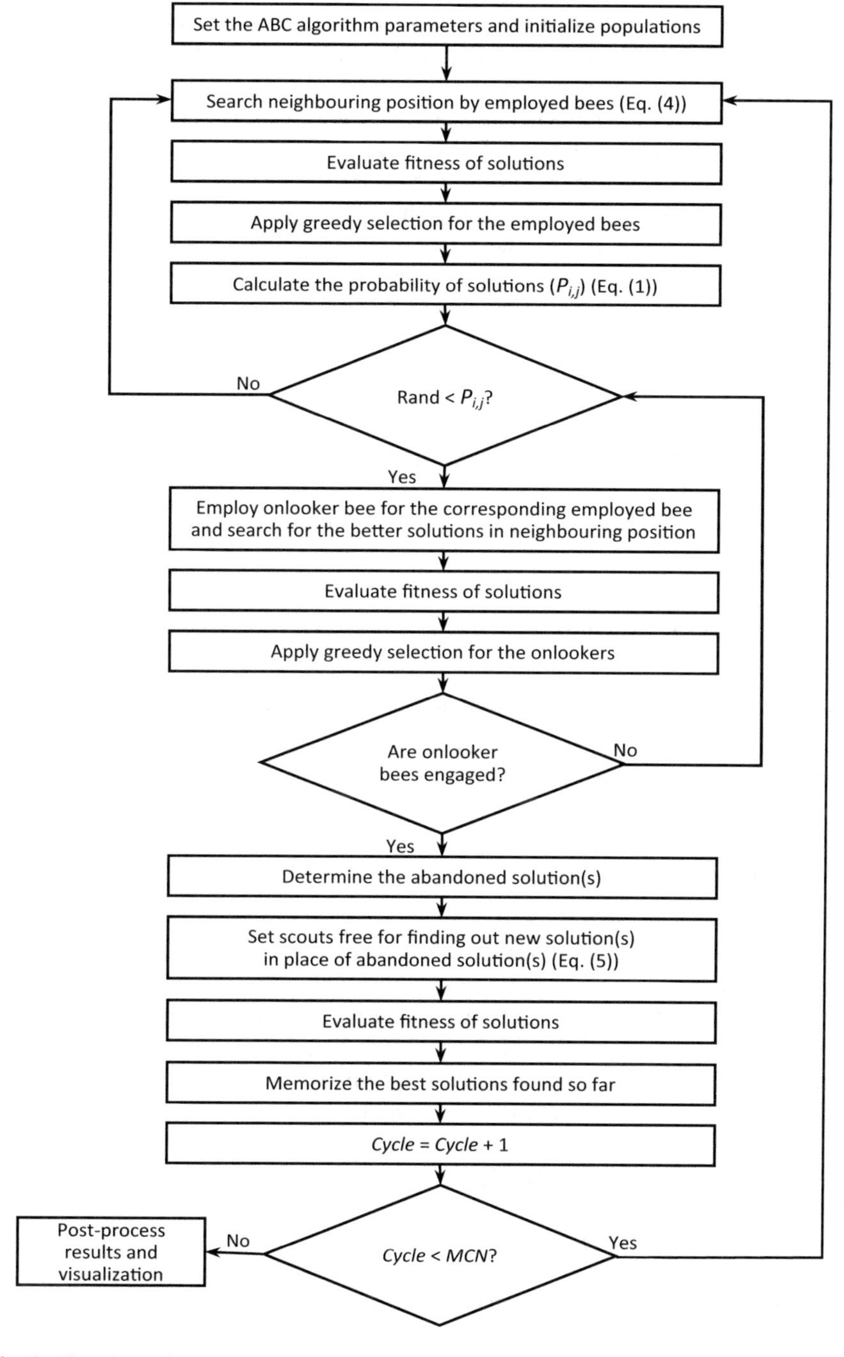

Fig. 1 Flowchart of ABC algorithm

The cultivation was performed in the Institute of Technical Chemistry, University of Hannover, Germany, during the collaboration work with the Institute of Biophysics and Biomedical Engineering, BAS, Bulgaria, granted by DFG [61]. The strain used for the fermentation process was *E. coli* MC4110. The fed-batch cultivation started at $t_0 = 6.68$ h. The process was carried out up to $t = 11.54$ h. The cultivation was performed in a 2 L bioreactor containing a mineral medium. The initial liquid volume was 1350 ml. The temperature was maintained at 35 °C and pH was controlled at 6.8. The aeration rate was kept at 275 l/h. Oxygen was controlled at around 35%. The glucose concentration in the feeding solution was $S_{in} = 100$ g/l. The initial concentrations of the two process variables (biomass and substrate) were as follows:

$$X(t_0) = 1.25 \text{ g/l and } S(t_0) = 0.8 \text{ g/l} \tag{9}$$

The detailed description of the cultivation condition and experimental data can be found in [61].

The objective function is presented as a minimization of a distance measure J between experimental data and model predicted values of the process variables:

$$J = \sum_{i=1}^{n} \left(X_{\text{mod}}(i) - X_{\exp}(i)\right)^2 + \sum_{i=1}^{n} \left(S_{\text{mod}}(i) - S_{\exp}(i)\right)^2 \rightarrow min \tag{10}$$

where n is the number of data for each process variable (X, S); $X_{\exp}$ and $S_{\exp}$ are the experimental data; X_{mod} and S_{mod} are model predictions with a given set of the parameters ($\mu_{\max}$, k_S and $Y_{S/X}$).

3 Numerical Results

Each optimization algorithm has its own parameters and functions that determine the algorithm performance in terms of solution quality and execution time. In order to increase the performance of the ABC algorithm, the adjustments of the parameters and functions, depending on the problem domain, are provided. Some of the parameters of the ABC algorithm are tuned on the basis of a series of pre-tests, according to the parameter identification problem, considered here.

All computations are performed using a PC/Intel Core i5-2320 CPU @ 3.00 GHz, 8 GB Memory (RAM), Windows 7 (64 bit) operating system and Matlab 7.5 environment.

Eight differently tuned ABC algorithms are investigated. In Table 1, the used distinct ABC algorithm parameters and functions are summarized. These settings are chosen on the basis of performed pre-test procedures.

The rest of ABC parameters are as follows:

- number of food sources $SN = NP/2$;
- number of parameters $D = 3$;
- *limit* = 100.

Table 1 ABC algorithm—parameters and functions

ABC algorithm	Parameters		Probability
	NP	MCN	
ABC1	60	500	Eq. (1)
ABC2	40	400	
ABC3	40	200	
ABC4	20	400	
ABC5	60	500	Eq. (3)
ABC6	40	400	
ABC7	40	200	
ABC8	20	400	

Table 2 Results from model parameters identification procedures

ABC algorithm		*J* value				*T*, s	
	Average	Best	Worst	STD[a]	Average	Best	Worst
ABC1	4.4149	4.3001	4.5076	0.0517	534.4448	530.4688	538.4688
ABC2	4.4568	4.3023	4.5688	0.0671	286.35574	277.3438	291.8906
ABC3	4.5094	4.3758	4.6248	0.0602	146.7448	141.5313	149.5156
ABC4	4.5124	4.3785	4.6297	0.0577	146.7922	143.2344	150.6406
ABC5	4.4563	4.3624	4.5329	0.0456	544.8828	536.3594	560.9688
ABC6	4.4682	4.3774	4.5491	0.0523	280.7047	276.7813	288.6406
ABC7	4.4907	4.3268	4.6032	0.0574	146.1297	141.0000	149.9219
ABC8	4.4921	4.3195	4.5981	0.0643	149.2307	147.5156	152.1719

[a]average results

Because of the stochastic characteristics of the applied algorithm, a series of 30 runs for each differently tuned ABC algorithm were performed.

The following upper and lower model parameters (Eqs. 6–8) bounds are considered:

$0 < \mu_{max} < 0.8$; $0 < k_S < 1$ and $Y_{S/X} < 30$.

For comparison of the results of the identification of model parameters, the average, the best and the worst results of the 30 runs, for the μ_{max}, k_S and $Y_{S/X}$, *J* value and execution time (*T*) were observed. The obtained results are summarized in Tables 2 and 3.

4 Discussion

The results, presented in Tables 2 and 3, show that the ABC1 algorithm finds the solution with the highest quality, e.g. the smallest objective function value ($J = 4.3001$). A similar result is shown by the ABC2 algorithm: $J = 4.3023$. These two ABC algorithms operate with large number of population and cycles, $NP = 60$,

Table 3 Identified best model parameters

ABC algorithm	Model parameters					
	μ_{max}		k_S		$Y_{S/X}$	
	Value	STD	Value	STD	Value	STD
ABC1	0.471	0.0076	0.009	0.0015	2.020	0.0016
ABC2	0.476	0.0099	0.009	0.0019	2.023	0.0022
ABC3	0.504	0.0102	0.014	0.0018	2.018	0.0019
ABC4	0.490	0.0110	0.011	0.0022	2.021	0.0019
ABC5	0.500	0.0078	0.014	0.0016	2.019	0.0017
ABC6	0.469	0.0093	0.009	0.0018	2.021	0.0018
ABC7	0.475	0.0098	0.010	0.0017	2.018	0.0021
ABC8	0.491	0.0093	0.012	0.0017	2.020	0.0017

$MCN = 500$ and $NP = 40$, $MCN = 400$, respectively for ABC1 and ABC2. Both algorithms use the form of calculation of the probability (Eq. 1), proposed in [36]. The other two ABC algorithms, ABC3 and ABC4, using the same form obtained practically identical results for J value: $J = 4.3758$ and $J = 4.3785$, as well as for computation time (see Table 2, best values). Such result shows that the ABC algorithm will produce very similar results for a fixed value of $NP \times MCN$.

The same tendencies are observed when the average results are analyzed. In this case, as expected, ABC1 obtains results with the lowest value of standard deviation (STD) compared to the algorithms ABC2, ABC3 and ABC4.

When the second form of calculation of the probability (Eq. 3) is considered, the worst results are observed in case of large number of population and cycles. In the case of $NP \times MCN$ 40×200 and 20×400, the average results are similar to the previous ones. As it can be seen, the best results of ABC7 and ABC8 are even better than the best results of ABC3 and ABC4. The results show the importance of the choice of appropriate functions and parameters of ABC algorithm. Using the probability formula Eq. (3), ABC5 and ABC6 obtained similar to the best results of ABC3 and ABC4. Moreover, the standard deviation of the average results of the objective function value is smaller than the observed for ABC1 algorithm. These are very interesting results. The probability presented as Eq. (1) gives better results in the case of larger number of population and cycles, while the probability presented as Eq. (3) gives better results in the case of smaller number of population and cycles. So, it may be a matter of choice what probability function to use - the first one (Eq. 1) in which we have almost 3.5 times bigger computation time, or the second one (Eq. 3), with much less computation time. The ABC1 algorithm found the solution for about 530 s, whereas the ABC7 or ABC8—for about 150 s. In other words, we can choose between algorithm ABC1 that finds the solution with accuracy of $J = 4.3$ (best value) for time $T = 534$ s or algorithm ABC8 that finds the same or near to this solution for time $T = 149$ s.

The results presented in Table 3 can facilitate the choice of the most appropriate ABC algorithm for the parameter identification problem considered here. Since the model parameters (Eqs. (6–8) have strict physical meaning, we can round the estimates and accept that there are two different estimated values of model parameter μ_{max}, 0.5 1/h and 0.47 1/h. In the case of the estimations of model parameters k_S and $Y_{S/X}$, practically there is one estimate, respectively, 0.01 g/l and 2.02 g/l. Such a representation is also supported by the small values observed for the standard deviation of the parameter estimates obtained by the investigated eight ABC algorithms. As it can be seen from Table 3, the standard deviation of k_S and $Y_{S/X}$ parameters estimates is very small, about 0.002, while the standard deviation of μ_{max} parameters estimates is larger, about 0.01, compared to the other two model parameters.

Based on this fact, aiming real application, the fastest ABC algorithm (ABC8) will be more appropriate and the obtained model parameters estimates will be accurate enough. Considering some scientific research and analysis, the more accurate ABC algorithm (ABC1) will be the better choice.

In order to show the distribution of the model parameters estimates obtained by the eight considered ABC algorithms (at a glance), an exploratory graphic as box plot can be used. The box plot is a standard technique for presenting a summary of the distribution of a dataset. The graphics are presented in Figs. 2, 3 and 4.

The results in the box plot related to model parameter μ_{max} (Fig. 2) show that the algorithms with similar performance could be classified in two groups—algorithms ABC1 and ABC5, and algorithms ABC2, ABC3, ABC4, ABC6, ABC7 and ABC8. The almost identical groups could be defined based on the results in the box plot related to model parameter k_S (Fig. 3). In the last box plot (Fig. 4), the algorithm ABC6 can be added to the group of algorithms ABC1 and ABC5. In the case of the box plot related to model parameter $Y_{S/X}$, some outliers occur. Such outliers may appear in a sample from a normally distributed population, but such a result would be of interest for future research.

The results from the considered model parameter identification problem presented here are compared to the already published results from the application of different population-based biologically inspired algorithms solving the same problem. The performance of the ABC algorithms investigated here are compared to the best known performance of ACO and CS algorithms [58], GA [65] and FA

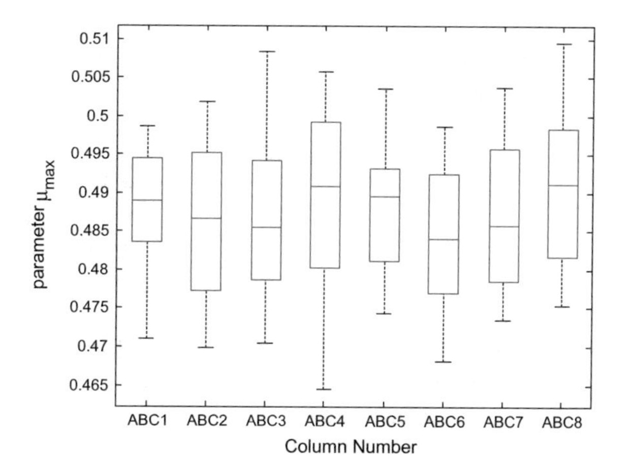

Fig. 2 Box plot for model parameter μ_{max}

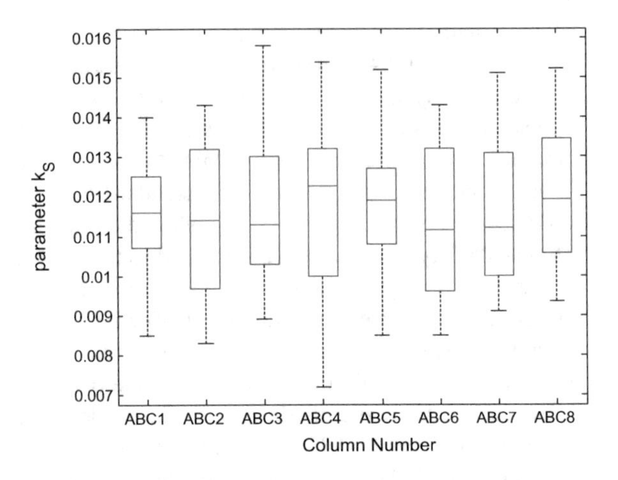

Fig. 3 Box plot for model parameter k_S

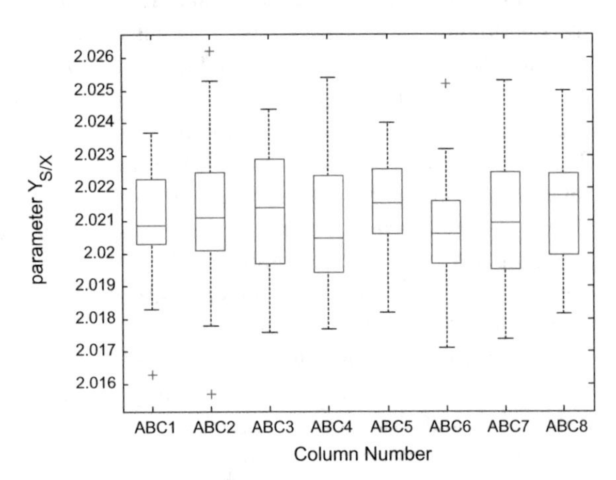

Fig. 4 Box plot for model parameter $Y_{S/X}$

Table 4 Comparison of the results obtained by ABC with those obtained by GA, FA, ACO and CS algorithms

Algorithms		*J* value		Research
	Average	Best	Worst	
ABC1	4.4149	4.3001	4.5076	The present research
ABC8	4.4921	4.3195	4.5981	
ABC4	4.5124	4.3785	4.6297	
FA	4.4581	4.3965	4.6269	[59]
GA	4.5341	4.4396	5.6920	[65]
CS	4.5662	4.4440	4.6641	[58]
ACO	5.2849	4.7402	6.2202	

[59]. The comparison of the two best performing ABC algorithms (ABC1 and ABC8) and the worst one (ABC4) is presented in Table 4. In Table 5, some of the parameters, namely number of population and number of generations of the regarded metaheuristics, are listed.

Table 5 Parameters and functions of GA, FA, ACO and CS algorithms

Algorithm	Parameters	
	Number of population	Number of generations
GA	180	200
FA	180	200
ACO	30	200
CS	30	200

As can be seen from Table 4 none of the considered metaheuristic algorithms obtain the solution with better precision than the precision obtained applying ABC1 or ABC8 algorithms. The FA solves the problem better than GA, SS and ACO algorithms. The average result of FA is better than ABC8 algorithm, although the FA best result is worst than ABC8 one. The GA and CS best results are practically the same. Considering the average results, the GA shows a bit better performance.

Having in mind that presented ACO and CS algorithms, unlike GA and FA, work with a small population, it would be fair to compare their results with the results of algorithms ABC3, ABC4 and ABC7, instead of algorithm ABC1. For example, the best result of CS $J = 4.4440$ to be compared to the results $J = 4.3758$, $J = 4.3785$ and $J = 4.3268$, respectively obtained by ABC3, ABC4 and ABC7 algorithms. Even with such a comparison the results show that ABC algorithms find the solution with higher quality. Moreover, considering the best obtained results none of the algorithms find better solution even than the solution of the worst performing ABC algorithm (ABC4). Compared to the ABC4 algorithm, FA obtains the same worst value of J ($J = 4.63$) and better average J value. A further investigation of the performance of FA with bigger number of population and/or number of generations would be of interest.

5 Conclusions

In this chapter, eight ABC metaheuristic algorithms, based on the foraging behaviour of honey bees, are investigated. The ABC algorithms are designed and applied to the parameter identification of an *E. coli* cultivation model. A system of nonlinear ordinary differential equations is used to model bacteria growth and substrate utilization. Model parameter identification was performed using a real experimental data set from an *E. coli* MC4110 fed-batch cultivation process.

Since the algorithm parameter settings may have a great influence on the computational results, an investigation of the ABC algorithm performance is performed. The eight ABC algorithms are examined using different values of the algorithm parameters, e.g. number of population, maximum cycle number and two different formulas for calculation of the probability value of the food source. The considered eight ABC algorithm are compared based on the obtained estimates of model parameters (maximum specific growth rate (μ_{max}), saturation constant (k_S)

and yield coefficient ($Y_{S/X}$), on the observed objective function value and the computation time. The achieved results show that with the appropriate choice of the algorithm settings, the accuracy of the decisions and the execution time can be optimized. Based on exhaustive analysis of the results, two ABC algorithms are determined as the best performing ones, regarding the optimization problem being solved. The most efficient algorithms are the ABC algorithms with number of population × maximum cycle number set at 60×500 (ABC1) and 20×400 (ABC8).

The proposed ABC algorithms are further compared to other nature-inspired population-based metaheuristics known in the literature. As competing algorithms, GA, FA, ACO and CS algorithms were chosen. Based on several conducted experimental studies, it is shown that the ABC algorithm outperforms all other competitor algorithms in all of the cases.

6 Future Research Directions

A possible future research direction is to consider various hybridizations of the ABC algorithm. It would be interesting to hybridize the basic ABC algorithm with another state-of-the-art algorithms. This is a growing research trend [77] with a wide range of potentialities for future research. In order to obtain the most of advantages of the nature inspired heuristic methods and to eliminate their disadvantages (premature convergence and computational time), hybridization could be performed. Particularly, hybridization of ABC may provide a platform for developing a meta-heuristic algorithm with better convergence speed and a better balance between exploration and exploitation capabilities [6, 35, 43, 47, 52, 69].

Acknowledgements Work presented here is dedicated to COMPSE 2016, First EAI International Conference on Computer Science and Engineering, November 11–12, 2016, Penang, Malaysia. Work is partially supported by the National Scientific Fund of Bulgaria under the Grant DN02/10 “New Instruments for Knowledge Discovery from Data, and their Modelling”.

References

1. AbdAllah, A., Essam, D., & Sarker, R. (2016). Genetic algorithms-based techniques for solving dynamic optimization problems with unknown active variables and boundaries. In *First EAI international conference on computer science and engineering*, November 11–12, 2016, Penang, Malaysia. http://eudl.eu/doi/10.4108/eai.27-2-2017.152266.
2. Abraham, A., Guo, H., & Liu, H. (2006). Swarm intelligence: foundations, perspectives and applications. In N. Nadjah, &. Mourelle (Eds.), *Swarm intelligent systems. studies in computational intelligence* 3–25 (2006).
3. Abualigah, L., Khader, A., & Al-Betar, M. (2016). A new hybridization strategy for krill herd algorithm and harmony search algorithm applied to improve the data clustering. In *First EAI*

International Conference on Computer Science and Engineering, November 11–12, 2016, Penang, Malaysia. http://eudl.eu/doi/10.4108/eai.27-2-2017.152255.
4. Akpinar, S., & Bayhan, G. M. (2011). A hybrid genetic aalgorithm for mixed model assembly line balancing problem with parallel workstations and zoning constraints. *Engineering Applications of Artificial Intelligence, 24*(3), 449–457.
5. Albayrak, G., & Özdemir, İ. (2017). A state of art review on metaheuristic methods in time-cost trade-off problems. *International Journal of Structural and Civil Engineering Research, 6*(1), 30–34. https://doi.org/10.18178/ijscer.6.1.30-34.
6. Ampellio, E., & Vassio, L. (2016). A hybrid ABC for expensive optimizations: CEC 2016 competition benchmark. In *2016 IEEE congress on evolutionary computation (CEC)*. https://doi.org/10.1109/CEC.2016.7743918.
7. Anh, T. N. L., Vo Ngoc, D., Ongsakul, W., Vasant, P., & Ganesan, T. (5015). Cuckoo optimization algorithm for optimal power flow. In: Handa, H., Ishibuchi, H., Ong, Y.S., Tan, K. (eds.), *Proceedings of the 18th Asia pacific symposium on intelligent and evolutionary systems, Vol. 1. Proceedings in Adaptation, Learning and Optimization* (Vol. 1, pp. 479–493). Springer.
8. Apostolopoulos, T., & Vlachos, A. (2011). Application of the firefly algorithm for solving the economic emissions load dispatch problem. *International journal of combinatorics*, Article ID 523806.
9. Atanassova, V., Fidanova, S., Popchev, I., & Chountas, P. (2012). Chapter 5. Generalized nets, ACO algorithms, and genetic algoriDimov, I. (eds.), *Monte carlo methods and applications, proceedings of the 8th IMACS seminar on monte carlo methods* (pp. 39–46). De Gruyter.
10. Benjamin, K. K., Ammanuel, A. N., David, A., & Benjamin, Y. K. (2008). Genetic algorithm using for a batch fermentation process identification. *Journal of Applied Sciences, 8*(12), 2272–2278.
11. Bolaji, A. L., Khader, A. T., Al-Betar, M. A., & Awadallah, M. A. (2013). Artificial bee colony algorithm, its variants and applications: a survey. *Journal of Theoretical and Applied Information Technology, 47*(2), 434–459.
12. Brenna, M., Foiadelli, F., & Longo, M. (2016). Application of genetic algorithms for driverless subway train energy optimization. *International Journal of Vehicular Technology, 2016*, Article ID 8073523. https://doi.org/10.1155/2016/8073523.
13. Brownlee, J. (2011). *Clever algorithms*. LuLu: Nature-Inspired Programming Recipes.
14. Chai-ead, N., Aungkulanon, P., & Luangpaiboon, P. (2011). Bees and firefly algorithms for noisy non-linear optimisation problems. *International multiconference of engineers and computer scientists* (Vol. 2, pp. 1449–1454).
15. Cronan, J. E. (2014). Escherichia coli as an experimental organism. In: eLS. Chichester: Wiley. https://doi.org/10.1002/9780470015902.a0002026.pub2.
16. da Silva, M. F. J., Perez, J. M. S., Pulido, J. A. G., & Rodriguez, M. A. V. (2010). AlineaGA—a genetic algorithm with local search optimization for multiple sequence alignment. *Applied Intelligence, 32,* 164–172.
17. Dorigo, M., & Stutzle, T. (2004). *Ant colony optimization*. MIT Press.
18. Ederer, M., Steinsiek, S., Stagge, S., Rolfe, M. D., Ter Beek, A., et al. (2014). A mathematical model of metabolism and regulation provides a systems-level view of how *Escherichia coli* responds to oxygen. *Frontiers in Microbiology*. https://doi.org/10.3389/fmicb.2014.00124.
19. Edgington, M., & Tindall, M. (2015). Understanding the link between single cell and population scale responses of *Escherichia coli* in differing ligand gradients. *Computational and Structural Biotechnology Journal, 13,* 528–538.
20. Fidanova, S, Marinov, P., & Alba, E. (2010a). ACO for optimal sensor layout. In J. Filipe, J. Kacprzyk (eds.), *Proceedings of international conference on evolutionary computing,* Valencia, Spain (pp. 5–9). SciTePress-Science and Technology Publications Portugal.
21. Fidanova, S., Alba, E., & Molina, G. (2010). Hybrid ACO algorithm for the GPS surveying problem. *Lecture Notes on Computer Science, 5910,* 318–325.

22. Fidanova, S., & Roeva, O. (2013). Metaheuristic techniques for optimization of an *E. coli* cultivation model. *Biotechnology and Biotechnological Equipment, 27*(3), 3870–3876.
23. Fidanova, S. (2010). An improvement of the grid-based hydrophobic-hydrophilic model. *International Journal Bioautomation, 14*(2), 147–156.
24. Fister, I., Fister, I., Jr., Yang, X.-S., & Brest, J. (2013). A comprehensive review of firefly algorithms. *Swarm and Evolutionary Computation, 13,* 34–46.
25. Ghaheri, A., Shoar, M., Naderan, M., & Hoseini, S. S. (2015). The applications of genetic algorithms in medicine. *Oman Medical Journal, 30*(6), 406–416.
26. Ghanem, W. (2016a). Hybridizing bat algorithm with modified pitch-adjustment operator for numerical optimization problems. In *First EAI international conference on computer science and engineering*, November 11–12, 2016, Penang, Malaysia. http://eudl.eu/doi/10.4108/eai.27-2-2017.152269.
27. Ghanem, W. (2016b). Hybridizing artificial bee colony with monarch butterfly optimization for numerical optimization problems. In *First EAI international conference on computer science and engineering*, November 11–12, 2016, Penang, Malaysia. http://eudl.eu/doi/10.4108/eai.27-2-2017.152257.
28. Goldberg, D. E. (2006). *Genetic algorithms in search, optimization and machine learning*. London: Addison Wesley Longman.
29. González, C. I., Castro, J. R., Melin, P., & Castillo, O. (2015). *Cuckoo search algorithm for the optimization of type-2 fuzzy image edge detection systems* (pp. 449–455). CEC.
30. Gu, W., Yu, Y., & Hu, W. (2017). Artificial bee colony algorithm-based parameter estimation of fractional-order chaotic system with time delay. *IEEE/CAA Journal of Automatica Sinica, 4* (1), 107–113.
31. Guerrero, M., Castillo, O., & Valdez, M.G. (2015a). Cuckoo search via lévy flights and a comparison with genetic algorithms. *Fuzzy Logic Augmentation of Nature-inspired Optimization Metaheuristics*, 91–103.
32. Guerrero, M., Castillo, O., & Valdez, M. G. (2015b). Study of parameter variations in the cuckoo search algorithm and the influence in its behavior. *Design of intelligent systems based on fuzzy logic, neural networks and nature-inspired optimization*, 199–210.
33. Holland, J. H. (1992). *Adaptation in Natural and Artificial Systems* (2nd ed.). Cambridge: MIT Press.
34. Ismail, M. M., Hezam, I. M., & El-Sharkawy, E. (2017). Enhanced cuckoo search algorithm with SPV rule for quadratic assignment problem. *International Journal of Computer Applications, 158*(4), 39–42.
35. Jadon, S. S., Tiwari, R., Sharma, H., & Bansal, J. C. (2017). Hybrid artificial bee colony algorithm with differential evolution. *Applied Soft Computing, 58,* 11–24.
36. Karaboga, D., & Akay, B. (2009). A comparative study of artificial bee colony algorithm. *Applied Mathematics and Computation, 214,* 108–132.
37. Karaboga, D., Gorkemli, B., Ozturk, C., & Karaboga, N. (2014). A comprehensive survey: Artificial bee colony (ABC) algorithm and applications. *Artificial Intelligence Review, 42*(1), 21–57.
38. Karaboga, D., & Ozturk, C. (2011). A novel clustering approach: Artificial bee colony (ABC) algorithm. *Applied Soft Computing, 11*(1), 652–657.
39. Karaboga, D. (2005). An idea based on honeybee swarm for numerical optimization. Technical Report TR06, Erciyes University, Engineering Faculty, Computer Engineering Department.
40. Kumar, S. M., Giriraj, R., Jain, N., Anantharaman, V., et al. (2008). Genetic algorithm based PID controller tuning for a model bioreactor. *Indian Chemical Engineer, 50*(3), 214–226.
41. Kwiecień, J., Filipowicz, B. (2017). Optimization of complex systems reliability by firefly algorithm. *Maintenance and Reliability 19*(2), 296–301. http://dx.doi.org/10.17531/ein.2017.2.18.
42. Le Dinh, L., Vo Ngoc, D., & Vasant, P. (2013). Artificial bee colony algorithm for solving optimal power flow problem. Hindawi Publishing Corporation. *The ScientificWorld Journal, 2013*, Article ID 159040. http://dx.doi.org/10.1155/2013/159040.

43. Li, Y., Wang, Y., & Li, B. (2013). A hybrid artificial bee colony assisted differential evolution algorithm for optimal reactive power flow. *International Journal of Electrical Power & Energy Systems, 52,* 25–33.
44. Li, Y., Zhou, C., & Zheng, X. (2014). The application of artificial bee colony algorithm in protein structure prediction. In L. Pan, G. Păun, M. J., Pérez-Jiménez, & T. Song, (Eds.), *Bio-inspired Computing—theories and applications. Communications in computer and information science* (Vol. 472, pp. 255–258). Berlin: Springer.
45. Maddala, V., Katta, & R. R. (2017). Adaptive ABC algorithm based PTS Scheme for PAPR reduction in MIMO-OFDM. *International Journal of Intelligent Engineering and Systems, 10* (2) (2017). http://dx.doi.org/10.22266/ijies2017.0430.06.
46. Majumder, A., & Laha, D. (2016). A new cuckoo search algorithm for 2-machine robotic cell scheduling problem with sequence-dependent setup times. *Swarm and Evolutionary Computation, 28,* 131–143.
47. Mao, L., Mao, Y., Zhou, C., Li, C., Wei, X., & Yang, H. (2016). Particle swarm and bacterial for aging inspired hybrid artificial bee colony algorithm for numerical function optimization. *Mathematical Problems in Engineering, 2016.* https://doi.org/10.1155/2016/9791060.
48. Mlakar, U., Fister Jr., I., & Fister, I. (2016). Hybrid self-adaptive cuckoo search for global optimization. Swarm and evolutionary computation. https://doi.org/10.1016/j.swevo. 2016. 03.001.
49. Mohamad, A. B., Zain, A. M., & Bazin, N. E. N. (2014). Cuckoo search algorithm for optimization problems—a literature review and its applications. *Applied Artificial Intelligence. An International Journal, 28*(5), 419–448.
50. Mucherino, A., Fidanova, S., & Ganzha, M. (2016). Introducing the environment in ant colony optimization. *Recent Advances in Computational Optimization, Studies in Computational Intelligence, 655,* 147–158.
51. Nasiri, B., & Meybodi, M. R. (2012). Speciation-based firefly algorithm for optimization in dynamic environments. *International Journal of Artificial Intelligence, 8*(S12), 118–132.
52. Nguyen, T. T., Pan, J. S., Dao, T. K., Kuo, M. Y., & Horng, M. F. (2014). Hybrid bat algorithm with artificial bee colony. In J. S. Pan, V. Snasel, E. Corchado, A. Abraham, S. L. Wang (Eds.), *Intelligent data analysis and its applications, Volume II. Advances in intelligent systems and computing,* (Vol. 298). Cham: Springer.
53. Ong, P. (2016). Performances of adaptive cuckoo search algorithm in engineering optimization. In P. Vasant, G.-W. Weber, D. Vo Ngoc (eds.), *Handbook of research on modern optimization algorithms and applications in engineering and economics*. https://doi.org/10.4018/978-1-4666-9644-0.ch026.
54. Pan, Q. K., Fatih Tasgetiren, M., Suganthan, P. N., & Chua, T. J. (2011). A discrete artificial bee colony algorithm for the lot-streaming flow shop scheduling problem. *Information Sciences, 181*(12), 2455–2468.
55. Paplinski, J. P. (2010). The genetic algorithm with simplex crossover for identification of time delays. *Intelligent Information Systems*, 337–346.
56. Petersen, C. M., Rifai, H. S., Villarreal, G. C., & Stein, R. (2011). Modeling *Escherichia coli* and its sources in an urban bayou with hydrologic simulation program—FORTRAN. *Journal of Environmental Engineering, 137*(6), 487–503.
57. Pham, D. T., Ghanbarzadeh, A., Koc, E., Otri, S., Rahim, S., & Zaidi, M. (2005). The Bees algorithm, technical report, manufacturing engineering centre, Cardiff University, UK.
58. Roeva, O., & Atanassova, V. (2016). Cuckoo search algorithm for model parameter identification. *International Journal Bioautomation, 20*(4), 483–492.
59. Roeva, O., Fidanova, S. (2014). Parameter identification of an *E. coli* cultivation process model using hybrid metaheuristics. *The International Journal of Metaheuristics*, *3*(2), 133–148.
60. Roeva, O., Fidanova, S., & Paprzycki, M. (2015). Influence on the genetic and ant algorithms performance in case of cultivation process modeling. *Recent Advances in Computational Optimization, Studies in Computational Intelligence, 580,* 107–120.

61. Roeva, O., Pencheva, T., Hitzmann, B., & Tzonkov, S. T. (2004). A genetic algorithms based approach for identification of escherichia coli fed-batch fermentation. *International Journal Bioautomation*, *1*, 30–41.
62. Roeva, O., & Slavov, T. S. (2011). Fed-batch cultivation control based on genetic algorithm PID controller tuning. *Lecture Notes on Computer Science*, *6046*, 289–296.
63. Roeva, O., & Slavov, T. S. (2012). Firefly algorithm tuning of PID controller for glucose concentration control during *E. coli* fed-batch cultivation process. In *Proceedings of the federated conference on computer science and information systems*, WCO, Poland (pp. 455–462).
64. Roeva, O. (2013). Chapter 21. A comparison of simulated annealing and genetic algorithm approaches for cultivation model identification. *Monte Carlo Methods and Applications*, 193–201 (2013).
65. Roeva, O. (2014). Genetic algorithm and firefly algorithm hybrid schemes for cultivation processes modelling. In R. Kowalczyk, A. Fred, & F. Joaquim (Eds.), *Transactions on computational collective intelligence XVII* (Vol. 8790, pp. 196–211). series Lecture Notes in Computer Science. Springer.
66. Roeva, O. (2012). Optimization of *E. coli* cultivation model parameters using firefly algorithm. *International Journal Bioautomation, 16*(1), 23–32 (2012).
67. Roeva, O. (2008). Parameter estimation of a monod-type model based on genetic algorithms and sensitivity analysis. *Lecture Notes on Computer Science, 4818,* 601–608.
68. Sörensen, K., Sevaux, M., & Glover, F. (2017). A history of metaheuristics, In R. Martí, P Pardalos, & M. Resende (Eds.), *Handbook of heuristics*. Springer. https://arxiv.org/pdf/1704.00853.pdf.
69. Thammano, A., & Phu-ang, A. (2013). A hybrid artificial bee colony algorithm with local search for flexible job-shop scheduling problem. *Procedia Computer Science, 20,* 96–101.
70. Tilahun, S. L., & Ngnotchouye, J. M. T. (2017). Firefly algorithm for discrete optimization problems: A survey. *Journal of Civil Engineering, 21*(2), 535–545.
71. Toimil, D., & Gómes, A. (2017). Review of metaheuristics applied to heat exchanger network design. *International Transactions in Operational Research, 24*(1–2), 7–26.
72. Tsai, P.-W., Pan, J.-S., Liao, B.-Y., & Chu, S.-C. (2009). Enhanced artificial bee colony optimization. *International Journal of Innovative, 5*(12B), 5081–5092.
73. Tumuluru, J. S., & McCulloch, R. (2016). Application of hybrid genetic algorithm routine in optimizing food and bioengineering processes. *Foods, 5.* https://doi.org/10.3390/foods5040076.
74. Vasant, P. (2015). *Handbook of research on artificial intelligence techniques and algorithms*. Hershey, PA: IGI-Global.
75. Vasant, P. (2013a). Hybrid linear search, genetic algorithms, and simulated annealing for fuzzy non-linear industrial production planning problems. In *Meta-Heuristics optimization algorithms in engineering, business, economics, and finance* (pp. 87–109). Hershey: Idea Group.
76. Vasant, P. (2013). *Meta-heuristics optimization algorithms in engineering, business, economics, and finance*. Hershey, PA: IGI-Global.
77. Vasant, P. (2014). Hybrid optimization techniques for industrial production planning: a review. In *Handbook of research on novel soft computing intelligent algorithms: theory and practical applications* (pp. 41–48). IGI Global.
78. Vazquez, R. A., & Garro, B. A. (2016). Crop classification using artificial bee colony (ABC) algorithm. In Y. Tan, Y. Shi, L. Li (Eds.), *Advances in swarm intelligence, ICSI 2016, Lecture notes in computer science* (Vol. 9713, pp. 171–178). Springer.
79. Wahdan, H. G., Kassem, S. S., & Abdelsalam, H. M. E. (2017). Product modularization using cuckoo search algorithm. In B. Vitoriano, G. H. Parlier (Eds.), *ICORES 2016, CCIS 695* (pp. 20–34).
80. Yang, X. S., & Deb, S. (2009). Cuckoo search via lévy flights. In *Proceeding of world congress on nature & biologically inspired computing (NaBIC 2009)* (pp. 210–214). USA: IEEE Publications.

81. Yang, X. S., & Deb, S. (2010). Engineering optimization by cuckoo search. *International Journal of Mathematical Modellling & Numerical Optimisation, 1*(4), 330–343.
82. Yang, X. S., & Deb, S. (2013). Multiobjective cuckoo search for design optimization. *Computers & Operations Research, 40*(6), 1616–1624.
83. Yang, X. S. (2005). Engineering optimizations via nature-inspired virtual bee algorithms. *Artificial Intelligence and Knowledge Engineering Applications: A Bioinspired Approach, Lecture Notes in Computer Science, 3562,* 317–323.
84. Yang, X. S. (2009). Firefly algorithm for multimodal optimization. *Lecture Notes in Computing Sciences, 5792,* 169–178.
85. Yang, X. S. (2010a) Firefly algorithm, levy flights and global optimization. *Research and development in intelligent systems XXVI* (pp. 209–218). London, UK: Springer.
86. Yang, X. S. (2010). Firefly algorithm, stochastic test functions and design optimization. *International Journal of Bio-inspired Computation, 2*(2), 78–84.
87. Yang, X. S. (2008). *Nature-inspired Meta-heuristic algorithms*. Beckington, UK: Luniver Press.
88. Yang, X. S. (2014). *Nature-inspired optimization algorithms*. London: Elsevier.
89. Yousif, A., Abdullah, A. H., Nor, S. M., & Abdelaziz, A. A. (2011). Scheduling jobs on grid computing using firefly algorithm. *Journal of Theoretical and Applied Information Technology, 33*(2), 155–164.
90. Zhang, C., Ouyang, D., & Ning, J. (2010). An artificial bee colony approach for clustering. *Expert Systems with Applications, 37*(7), 4761–4767.

A Novel Weighting Scheme Applied to Improve the Text Document Clustering Techniques

Laith Mohammad Abualigah, Ahamad Tajudin Khader and Essam Said Hanandeh

Abstract Text clustering is an efficient analysis technique used in the domain of the text mining to arrange a huge of unorganized text documents into a subset of coherent clusters. Where, the similar documents in the same cluster. In this paper, we proposed a novel term weighting scheme, namely, length feature weight (LFW), to improve the text document clustering algorithms based on new factors. The proposed scheme assigns a favorable term weight according to the obtained information from the documents collection. It recognizes the terms which are particular to each cluster and enhances their weights based on the proposed factors at the level of the document. β-hill climbing technique is used to validate the proposed scheme in the text clustering. The proposed weight scheme is compared with the existing weight scheme (TF-IDF) to validate its results in that domain. Experiments are conducted on eight standard benchmark text datasets taken from the Laboratory of Computational Intelligence (LABIC). The results proved that the proposed weighting scheme LFW overcomes the existing weighting scheme and enhances the result of text document clustering technique in terms of the F-measure, precision, and recall.

Keywords Text document clustering · β-hill climbing technique
Length feature weight scheme

L.M. Abualigah (✉) · A.T. Khader
School of Computer Science, Universiti Sains Malaysia, George Town, Penang, Malaysia
e-mail: lmqa15_com072@student.usm.my; laythdyabat@ymail.com

A.T. Khader
e-mail: tajudin@cs.usm.my

E.S. Hanandeh
Department of Computer Information System, Zarqa University, Zarqa 13132 Jordan
e-mail: Hanandeh@zu.edu.jo

I. Zelinka et al. (eds.), *Innovative Computing, Optimization and Its Applications*, Studies in Computational Intelligence 741,
https://doi.org/10.1007/978-3-319-66984-7_18

1 Introduction

Nowadays, the main important thing the domain of the text analysis is how to reproduce an enormous amount of text information in an accessible form, which means how to display the documents as groups. Although, all internet web pages and most of the advanced applications contain an enormous amount of text information which is needed by users to be in tidy form [1–3]. Text clustering is one of the most efficient unsupervised learning technique, it is used to solve the problem of partition many documents into a subset of clusters with foreknowledge the numbers of groups (clusters). This method is much use in the area of text mining to perform comprehensive analysis for all the text information include: data clustering, detection and disease clustering, open source clustering software, text information retrieval, clustering the results of the search engine, time series clustering and wireless sensor clustering [4].

Several challenges facing the text analysis techniques in all domains of text mining area and especially in the domain of the text document clustering are the content. Which means that the text document contains many informative and uninformative features [2, 5]. These uninformative features mislead the clustering algorithms or techniques and reduced its performance. In this research, the problem of uninformative features is relatively solved by giving a favorable weight to each term or feature according to some factors that affect the weight values indeed [6]. These classifications of text features affect effectiveness and performance of the text document clustering procedures, where the uninformative features are unnecessary, unrelated, and noisy features. Hence, the clustering method needs a powerful decision technique to improve the clustering process through the portion smaller document together in the same clusters [7].

Text document clustering is an unsupervised learning technique, which does not give the information about the given class label of the data to the clustering algorithm or technique. It seeks to find the unknown information (class label) in the collection by itself. It means that only specific terms selected from a text document are used for identifying a text document within the documents collection [1, 4, 8, 9].

Term weighting schemes are utilized to identify the significance and importance of each feature or term at the level of the documents collection. On the other hand, it assigns weight score to them according to some factors. These factors assist in calculating the weight value or score such as term frequency, document frequency, the number of terms in the collection and so on. Document clustering utilizes the term weight schemes to calculate the similarity between each document with all clusters centroids. Numerous weighting schemes are in use in our current time, but none of them is unique to the text document clustering problem [2, 10, 11].

Recently, many researchers have suggested various text clustering methods in order to solve the difficulties which faced the text clustering process. Local search is one of the robust clustering techniques easily used to generate a subset of document clusters. By using this technique, the information view became easier and the user time became less as well [4, 11, 12].

Generally, text document clustering defined as an optimization problem in terms of maximizing or minimizing the performance of the clustering algorithm [13]. In terms of minimizing, find the minimize distance value between the text document with the clusters centroid. However, In terms of maximizing, find the maximize similarity value between any text document with clusters centroids. The text clustering has been successfully used in many domains include ontology-based text clustering, text mining, text feature selection, text classification, automated clustering of newspapers, image clustering, and text categorization [4, 14].

Vector Space Model (VSM) is a popular pattern (representation way) used in the area of the text mining, especially in the text document clustering and text feature selection to facilitate the analysis process [6, 7]. This pattern represents the component of each document as a row (vector) of terms frequency; each term frequency represented as one position (dimension space).Then, based on the terms frequency, VSM generate new vectors that contain the features (terms) weighting. These vectors of weighing are used during the text analysis process to find the characteristic of each term, document, cluster, collection and so on. Therefore, the performance of the β-hill climbing text clustering technique affected positively if the number of represented feature is small with more accurate weight consideration [1, 10, 15].

β-hill climbing is an optimization technique introduced in 2016 by [16]. It can produce a search path in the available search space until moving to the local optimal solution and it obtained a superior results in comparison with the other comparative algorithms. This technique has several extensions to overcome such problems such as Tabu Search and Simulated Annealing. One of the primary features of the β-hill climbing is it leads to escape stuck in local optima.

In this paper, we proposed a novel term weight scheme, namely, length feature weight (LFW) to improve the text document clustering by giving an appropriate terms weights for the documents. The proposed weighting scheme (LFW) utilized the β-hell climbing technique for the text document clustering. The primary idea in this method is applying LFW to find more accurate weighing for each term to improve the clustering problem. The text clustering is applied using β-hell climbing technique to find more related and coherent clusters. The proposed method seeks to make the performance of the text clustering higher in terms of clusters accuracy. Experiments results were conducted on eight standard benchmark text datasets taken with varying characteristics in order to test the proposed scheme. The results proved that the proposed LFW for the text clustering obtained better results in comparison with the existing weight scheme using β-hill climbing technique measured by F-measure, precision, recall, and accuracy. As well, LFW improved the text clustering algorithm by dealing with a large number of clusters.

The remainder of this paper prepared as follows: Sect. 2??? presents the related works in the domain of the text document clustering. Section 3??? reviews the proposed length feature weighing scheme, β-hill climbing technique and the text document clustering problem. Section 4??? shows the evaluation measures. Results and discussion are given in Sect. 5???. Finally, Sect. 6 provided the conclusion.

2 Related Works

This section shows the most related works in the domain of text document clustering, β-hill climbing, and weighing schemes.

Recently, the text document clustering is a useful technique for partitioning an extensive amount of text information into associated clusters [15, 17–20]. Therefore, one of the fundamental problems that affect any clustering technique is the appearance many uninformative and sparse features in the texts. Unsupervised feature selection (FS) is an essential technique for eliminating possible uninformative features to support the text clustering method. In this paper [7], the harmony search (HS) algorithm is proposed, namely, feature selection based on harmony search algorithm for text clustering (FSHSTC), to solve the text feature selection problem. Eventually, FSHSTC is done to enhance the text clustering by getting a new subset of informative text features. Experiments were carried out on four text benchmark datasets. The results prove that the proposed method is enhanced the effectiveness and performance of the text clustering algorithm (i.e., k-mean algorithm) in terms of F-measure and accuracy.

A new four term-weighting schemes are introduced for the text document clustering problem [21]. The authors used the k-mean clustering algorithm to validate the proposed schemes. Experiments conducted on text document datasets, and the results showed that the proposed weighted pair group method with arithmetic mean improved the results in comparison with the other term-weighting schemes for the text clustering. The results are evaluated in terms of the entropy, purity, and F-measure.

The meta-heuristic optimization algorithms are actively employed to solve several complex optimization problems [22–26]. In this paper [6], genetic algorithm is proposed, namely, feature selection based on genetic algorithm for text clustering (FSGATC), to solve the text feature selection problem. Finally, FSGATC is proposed to enhance the text clustering by creating a new subset of more informative text features. Experiments were carried out on four text benchmark datasets and compared with another well-known algorithm in the same domain. The results proved that the proposed method (FSGATC) got better results according to the effectiveness and performance of the text clustering algorithm (i.e., k-mean algorithm) in comparison with k-mean and HS algorithm according to the F-measure and accuracy values.

A new weighing scheme for text clustering, namely, Cluster-Based Term weighting scheme (CBT), is proposed to enhance the performance of the clustering algorithm [11]. The proposed scheme is based on two common factors, which are term frequency and inverse document frequency (TF-IDF). The authors work to find a new scheme to assign the weighting values according to the information obtained from the produced clusters. On the other hand, they consider the proposed scheme as a specific scheme to the clustering technique in order to enhance the term weighting values based on their interest. The experimental results were done using the k-mean clustering algorithm and it is compared with other existing three weighing schemes. The results showed that the proposed scheme outweighs the existing weighting schemes and develops the product of the clustering algorithm.

Firefly algorithm (FA) is utilized to improve the solution diversity by introducing a new weighting scheme [27]. The authors applied FA for the text document clustering using the Reuters-21578 text dataset. The results showed that the proposed weighting scheme based on FA is competitive in the area of text. The results are evaluated by two standard measures in the domain of the text mining, namely, purity and F-measure.

Due to the tremendous growth of web pages, and modern applications, the text clustering has developed as a vital task to deal with many text documents. Unusual, web pages are simply browsed and tidily shown via applying the clustering method in order to distribution the documents into a subset of similar clusters. In this paper [4], the authors proposed two novel text document clustering methods based on krill herd algorithm to develop and enhance the text documents clustering. In the first method, the basic krill herd algorithm utilizes all genetic operators. While in the second method, the basic krill herd algorithm utilizes without all genetic operators. Experiments conducted on four standard text datasets. The results revealed that the proposed krill herd algorithm overcomes the k-mean text clustering algorithm in term of the clusters accuracy (i.e., purity and entropy).

One of the popular unsupervised text mining tools is text documents clustering. In text clustering algorithm, the correct decision for any document distribution is made using an objective function (i.e., similarity measurements or distance measurements). Text clustering algorithms work very poorly when the form of the objective function is not valid and complete. Hence, the authors proposed multi-objective function, namely, aggregation Euclidian distance and Cosine similarity measurements for adjusting the text clustering distribution [1, 28]. The multi-objective function has been by combined two evaluating functions which rise as an efficient alternative in numerous clustering situations. In particular, the multi-objective function is not a popular in the domain of the text clustering. The authors said that the multi-objective function is a core problem that reduces the performance of the k-mean text clustering algorithm. Experiments conducted on seven standard benchmark text datasets, which is common in the domain of the text clustering. The results revealed that the proposed multi-objective function outperforms the other measure standalone in term of the achievement of the k-mean text clustering. The results evaluated by using two well-known clustering measures, namely, accuracy and F-measure.

A new version of hill climbing technique, namely, β-hill climbing, has been proposed to solve benchmark optimization problems [16]. The authors add a new stochastic operator in hill climbing, called, β operator, to establish a balance between the exploitation (i.e., intensification) and exploration (i.e., diversification) during the search. Experiments conducted on IEEE-CEC2005 global optimization functions. The results reveal that the proposed β-hill climbing obtained better results to the hill climbing providing reliable results when it compares with other comparative methods using the same IEEE-CEC2005 global optimization functions.

3 Text Clustering Method

A. *Text documents preprocess*

- In the first stage (Tokenization), it is the process of cutting continues letters of documents into token (i.e., words, phrases, symbols, and important elements). On the other hand, the process of removing an empty sequence by taking each token from the first letter to the last letter. This process saved the memory [6, 29, 30].
- In the second stage (Removal the stop words), it is the process of removing all the common words in the documents (i.e., “a”, “against”, “about”, “am”, “all”, “above”, “after”, “and”, “again”, “any”, “an” and so on). These stop words are available at http://www.unine.ch/Info/clef/. Available list of the stop-words contains 571 stop-words [6, 29].
- In the third stages (Stemming), the process of dismantling some of the related words in terms of structure and meaning to be in the same form. Which means that every some of the related words will represent by one root (i.e., feature). Usually, this process is done by using the Porter Stemmer. It is available at http://tartarus.org/martin/PorterStemmer/. Porter Stemmer gets rid of some of the parties such as eliminating the prefixes and suffixes of each term (i.e., “ed”, “ly”, “ing”, and so on). For example, “connection”, “connective”, “connections”, “connecting”, and “connected” all these words or terms have the common root “connect”. This root after these some preprocessing will call feature [6, 29, 31].

B. *Calculating the term weighting*

(a) Term frequency–invers document frequency (TF-IDF)

After finish the preprocess steps, we moved to calculate the term weighting for each feature. In the area of text mining particularly in the domain of the text clustering, the term weighting scheme, namely, term frequency-inverse document frequency (TF-IDF), is used to calculate the weighting score for each term (feature) by Eq. (1) [1, 4, 32, 33]. In this paper, this scheme is used in order to make a comparison with the proposed scheme to distinguish between the document terms (classification).

$$d_i = \left(w_{i,1}, w_{i,2}, \ldots, w_{i,j}, \ldots, w_{i,t}\right), \quad (1)$$

where, d_i is the vector of the document i, each document in the dataset is represented as a row (vector) of terms weighting, $w_{i,j}$ is the weight value of the terms j in document number i, and t is the number of all unique terms in the documents. Equation (2) is used to calculate the weighting value of the terms j in document number i.

$$w_{i,j} = ft(i.j) * \log\left(\frac{n}{df(j)}\right), \tag{2}$$

where, *tf(i,j)* is the terms frequency of the terms number *j* in document number *I*, *n* is the number of all documents, and *df(j)* is the document frequency of the feature number *j*. The following matrix shows the documents in format of the vector space model [7, 34].

$$VSM = \begin{bmatrix} w_{1,1} & \cdots & \cdots \cdots & w_{1,t-1} & w_{1,t} \\ \cdots & \cdots & \cdots \cdots & \cdots & \cdots \\ w_{n-1,1} & \cdots & \cdots \cdots & \cdots & w_{n-1,t} \\ w_{n,1} & \cdots & \cdots \cdots & w_{n,t-1} & w_{n,t} \end{bmatrix} \tag{3}$$

(b) The proposed length feature weigh scheme (LFW)

In text mining domains, the term weighting schemes are used to assign an appropriate weight score for the document's terms (features) in order to improve terms classification (discrimination).

A weighting scheme, namely, length feature weight (LFW), is proposed in order to obtain a better term weighting (feature score) to facilitate and improve the feature selection process by distinguishing between informative and uninformative text features more efficiently. As known in the literature of the term weighting schemes, TF-IDF is the common weight scheme. In this research, we focus on improving the current TF-IDF's weakness that affects the assessment of the term weighing of each document.

Three main factors have been developed to improve the effectiveness of the weighting scheme for the unsupervised text feature selection technique as follows.

Firstly, the document frequency (*df*) value of each term is not considered as a major factor to magnified the term weighing in TF-IDF. Thus, we added a variable, namely, document frequency, to take into account that how many time exactly the term appears in all the documents. This variable affects the importance by increasing the weight value (score) if it appears in a lot of documents.

Secondly, the number of all terms in the document has not been inserted to the common weight schemes so far. One of the major objectives to add this variable in the LFW is to facilitate the unsupervised feature selection. In a situation, the document has many numbers of features will eliminate many uninformative features, in a situation, the document has small numbers of features will preserve it. Note, this factor was not used previously in any weighing scheme.

Third and final, the max term frequency (*maxtf(i)*) is an important factor plays an essential role to assign a better term weight score. This factor is added to the LFW in order to moderate the importance of terms vice versa. Generally, if the document contains a large number of terms that mean the importance of the terms is smaller than the small number of terms. The max term frequency increases the term weighting in case the max term frequency is small in comparison with the other documents. Note, this factor was not used previously in any weighing scheme. LFW is formulated as Eq. (4).

$$LFW_{i,j} = \left(\frac{tf(i,j) * df(j)}{\max(i)}\right) * \log\left(\frac{n}{df(j)}\right), \tag{4}$$

where *tf(i, j)* is the term frequency of the term *j*th in document *i*, *df(j)* is the number of documents which contain feature *i*, a_i is the number of the new selected features for the document *i*, *maxtf(i)* is maximum term frequency in the document *i*th, and n is the number of all documents in the given dataset.

C. *β-hill climbing technique for the text clustering*

In this section, β-hill climbing technique is used for the text clustering based on two weighing schemes [1, 4, 35].

(a) *Text clustering notations*

The text clustering problem is formulated as an optimization problem to create a subset of documents clusters.

Definition 1 Let *D* a set of documents, where d_i presents the document number *i* as a vector of terms weights $d_i = (w_{i,1}, w_{i,2}, \ldots, w_{i,j}, \ldots, w_{i,t})$ [36]. Where, $w_{i,j}$ denote the weight value of term *j* in the document *i*, *t* is the number of all features in dataset.

Definition 2 let *D* is a collection of text documents portion into *K* is the number of all clusters. Equation (3) shows documents representation in the dataset. *n* is the number of all documents, where d_i ε {1, *n*}, ($i = 1, 2, \ldots, n$). Clusters centroids are represented as a vector like a document $C = (c_1, c_2, \ldots, c_k, \ldots, c_K)$, where each cluster has one centroid such as c_k, which is the centroid of the cluster *k*.

Definition 3 Let *D* is a collection of documents contain $n = 100$ documents, where *n* is number of all documents. Hence, the search space of each solution will be equal *K*, where *d* ε {1…*K*}.

(b) *Solution representation*

In this paper, document clustering problem utilized the β-hill climbing technique, which begins with a random initial solution and seek to improve its solution by reaching a globally optimal solution. Each single document in the dataset reflects as a dimension in the search space. Figure 1 presents the solution of the text clustering problem [4].

Fig. 1 Text clustering solution representation

X represents the solution for solving the text clustering problem as the vector which provided in Fig. 1. In this case, the value of position number i is equal 5, which means that the ith document belongs to the cluster number 5 etc.

(c) *Distance measure*

Euclidean is a standard distance measure used in the domain of the text clustering to compute the dissimilarity (distance) score between each document with clusters centroids. This paper uses the Euclidean distance measure as the objective functions by Eq. (5). Normally, distance values are between (0, 1), although it is unlike the cosine similarity measure. Where, if the distance value close to 0, that means it is the best value [8, 28, 36].

$$Dis_{(d4,c2)} = \left(\sum_{j=1}^{t} \left| w_{d,j} - w_{c,j} \right|^2 \right)^{1/2}, \tag{5}$$

Equation (5) presents the distance between the document number 4 and the cluster centroid number 2. Where, $w_{d,j}$ is the weight of term j in document number 4, and $w_{c,j}$ is the weight of term j in cluster centroid number 2.

(d) *Fitness function*

The fitness function (F) is a class of the evaluation measure employed to evaluate the solution. Iteratively, the fitness function of each solution calculated. Finally, the solution, which has a greater fitness value is the optimal solution [4, 37, 38]. The proposed method used by the average distance of documents to the cluster centroid (ADDC) as Eq. (6).

$$ADDC = \left[\frac{\sum_{j=1}^{K} \frac{\left(\sum_{i=1}^{n} Dis(d_j, c_i) \right)}{r_j}}{K} \right] \tag{6}$$

where K is the number of clusters, r_j is the number of documents that belong to the cluster number j, and $Dis(d_i, c_i)$ is the distance measure between the document number i and the cluster centroid number j. The cluster centroid computed by Eq. (7).

$$c_j = \frac{1}{n_i} \sum_{d_i \in c_j} d_i \tag{7}$$

where d_i shows that the document i belongs to the ith centroid, and n_i represents the number of the documents that belong to cluster i.

(e) *β-hill climbing technique*

The β-hill climbing (see Algorithm I) begins with a random solution as $X = (x_1, x_2, \ldots, x_i, \ldots, x_n)$. It iteratively produces a new solution $X' = (x'_1, x'_2, \ldots, x'_i, \ldots, x'_n)$

using two operators: (i) β operator and (ii) neighborhood navigation. In the neighborhood navigation stage, the function improve the solution by using the acceptance rule where iteratively a random neighboring solution of the solution X is adopted as Eq. (8) [16].

$$x_i = (1 + rand)\ mod\ K, \tag{8}$$

Algorithm I

Pseudo-code of the β-hill climbing technique for the text document clustering problem

1: **Input**: A collection of documents D.
2: **Output**: Generate a new subset of clusters K.
3: **Termination criteria**
4: X^I = improve (X) by neighborhood navigation.
5: **for** $i = 1, \ldots, n$ **do** t> Note, n is the number of documents
6: **if** $rand \leq \beta$ **then**
7: x^I = 1+ rand mod K,
8: **end if**
9: ***end for***
10: ***if*** $F(X^I) \leq F(X)$ ***then*** *t> Note, F is the ADDC*
11: $X = X^I$
12: ***end if***
13: *Assign the final solution as a subset of documents clutters*
14: ***End***

In β operator stage, the positions of the new solution are selected values by one way of these two way: (i) according to the current values of the current solution (ii) randomly from possible search space (i.e., binary). This way based on a probability of β where $\beta \in [0, 1]$ as Eq. (9) [16]. β value is fixed (0.1).

$$x_i' \leftarrow \begin{cases} x_p & \text{if } rand \leq \beta \\ x_i & \text{otherwise} \end{cases}, \tag{9}$$

where $xp \in X$ is the possible region for the decision variable xi and *rand* presents a random number either one or zero.

4 Evaluation Measures

The comparative evaluations are done using the three common evaluation measures: accuracy (A) precision (P), recall (R) and F-measure (F). These measures are standard criteria utilized in that domain to evaluate the clusters accuracy [1, 6, 7, 39].

F-measure (F) is a standard evaluation criteria used to calculate the percentage of the truly distributed document in each cluster by using Eq. (10) [1, 6, 10].

$$F(j) = \frac{2 * P(i,j) * R(i,j)}{P(i,j) + R(i,j)} \tag{10}$$

In Eq. (10), two measures employed to find the F-measure value: precision (P) and recall (R), which are calculated by Eqs. (11) and (12), respectively.

$$P(i,j) = \frac{n_{i,j}}{n_j} \tag{11}$$

$$R(i,j) = \frac{n_{i,j}}{n_i}, \tag{12}$$

where *P(i, j)* is the precision of class *i* in cluster number *j*, *R(i, j)* is the recall of class *i* in cluster number *j*. Where, $n_{i,j}$ is the number of correct members of the class *i* in cluster *j*, n_i is the total number of members of the cluster *j*, and n_j is the number of original members of the class *j* as the class labels given in the dataset. F-measure value for all clusters calculated by find the average F-measure value of all clusters by Eq. (13).

$$F = \frac{\sum_{j=1}^{K} F(j)}{K} \tag{13}$$

The accuracy (A) measurement is one of the main external measurements used to compute the percentage of correct assigned documents to each cluster by using Eq. (14) [6, 39, 40].

$$A = \sum_{i=1}^{K} \frac{1}{n} P(i,i) \tag{14}$$

where, P(i, i) is the precision value for class *i* in cluster *i*, *K* is the number of all clusters, and *n* is the number of all documents.

5 Experimental Results

We applied the proposed text clustering method on variant weighing schemes using MATLAB (R), version 8.3.0.532, and 64-bit. Text clustering techniques run 20 times, each run 1000 iterations. Table 1 displays eight text datasets taken from Laboratory of Computational Intelligence LABIC at http://sites.labic.icmc.usp.br/text_collections/. These datasets are used to test and compared the performance of the proposed LFW in the domain of the text clustering.

A. *Results and discussion*

This section examines the proposed weighting scheme (i.e., LFW) in comparison with the common existing weighting scheme (i.e., TF-IDF) for the text document clustering problem. Table 2 shows the performance of the β-hill climbing in solving the text document clustering problem using two weighting schemes.

Table 1 Characteristics of datasets

Dataset	# of documents	# of terms	# of clusters
DS1	299	1725	4
DS2	333	4339	4
DS3	204	5832	6
DS4	313	5804	8
DS5	414	6429	9
DS6	878	7454	10
DS7	913	3100	10
DS8	18828	45433	20

Table 2 Clusters quality

Datasets	Measure	TF-IDF	LFW
DS1	Accuracy	0.5339	0.5419
	Precision	0.4305	0.4451
	Recall	0.5096	0.5265
	F-measure	0.5152	0.5210
DS2	Accuracy	0.7436	0.7534
	Precision	0.7056	0.7269
	Recall	0.7295	0.7398
	F-measure	0.7171	0.7214
DS3	Accuracy	0.3997	0.3905
	Precision	0.3617	0.3456
	Recall	0.3587	0.3754
	F-measure	0.3583	0.3664
DS4	Accuracy	0.5582	0.5596
	Precision	0.5224	0.5301
	Recall	0.5311	0.5518
	F-measure	0.5285	0.5412
DS5	Accuracy	0.5679	0.5695
	Precision	0.4809	0.4740
	Recall	0.4670	0.4621
	F-measure	0.4731	0.4651
DS6	Accuracy	0.5772	0.5697
	Precision	0.4685	0.4755
	Recall	0.5043	0.4911
	F-measure	0.4810	0.4901
DS7	Accuracy	0.5202	0.5334
	Precision	0.4462	0.4564
	Recall	0.4498	0.4705
	F-measure	0.4480	0.4685
DS8	Accuracy	0.2117	0.2216
	Precision	0.2014	0.2248
	Recall	0.1954	0.2156
	F-measure	0.2019	0.2200

The proposed text clustering methods using LFW obtained the best results in comparison with TF-IDF almost in all datasets. According to the accuracy measure, LWF obtained the best results in six out of eight datasets (i.e., DS1, DS2, DS4, DS5, DS7, and DS8). According to the precision measure, LWF obtained the best results in six out of eight datasets (i.e., DS1, DS2, DS4, DS6, DS7, and DS8). According to the recall measure, LWF obtained the best results in six out of eight datasets (i.e., DS1, DS2, DS43 DS4, DS7, and DS8). Finally, according to the F-measure evaluation criteria, which is the common evaluation measure used in the domain of the text clustering to evaluation the clusters accuracy. LWF obtained the best results in seven out of eight datasets (i.e., DS1, DS2, DS3, DS4, DS6, DS7, and DS8).

These results are a big proof to prove that the proposed LFW overcome the TF-IDF in the domain of the text document clustering (see Table 2). We conclude based on the obtained results that the factors which have been added to the LFW scheme were effective in terms of improving the performance of the clustering technique through the dealing with the document features in more consideration for the features at the level of each document.

Table 3 shows the statistical analysis (Friedman test ranking) based on the four evaluation criteria (i.e., accuracy, precision, recall, and F-measure). These measures are considered as a standard and active measures in the area of text mining and particularly in the domain of the text clustering. The proposed LFW obtained the higher ranking (i.e., 1) followed by TF-IDF obtained the second ranking (i.e., 2). Particularly, LFW obtained the best results according the all evaluation measures in five of eight datasets (i.e., DS1, DS2, DS4, DS67 and DS8). Thus, we conclude that the LFW is effective weight scheme to solve the text clustering problem.

Table 3 Statistical analysis

Dataset	Schemes	
	TF-IDF	LFW
DS1	0	4
DS2	0	4
DS3	2	2
DS4	0	4
DS5	3	1
DS6	2	2
DS7	0	4
DS8	0	4
Summation	07	25
Mean rank	0.875	3.125
Final ranking	2	**1**

6 Conclusion

In this paper, a novel weighting scheme, namely, length feature weight (LFW), is used for solving the text document clustering problem. This scheme improved the text documents clustering by giving a favorable weighting values for the most informative features at the level of each document. The β-hill climbing technique got better results using LFW according to all evaluation measures that utilized in the experiments. The performance of the text clustering is improved by adding the new three factors. For future work, can investigate the effectiveness of the proposed weight scheme by applying to solve other problem in the area of the text analysis u another techniques.

Acknowledgements The authors would like to thank the editors, reviewers for their helpful comments and EAI COMPSE 2016.

References

1. Abualigah, L. M., Khader, A. T., & Al-Betar, M. A. (2016, July). Multi-objectives-based text clustering technique using K-mean algorithm. In *7th International Conference on Computer Science and Information Technology (CSIT)* (pp. 1–6). IEEE.
2. Makki, S., Yaakob, R., Mustapha, N., & Ibrahim, H. (2015). Advances in document clustering with evolutionary-based algorithms. *American Journal of Applied Sciences, 12*(10), 689.
3. Tang, B., Shepherd, M., Milios, E., & Heywood, M. I. (2005, April). Comparing and combining dimension reduction techniques for efficient text clustering. In *International Workshop on Feature Selection for Data Mining*, 39 (pp. 81–88).
4. Abualigah, L. M., Khader, A. T., Al-Betar, M. A., & Awadallah, M. A. (2016, May). A krill herd algorithm for efficient text documents clustering. In *IEEE Symposium on Computer Applications & Industrial Electronics (ISCAIE)* (pp. 67–72). IEEE.
5. Bharti, K. K., & Singh, P. K. (2015). Hybrid dimension reduction by integrating feature selection with feature extraction method for text clustering. *Expert Systems with Applications, 42*(6), 3105–3114.
6. Abualigah, L. M., Khader, A. T., & Al-Betar, M. A. (2016, July). Unsupervised feature selection technique based on genetic algorithm for improving the Text Clustering. In*7th International Conference on Computer Science and Information Technology (CSIT)* (pp. 1–6). IEEE.
7. Abualigah, L. M., Khader, A. T., & Al-Betar, M. A. (2016, July). Unsupervised feature selection technique based on harmony search algorithm for improving the Text Clustering. In *7th International Conference on Computer Science and Information Technology (CSIT) 2016* (pp. 1–6). IEEE.
8. Aggarwal, C. C., & Zhai, C. (2012). A survey of text clustering algorithms. In *Mining text data* (pp. 77–128). US: Springer.
9. Mahdavi, M., Chehreghani, M. H., Abolhassani, H., & Forsati, R. (2008). Novel meta-heuristic algorithms for clustering web documents. *Applied Mathematics and Computation, 201*(1), 441–451.
10. Abualigah, L. M. Q., & Hanandeh, E. S. (2015). Applying genetic algorithms to information retrieval using vector space model. *International Journal of Computer Science, Engineering and Applications, 5*(1), 19.

11. Murugesan, A. K., & Zhang, B. J. (2011). A new term weighting scheme for document clustering. In *7th International Conference Data Min. (DMIN 2011-WORLDCOMP 2011, Las Vegas, Nevada, USA*.
12. Cui, X., Potok, T. E., & Palathingal, P. (2005, June). Document clustering using particle swarm optimization. In *Swarm Intelligence Symposium, 2005. SIS 2005. Proceedings 2005 IEEE* (pp. 185–191). IEEE.
13. Jensi, R., & Jiji, D. G. W. (2014). A survey on optimization approaches to text document clustering. arXiv:1401.2229.
14. Bolaji, A. L. A., Al-Betar, M. A., Awadallah, M. A., Khader, A. T., & Abualigah, L. M. (2016). A comprehensive review: Krill Herd algorithm (KH) and its applications. *Applied Soft Computing, 49,* 437–446.
15. Hanandeh, E., & Maabreh, K. (2015). Effective information retrieval method based on matching adaptive genetic algorithm. *Journal of Theoretical and Applied Information Technology, 81*(3), 446.
16. Abualigah, L. M., Khader, A. T., Al-Betar, M. A., Alyasseri Z. A., Alomari, O. A., & Hanandeh, E. S. (2017). Feature Selection with β-hill climbing Search for Text Clustering Application. In *Second Palestinian International Conference on Information and Communication Technology*. IEEE.
17. Yeh, W. C., Lai, C. M., & Chang, K. H. (2016). A novel hybrid clustering approach based on K-harmonic means using robust design. *Neurocomputing, 173,* 1720–1732.
18. Chandran, T. R., Reddy, A. V., & Janet, B. (2017). Text Clustering Quality Improvement using a hybrid Social spider optimization. *International Journal of Applied Engineering Research, 12*(6), 995–1008.
19. Tunali, V., Bilgin, T., & Camurcu, A. (2016). An improved clustering algorithm for text mining: multi-cluster spherical k-means. *International Arab Journal of Information Technology, 13*(1), 12–19.
20. Kohli, S., & Mehrotra, S. (2016). A clustering approach for optimization of search result. *Journal of Images and Graphics, 4*(1), 63–66.
21. Prakash, B. R., Hanumanthappa, M., & Mamatha, M. (2014). Cluster based term weighting model for web document clustering. In *Proceedings of the Third International Conference on Soft Computing for Problem Solving* (pp. 815–822). India: Springer.
22. Vahdani, B., Behzadi, S. S., Mousavi, S. M., & Shahriari, M. R. (2016). A dynamic virtual air hub location problem with balancing requirements via robust optimization: Mathematical modeling and solution methods. *Journal of Intelligent & Fuzzy Systems, 31*(3), 1521–1534.
23. Vasant, P. (2015). *Handbook of Research on Artificial Intelligence Techniques and Algorithms, 2 Volumes*. Information Science Reference-Imprint of IGI Publishing.
24. Vasant, P. (Ed.). (2013). *Handbook of research on novel soft computing intelligent algorithms: Theory and practical applications*. IGI Global.
25. Vasant, P. (Ed.). (2011). *Innovation in power, control, and optimization: Emerging energy technologies: Emerging energy technologies*. IGI Global.
26. Vasant, P. (Ed.). (2016). *Handbook of research on modern optimization algorithms and applications in engineering and economics*. IGI Global.
27. Mohammed, A. J., Yusof, Y., & Husni, H. (2014). Weight-based Firefly algorithm for document clustering. In *Proceedings of the First International Conference on Advanced Data and Information Engineering (DaEng-2013)* (pp. 259–266). Singapore: Springer.
28. Punitha, S. C., & Punithavalli, M. (2012). Performance evaluation of semantic based and ontology based text document clustering techniques. *Procedia Engineering, 30,* 100–106.
29. Liu, W., & Wong, W. (2009). Web service clustering using text mining techniques. *International Journal of Agent-Oriented Software Engineering, 3*(1), 6–26.
30. Abualigah, L. M., Khader, A. T., Al-Betar, M. A., & Hanandeh, E. S. A new hybridization strategy for krill herd algorithm and harmony search algorithm applied to improve the data clustering. *management*, *9*, 11.

31. Abualigah, L. M., Khader, A. T., Al-Betar, M. A., & Alomari, O. A. (2017). Text feature selection with a robust weight scheme and dynamic dimension reduction to text document clustering. *Expert Systems with Applications*.
32. Rangrej, A., Kulkarni, S., & Tendulkar, A. V. (2011, March). Comparative study of clustering techniques for short text documents. In *Proceedings of the 20th International Conference Companion on World wide web* (pp. 111–112). ACM.
33. Abualigah, L. M., & Khader, A. T. (2017). Unsupervised text feature selection technique based on hybrid particle swarm optimization algorithm with genetic operators for the text clustering. *The Journal of Supercomputing*, 1–23.
34. Abualigah, L. M., Khader, A. T., AlBetar, M. A., & Hanandeh, E. S. (2017). Unsupervised text feature selection technique based on particle swarm optimization algorithm for improving the text clustering.
35. Sharma, S., & Gupta, V. (2012). Recent developments in text clustering techniques. *Recent Developments in Text Clustering Techniques*, *37*(6).
36. Huang, A. (2008, April). Similarity measures for text document clustering. In *Proceedings of the Sixth New Zealand Computer Science Research Student Conference (NZCSRSC2008)* (pp. 49–56), *Christchurch, New Zealand.*
37. Zaw, M. M., & Mon, E. E. (2013). Web document clustering using cuckoo search clustering algorithm based on levy flight. *International Journal of Innovation and Applied Studies, 4*(1), 182–188.
38. Forsati, R., Mahdavi, M., Shamsfard, M., & Meybodi, M. R. (2013). Efficient stochastic algorithms for document clustering. *Information Sciences, 220,* 269–291.
39. Karol, S., & Mangat, V. (2013). Evaluation of text document clustering approach based on particle swarm optimization. *Open Computer Science, 3*(2), 69–90.
40. Boyack, K. W., Small, H., & Klavans, R. (2013). Improving the accuracy of co-citation clustering using full text. *Journal of the American Society for Information Science and Technology, 64*(9), 1759–1767.

A Methodological Framework to Emulate the Visual of Malaysian Shadow Play With Computer-Generated Imagery

Kheng-Kia Khor

Abstract Wayang kulit Kelantan, a preeminent shadow play in Malaysia is currently threatened with imminent extinction. At the moment of this writing, the existing efforts made to preserve this traditional cultural heritage are apparently insufficient. Therefore it is clear that with the current situation in Malaysia and the level of official support, wayang kulit Kelantan is unlikely to last long. To overcome this problem, on one hand, some researchers suggested to digitise wayang kulit Kelantan as a mean of preservation. On the other hand, there are also researchers who attested that there is a dire need for Malaysians to promote this art form by using modern technology. This paper presents a methodology framework of using Computer Generated Imagery (CGI) to emulate the visual of wayang kulit Kelantan. Experiments have been carried out to vindicate the veracity of the framework, and the results show that realistic and plausible visual of wayang kulit Kelantan can be emulated by using the proposed framework.

1 Background

In one of its attempts to preserve and safeguard the unique traditional wayang kulit (shadow play), the United Nations Educational, Scientific and Cultural Organization (UNESCO) has designated Indonesian wayang kulit purwa as a Masterpiece of Oral and Intangible Heritage of Humanity on 7th November 2003. The term wayang kulit combines two important words: "wayang", which can be translated as shadow, and "kulit", which means leather. Thus wayang kulit indicates a form of shadow theatre performed with leather puppets [1].

Today, the preservation of Indonesian wayang kulit purwa has been executed in a very promising way but Malaysian wayang kulit Kelantan is threatened with imminent extinction [2]. This dying art form was one time fairly widespread and extremely popular in the state of Kelantan in Malaysia. The statistic showed that in

K.-K. Khor (✉)
Universiti Tunku Abdul Rahman, Kampar, Malaysia
e-mail: khengkia@yahoo.com

I. Zelinka et al. (eds.), *Innovative Computing, Optimization and Its Applications*, Studies in Computational Intelligence 741,
https://doi.org/10.1007/978-3-319-66984-7_19

the 1960s, there were more than 300 puppeteers in Kelantan but now there are less than six [3].

Existing evidences have indicated the origin of wayang kulit Kelantan to a proto-Javanese wayang kulit. However, after substantial adaptation and localisation, wayang kulit Kelantan has successfully developed its idiosyncratic characteristics and styles [4]. Although both Indonesian wayang kulit purwa and Malaysian wayang kulit Kelantan use the same source of repertoire (Ramayana) with the same principal characters (with slight differences in their names); there are distinctively different in terms of their visuals [5].

Wayang kulit Kelantan shows a greater level of shadow distortion compared to wayang kulit purwa. It is due to the differences in the size of *kelir* (screen), distances and positions of light source. Moreover, the puppets of wayang kulit Kelantan cast colourful shadows on the screen whereas the puppets of wayang kulit purwa are opaque and only cast black shadows. Figure 1 depicts the visual of wayang kulit Kelantan during a performance.

With the advent of modern entertainment technology and online media such as television, cinema and Internet, wayang kulit Kelantan is facing the predicament of imminent extinction. In order to survive in this era of digitalisation and globalisation, practitioners and researchers asserted that it needs a new alternative media output, to be digitalised into cyber world and to be watched on the computer screen. Tan, Abdullah and Osman from Universiti Sains Malaysia (USM) pointed out that there is a need for Malaysians to promote and provide greater accessibility to the dying art form by using modern technology and digital media [6]. Che. Mohd. Nasir Yussof, a professional puppeteer who helms the wayang kulit Kelantan groups in National Arts Academy (ASWARA) affirmed that the digitalisation of Wayang Kulit Kelantan will be very helpful in delivering the related information and thus promoting this art form through internet or online social media to a larger group of audience [7]. Kaplin believes the digital form of puppetry will not mean the "death" of traditional form of puppetry, but will probably lead them to be "preserved" for their historic, spiritual or folkloric value, like endangered species on a game preserve. Digital puppetry is a revolutionary idea, for it expands the realm of puppetry beyond all definitions that center upon the materiality of the puppets [8].

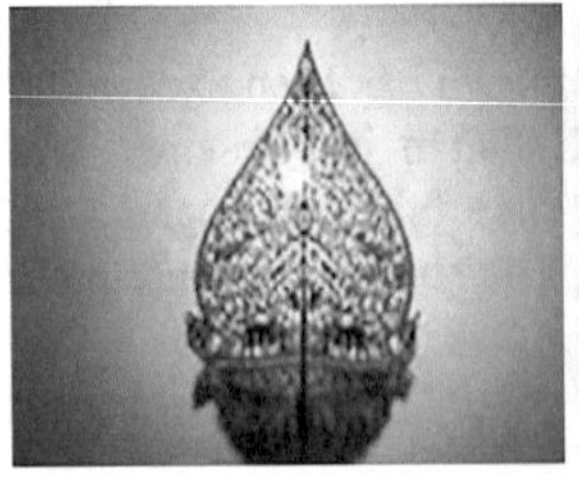
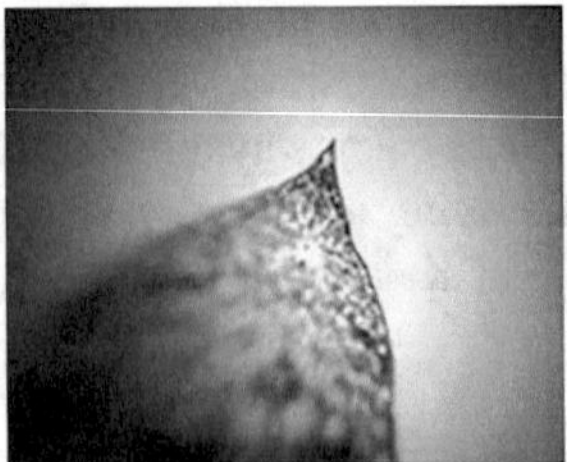
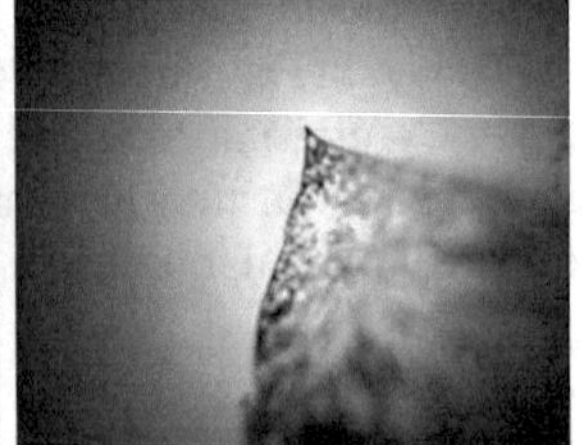

Fig. 1 The visual of wayang kulit Kelantan

2 Theories of Remediation

Remediation refers to processes whereby an old form of media blends or fuses with a new one; or vice versa. It has existed for centuries that can date back to at least the Renaissance paintings with linear perspectives. The renowned media theorist, Marshall McLuhan has mentioned: "The electric light is pure information. It is a medium without message, as it were, unless it is used to spell out some verbal advertisements or names [9]. This fact, characteristics of all media, means that the "content" of any medium is always another medium." In this paper, the author adopted the remediation theory of Marshall McLuhan, with his dictum "the medium is the message" to propose a methodological framework to remediate the visual of wayang kulit Kelantan on the tangible screen onto the virtual computer screen without losing its authenticity and identity. As what McLuhan believes, the introduction of any new form of media in a given culture can radically alters the way that members of that culture mediate between the real world and also the given values available to them.

The logics of remediation bifurcate into immediacy and hypermediacy [10]. The first logic, namely immediacy refers to the remediation that aims to erase and eliminate all the trademarks or hints of the medium and thus making that medium "disappear" or "transparent". The second logic of remediation refers to the remediation that do not aim to make the medium disappear/transparent, but rather to refashion the older medium in order to create a sense of multiplicity or hypermediacy. This logic of remediation is best expressed in web pages, desktop interface, multimedia applications, interactive programme, video games, etc., whereby videos, images, texts, animations, sound, music, etc. are represented and combined in ways to allow the audience to gain random access within the multiple media. The finished works of hypermediacy uses fragmentation, indeterminacy and heterogeneity to offer "random access" that cannot be achieved in the real world.

In both logics, the transparency is vital and important as it will enable the audience to know the remediated objects directly without being distracted or disturbed by the media. Simply put, a remediated work with high level of transparency is the one that is able to induce authentic or realistic feeling or experience from the audience, making them believe that the mediated objects or visuals presented to him are real and authentic. In the following section, the author presents the various works on the remediation of wayang kulit Kelantan and their appeal to the authenticity of the original art form.

3 Related Works on the Digital Remediation of Wayang Kulit Kelantan

The remediation of wayang kulit Kelantan can be traced back to about three decades ago, when several Malaysian researchers started to combine or apply the modern technology with the art of shadow play. For example, in 1996, under a short-term grant awarded by Universiti Malaysia Sarawak (UNIMAS), a research project entitled "Wayang Virtual" was carried out as an experimental version of the traditional shadow play. The virtual version of the traditional shadow puppets were staged together with a virtual 3D animated figure which was controlled by a human puppeteer using a computer mouse while the visuals were juxtaposed and projected on a white screen [11].

In 2011, Dahlan Abdul Ghani from Universiti Kuala Lumpur (UniKL) has developed a prototype design on wayang kulit in computer-generated environment. However, the result from his survey showed that the Malaysian audiences prefer wayang kulit in traditional manner compared to his computer-generated one. The reason is may be because the virtual puppet created by him lacks the visual aesthetics of wayang kulit Kelantan even though his aim was to preserve this traditional shadow puppet play [12].

In 2012, a group of researchers in Universiti Utara Malaysia (UUM) has invented a prototype of an interactive digital puppetry called e-WayCool to help students to learn mathematics in Malaysian Primary School Standard Curriculum. One of the objectives of this research is to preserve the wayang kulit Kelantan. However, the virtual puppet used in e-WayCool apparently denoted to Indonesian wayang kulit purwa rather than Malaysian wayang kulit Kelantan [13].

Since July 2012, an award-winning character designer-cum-founder of the Action Tintoy Studio in Malaysia, Tintoy Chua, has been collaborating with his friend, Take Huat to redesign and create a series of *Star Wars* puppets with the visual elements of wayang kulit Kelantan. They managed to invite a renowned puppeteer of wayang kulit Kelantan, Pak Dain, to perform a 25-min preview of "*Peperangan Bintang*" (Malay for "Star Wars") in October 2013 [14]. In this Wayang Fusion project, Chua and Take Huat have revamped and reinterpreted the famous characters in the popular 1977 Western science fiction blockbuster movie namely *Star Wars Episode IV: A New Hope*, which was written and directed by George Lucas. Based on Chua, one of the main objectives of this project is to promote the traditional wayang kulit Kelantan to the public, especially the younger generation who generally has lukewarm interest in watching the long-winded and slow-paced traditional performances. Nevertheless, this project seems to focus more on *Star Wars* rather than the traditional wayang kulit Kelantan, whereby the original repertoire and characters of the traditional art form have been completely abandoned. The only ostensible traditional vestige in this project is the visual elements of puppets that have been used to construct and compliment the characters of *Star Wars*.

In regard to the remediation of Malaysian wayang kulit by using digital technology and media, it is very obvious that most of the existing works paid no heed to its original visual aesthetics. Most of them depicted the visual of Indonesian wayang kulit which is very much different compared to the wayang kulit Kelantan.

Apparently the mentioned remediated works lack the logic of transparent immediacy as mentioned in the Sect. 2 of this paper. Neither can they be classified under the category of immediacy nor hypermediacy as they fail to recapitulate the visual and vibe of wayang kulit Kelantan and therefore the audience can hardly related them to the original art form, let alone induce their interest on it.

Therefore, in order to offer a solution to this conundrum as well as filling up the cap in the corpus of this body of knowledge, this paper presents a methodological framework to emulate the identical visual of wayang kulit Kelantan by using Computer-Generated Imagery (CGI).

4 Methodological Framework

Computer-generated imagery (CGI) is the application of the field of computer graphics or, more specifically, 3-Dimensional (3D) computer graphics to create special effects in animations, video games, films, television programs, commercials, simulation and printed media.

The objective of the experiments is to propose a methodological framework to emulate the plausible visual of wayang kulit Kelantan. To achieve this objective, there are three important elements to be considered; (i) puppet design, (ii) screen and (iii) light source. Similarly to an actual wayang kulit Kelantan performance, these elements must be established and ready prior to the performances. Moreover, experienced puppeteer(s) is needed to animate the puppets. For the experiment, first, we have to create the virtual puppet(s), a virtual screen and a virtual light source. Upon the completion of these elements, then only can a computer animator animate these elements in ways like an actual puppeteer, by using computer input devices such as a computer mouse.

As shown in Fig. 2, the task of emulating the visual of wayang kulit Kelantan can be divided into four phases. The first phase involves the creation of virtual

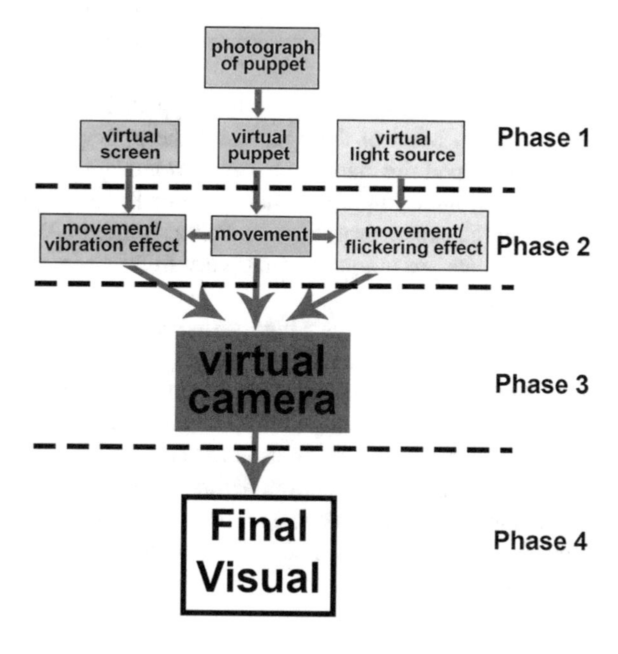

Fig. 2 The methodological framework of the experiments

puppet(s), a virtual light source and a virtual screen. The tangible puppets have to be photographed, digitalized and touched-up before being transferred (or mapped) onto the virtual objects with identical shapes and sizes. The second phase is to animate these virtual objects (puppets, light source and screen). The third phase is to create a virtual camera to capture the screen's movement/vibration effect, puppets' movement and light source's movement/flickering effect. The final phrase involves exportation of these visual elements into a final rendered video. The author presents the detailed processes and methodologies of these phases in the following sections.

5 Experiments

The puppets of traditional wayang kulit Kelantan depict near-naturalistic appearance and configurations, which is very different than the disproportioned long necks, exorbitant long arm, thin and pointed noises puppets of Indonesian shadow puppets. The existence of this traditional shadow play can be traced back to as early as 14th century, during the Majapahit period and is possibly a developed art form due to the spread of a proto-Javanese wayang kulit purwa shadow play to the northern part of the Malay Peninsula [15].

Puppets of traditional wayang kulit Kelantan are made by properly treated cowhide and mounted on bamboo sticks. They are vary in size; the shortest puppets stand about 10 cm in height while the tallest ones can be over 85. Painted with marker pen ink, the puppets of wayang kulit Kelantan today are translucent. In the performances, the shadows of these puppets appear in colors instead of solid dark that were the norm in the 1960s [16]. This is one of the most noticeable and distinctive features that differentiate the puppets of wayang kulit Kelantan from other shadow plays.

There are approximately sixty puppets in a basic traditional wayang kulit Kelantan set. Most of them represent the principal and secondary characters from the main repertoire namely *Hikayat Maharaja Wana*. Through meticulous considerations, three important wayang kulit Kelantan puppets have been selected to carry out the experiment; they are Pohon Beringin, Wak Long and Siti Dewi. These three puppets are unique and are very different in terms of their designs, sizes as well as apparatus.

The Pohon Beringin is a highly elegant tree or leaf-shaped puppet shown during the opening and closing of all wayang kulit Kelantan performances. It is unanimously perceived as the most important puppet among all. There are two types of Pohon Beringin in wayang kulit Kelantan; the first type is filled with the arabesques design of vegetal patterns on its entire surface (Fig. 3). The second type is designed and crated with the motifs from natural environments such as birds, fish, crocodiles, elephants, monkeys, snakes, tree, branches, flowers, etc. on a shape of a large leaf, tree or mountain. Both sides of the Pohon Beringin are a mirror image of the opposite.

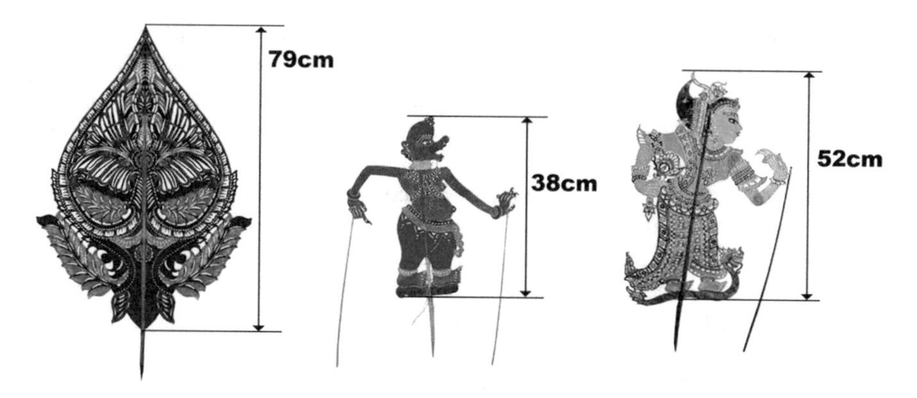

Fig. 3 The main puppets of wayang kulit Kelatan; from left: Pohon Beringin, Wak Long and Siti Dewi

The Pohon Beringin puppet comes in one piece with the height of approximately 79 cm and there is no extra joint or parts within. In wayang kulit Kelantan performances, the Pohon Beringin puppet represents the World Tree or Cosmic Mountain which connects the earth and sky. It also symbolises the cosmos, with different levels of creation manifested in it. It encapsulates the essence of all things in the performances, reflecting all phenomena and objects in the natural world. In the performances of wayang kulit Kelantan, the shape of the puppet was distorted when the puppeteer presses it towards the screen, move or shake it swiftly.

Wak Long is one of the two main clown characters in wayang kulit Kelantan performances, which is also an important comic characters not derived from the Hindu Ramayana epic [17]. With the height of 18 cm, the Wak Long puppet is always painted red and displays a comical characteristic with his broad, bulbous nose, round eye, pot-belly, large backside and sarong-clad torsos. Among all wayang kulit Kelantan puppets, Wak Long puppet contains the most sophisticate joints with two movable articulated arms, movable mouth and eyebrow.

Siti Dewi is the main heroine character in wayang kulit Kelantan. Due to her nobility and lofty princess status in the repertoire, the Siti Dewi puppet can be considered as one of the most refined puppets in wayang kulit Kelantan. Although her skin is unpainted (except for the details of her face features), she is usually clothed with highly filigreed costume, decorated with flowers and details been highlighted with gold paint. She only has one single movable hand. Her other fixed hand holds a symbol of flower or fan. She has extended, backward bent finger nails with a ring on her articulated arm. Like most of the important characters in wayang kulit Kelantan, the Siti Dewi puppet stands on a boat-or dragon-shaped vehicle with her body being positioned frontally with both feet point in the same direction.

The photographs of these three puppets were digitised, retouched and cleaned up by using computer graphic software. For the Wak Long puppet, the different joints and movable parts such as jaw and limbs need to be separated into different layers. In addition to it, the opacity map (grey-scale) files were also been created to specify the areas of opacity and transparency of the puppets. The pure white areas of the opacity map allow the corresponding areas of the puppets to be totally visible, while pure black areas cause the corresponding areas of the puppets to be completely transparent. The level of transparency in the materials is determined by the level of darkness in the opacity map.

These three puppets were modeled in 3D computer graphic application (in this case the author use Autodesk 3D Studio Max 2016). The digital photos of both puppets were mapped into the virtual puppets accordingly with their opacity files.

Lastly in the modeling stage, a virtual screen was created with its identical setting in the real world (250 cm width x 180 cm height with 77° titling forward).

The traditional Wayang Kulit Kelantan uses paraffin oil lamp as its light source. The flickering of the lamp sets the mystical atmosphere during the performance. The shimmering soft light of the lamp casts a mysterious atmosphere over the performance area, and because of the flickering, the puppets were seen from the shadow side often appears as if they were alive and breathing. In the recent years the use of the electric light bulb for this purpose often resulted in a loss of this natural quality. In order to simulate the visual of the modern wayang kulit Kelantan performance, a virtual light bulb was added in the virtual setting. It was animated to create the swinging effect just like how a puppeteer moves the hanging light bulb during the actual performances (Fig. 4).

Apart from the virtual light bulb, the author also replaced it with a flickering fire effect in another experiment. This is to emulate the fire effects of a paraffin oil lamp. The intensity of the virtual light source was animated and artificial flickering flame effect has been created and superimposed digitally onto the final visual (Fig. 5).

Next, a series of dummy objects were created in order to control the different parts of the Pohon Beringin puppet. These dummy objects are invisible on the

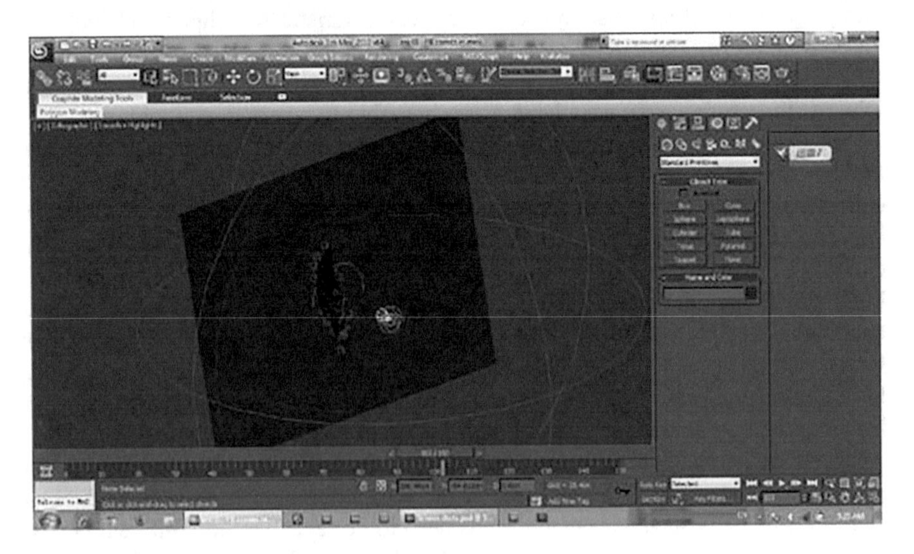

Fig. 4 The setting of virtual wayang kulit Kelantan

Fig. 5 The computer-generated fire effects as the light source

rendering/output and were linked with proper hierarchy in order to control the distortion of specific parts of the puppet (Fig. 6).

The Pohon Beringin puppet was animated with computer-generated tweening technique to perform an arc movement from right to left and touches the centre of the screen during the half-way of the movement. The term tweening in CGI software means the animator creates the keyframes (important frames of a sequence), such as the starting and ending position of the Pohon Beringin puppet, then the computer software will smoothly translate the object from the starting to the ending point. The animator can amend the movement at any point by shifting keyframes back and forth to improve the timing and dynamics of a movement; or insert additional keyframes into the 'in between' to further refine and improve the movement.

When the Pohon Beringin puppet touches the screen, the top part of the puppet was distorted and this effect was achieved by moving and rotating the dummy object on the top part of the puppet. Throughout the arc movement of the puppet, different parts of it have been distorted by manipulating the relevant dummy objects.

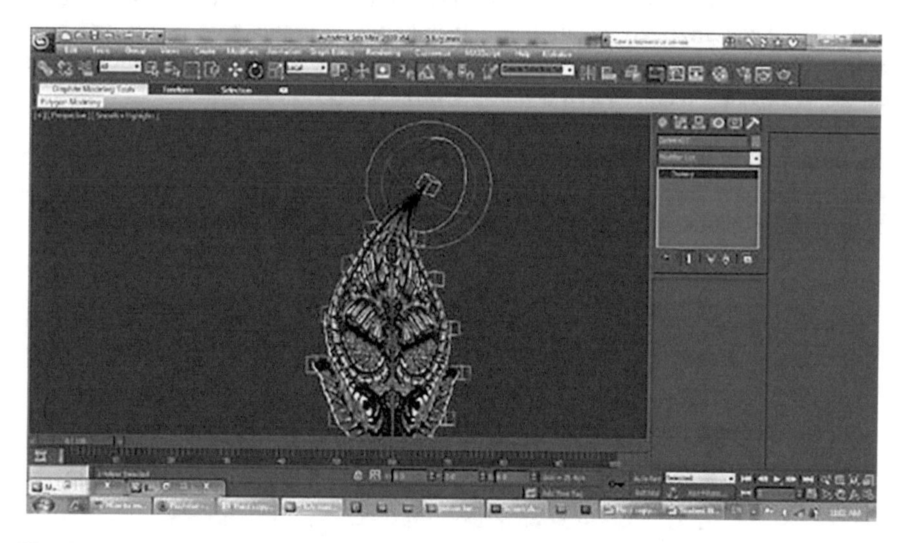

Fig. 6 The virtual Pohon Beringin with attached dummy objects

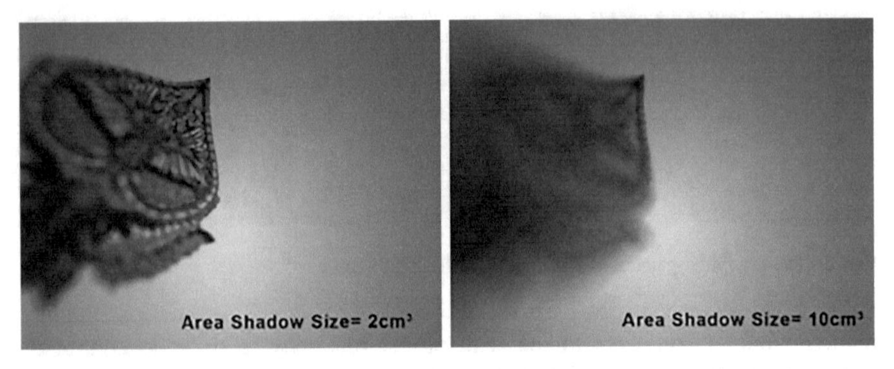

Fig. 7 Area shadow size determines the softness of the edge of the shadow

Generally the computer-generated shadow has two important parameters to control the transparency of shadow and the softness of its edges. The area shadow function creates a soft edge that becomes more noticeable as the distance between the object and the shadow increases. The softness of the shadow's edges was determined by the sizes of area shadow. The bigger size of the area shadow will make the softness of the shadow's edge become more distinctive (Fig. 7).

Both of the parameters have been activated and adjusted in this experiment. In addition to it, the cast colourful shadows function in the materials of the puppets also need to be activated.

The Wak Long puppet consists of two moveable articulated arms. The puppeteer controls and moves the arms of the puppet by manipulating the sticks connected to the end of each articulated arm. Each arm of the Wak Long puppet consists of two movable parts. In virtual wayang kulit Kelantan environment, all of these moveable parts were modeled and texture-mapped using the methods discussed earlier. They were later been arranged, positioned and linked with proper axis points and hierarchy. These parts were later being rotated and animated to mimic the movements of the Wak Long puppet in an actual performance (Fig. 8).

The movements of eyebrow and mouth of a tangible Wak Long puppet were controlled by a string. The moveable eyebrow of Wak Long puppet is actually a bent bamboo attached to a string. By pulling the string downward and releasing it to its original position, the puppet's eyebrow and mouth will move simultaneously.

Fig. 8 The movements of the virtual Wak Long puppet

To achieve the above mentioned visual, a virtual bent thin cylinder was modeled to operate as the eye brow of the puppet with a virtual string attached to it. The elasticity of the stick was manipulated and animated synchronously with the movement of its mouth and string; creating the visual of moveable mouth, eyebrow and string (Fig. 9).

To further investigate the capabilities of computer technology in animating the facial expression of puppets, another computer experiment has been conducted on Siti Dewi puppet. Like most of the puppets of wayang kulit Kelantan, the Siti Dewi puppet is craved with her face in profile. The structure and design of Siti Dewi puppet is very much constant and standard with a small nose, red mouth, almond-shaped eye and forehead in a slightly concaved line in profile of her configuration. However, unlike Wak Long puppets which have two moveable mouth pieces, such apparatus does not exist in Siti Dewi puppet. Therefore it is impossible in any actual wayang kulit Kelantan performances to have Siti Dewi puppet's facial expression being animated.

Nevertheless, in the virtual wayang kulit Kelantan environment, the mouth movements and facial expressions of Siti Dewi puppet can be achieved by utilizing the morphing function in computer animation software. A set of virtual Siti Dewi puppets with different mouth shapes and facial expressions (i.e. closed eye, wide-opened eye, aggressive eye, etc.) must be created prior to the animation process. With a principal virtual Siti Dewi puppet being selected, the morph modifier was applied onto it. Under the parameters of the morph modifier, the remaining virtual Siti Dewi puppets have been selected as its morph targets. Thenceforth, by animating the percentage of the morph targets, the authors managed to generate the facial expressions of a virtual Siti Dewi puppet as shown in Fig. 10.

Fig. 9 The movements of the mouths, eyebrow and string of Virtual Wak Long Puppet

Fig. 10 The facial expression of Siti Dewi puppet

One of the flies in the ointment of these experiments is that all the animations were not done in "real time". "Real-time" in virtual shadow play refers to a synchronicity between the puppeteer's control and the puppet's resultant movements [18]. Even though the animation process in virtual wayang kulit Kelantan eliminates the human errors, it also disconnects or halts the interaction between the puppeteer (in this case, animator) and the audience. Therefore in the final experiment, human agent in the animation process has been removed with the use of a motion capture devise. A total of five reflective markers have been attached on the various parts of the Pohon Beringin puppet while the puppeteer wore the standard motion capture suit with reflective markers attached on the different parts of the suit. OptiTrack Arena capture system was used in this experiment. It is an optical motion capture system with sixteen cameras set up, easy skeleton creation, multiple actors/props tracking, real-time solving and streaming, editing tools and with flexible data export options for full human body motion tracking function integrated with real-time Skeleton Solver software. This optical motion capture system uses real lights, multiple cameras and various reflective dots to determine and capture three-dimensional position and it gives freedom to the puppeteer to perform any motion with his puppet.

A virtual biped humanoid structure was firstly created to enable the transfer and manipulation of data obtained from the motion capture sensors. The movements of both puppeteer and Pohon Beringin puppet were then captured and transformed into a virtual biped humanoid character to achieve the absolute synchronicity in human movements and puppet's distortion. The author then rendered the view of a virtual camera located on the opposite side of the virtual screen (Fig. 11).

With the use of motion capture device, there is no need to acquire skillful computer animator to animate the virtual puppets. It produces greater immediacy

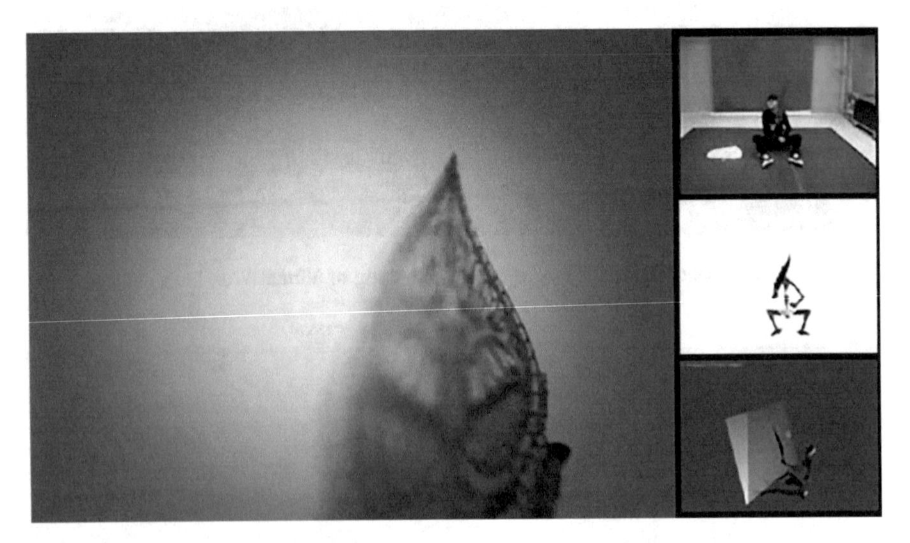

Fig. 11 Using motion capture device to preserve the movements of puppeteer and puppet's distortion

and transparency in puppet's movements and distortions and thus making the audience to believe that the computer-generated visual of wayang kulit Kelantan as authentic and real.

6 Results and Analysis

The author used the Interpretative Phenomenological Analysis (IPA) to analysis and validate the results of experiments. It is a widely-used and popular approach to psychological research that involves a close examination of the experiences and meaning-making activities of a small number of participants who have been carefully culled via the purposive sampling method [19].

With the use of this IPA's research methodology, two renown and well-established researchers in the related field—Prof. Dr. Ghulam-Sarwar Yousof and Prof. Dr. Patricia Matusky as well as three professional puppeteers of wayang kulit Kelantan—Che. Mohd. Nasir Yussof, Mohd. Kamarulbahri bin Hussin and Rahim bin Hamzah have been chosen to answer a series of semi-structure questions during interviews. To facilitate the necessary reflexivity, the author has conducted formal and informal interviews with them separately from time to time since 2009. Besides, in order to build rapport with them and to ensure the collected data are rich, the author has accompanied the mentioned academic researcher(s) to attend several wayang kulit Kelantan performances performed by these puppeteers in Kuala Lumpur and Kelantan ever since this research started in 2009. These data were collected and analyzed systematically adhered to the rigorous procedures of IPA.

The results of the analysis showed that the above-mentioned experiments using the proposed methodological framework had successfully produced plausible and convincing visual of wayang kulit Kelantan. The differences between the visual of computer-generated wayang kulit Kelantan with the actual one are indistinguishable (or very subtle). These computer-generated visual elements comprise distinctive shadow's colours, distortions, softness of edges, generated abreast with the movements of the virtual puppets and light source.

7 Discussion

In addition to the capabilities of simulating the plausible visual of wayang kulit Kelantan, the author has also found out that all computer-generated visual elements in the above-mentioned experiments can be adjusted and animated. These visual elements include the colour and intensity of light source, the softness of the edges of shadows and the transparency of the screen. It is very difficult (or maybe impossible) for puppeteers to control these elements in any actual performances. This could be another heuristic discovery which can be quite useful in the conservation of the visual of wayang kulit Kelantan.

8 Conclusion

The objective of this paper is to propose a methodological framework of using CGI to emulate the visual of wayang kulit Kelantan. The results of the experiments have shown a high degree of verisimilitude in terms of puppets' movements, colours of shadows (translucency of puppets), movements of light source, distortions of puppets, softness of the edges of shadows, movements of eyebrow, mouth and string which was attached to the puppets. Moreover, one of the experiments also corroborated the capability of computer-graphics in simulating the facial expressions of puppet, which the author believes is very conducive towards the creation of digitally animated wayang kulit Kelantan.

The author hopes the experiment results will boost more creation and adaptation of wayang kulit Kelantan's visual in movies, animations, video games and films. It is also hoped that with the advent of computer technology, the simulation of virtual wayang kulit Kelantan will become more user-friendly and accessible. Last but not least, the author hopes to see more efforts in the creation of virtual wayang kulit Kelantan as both of them share the similarities in their visuals on screen and the author believes they can co-exist and benefit each other.

In summary, the work presented here focuses on the emulation of the visual of wayang kulit Kelantan with the use of CGI. On one hand, the work by Ghani [12] considers only the 3D puppet modeling without referring to the actual setting, materials, techniques and visual of wayang kulit Kelantan. On the other hand, the work by Tan et al. [6] takes a different approach based on the visual simulation and interactive animation of a Javanese shadow play puppet by using OpenGL technique. While the present study is related to the diminishing wayang kulit Kelantan, it has successfully remediated the visual of wayang kulit Kelantan from tangible muslin screen to digital screen without compromising on its immediacy and transparency. This is an aspect that was not considered, or has been ignored, in other earlier works.

9 Future Research Directions

It is well known that we are living in a world which is full of uncertainty especially on the potentials and capabilities related to computer technologies [20]. The rapid development in information technologies has created many new research fields and one of them is Computer-generated Imagery (CGI) [21]. It is a new body of knowledge with less than six decades of history which encapsulates a broad and ever growing spectrum of practices and methodologies. Hence it is impossible for the author to cover all capabilities and possibilities of CGI and relate them with Malaysian wayang kulit in this paper. Topics such as virtual/augmented reality, immersive technology, inverse kinematics, character rigging and interactive methods related to the remediation of wayang kulit are still very limited and are worthy to be studied; therefore the author would recommend further research to be conducted, but not limited, on these topics.

Acknowledgements This paper is a revised and expanded version of my paper entitled "A Framework to Emulate the Visual of Malaysian Shadow Play with Computer-generated Imagery (CGI)" presented in the First EAI International Conference on Computer Science and Engineering, NOVEMBER 11–12, 2016, PENANG, MALAYSIA.

References

1. Van Ness, C. E., & Prawirohardjo, S. (1980). *Javanese wayang kulit*. Malaysia: Oxford University Press.
2. Yousof, G. S. (2004). *The Encyclopedia of Malaysia: Performing arts*. Singapore: Archipelago Press.
3. Sweeney, A. (1972). *The Ramayana and the Malay shadow-play*. Kuala Lumpur: Penerbit Universiti Kebangsaan Malaysia.
4. Osnes, M. B. (2010). *The shadow puppet theatre of Malaysia—A study of wayang kulit with performance scripts and puppet design*. Carolina: McFarland & Company Inc.
5. Khor, K. K., & Yuen, M. C. (2009). A study on the visual styles of wayang kulit kelantan and its capturing methods. In *Computer Graphic, Imaging & Visualisation (CGIV), 6th International Conference on Computer Graphics, Imaging and Visualization* (pp. 423–428). China: Tianjin University.
6. Tan, K. L., Talib, A. Z., & Osman, M. A. (2008). Real-time visual simulation and interactive animation of shadow play puppets using OpenGL. In *Proceedings of World Academy of Science, Engineering and Technology* (Vol. 35).
7. Lugiman, F. A. (2006). *Performing art traditional heritage: implementation of website to convey information of wayang kulit*. Shah Alam: Universiti Teknologi MARA (UiTM).
8. Kaplin, S. (1994). Puppetry into the next millennium. *In Pupptry International, 1,* 37–39.
9. McLuhan, M. (1964). *Understanding media: The extensions of man*. New York: McGraw-Hill.
10. Bolter, J. D., & Grusin, R. (2000). *Remediation: Understanding New Media*. Ann Arbor: MIT Press.
11. Abdul Wahid, H. (2007). The integration and an experiment of a traditional wayang kulit performance and electroacoustic music. In *The languange of electroacoustic music* (pp. 1–7).
12. Abdul Ghani, D. (2011). wayang kulit: Digital puppetry character rigging using Maya MEL language. In *4th International Conference on Modeling, Simulation and Applied Optimization (ICMSAO)* (pp. 1–4). Kuala Lumpur: IEEE.
13. Jasni, A., Zulikha, J., & Mohd-Amran, M. A. (2012). eWayCool: Utilizing the digital wayang kulit for mathematic learning. In *AWER Procedia Information Technology & Computer Science*.
14. Mayberry, K. (2014). Star wars meets Malaysia's arts renaissance. Retrieved January 12, 2015, from Al Jazeera Live News: http://www.aljazeera.com/indepth/features/2014/09/star-wars-meets-malaysia-arts-renaissance-201492261333584683.html.
15. Yousof, G. S. (2004). *Panggung Inu: Essays on traditional Malay theatre*. Singapore: National University of Singapore.
16. Yousof, G. S., & Khor, K. K. (2017). wayang kulit Kelantan: A study of characterization and puppets. *Asian Theatre Journal*, *34*(1), 1–25 (2017).
17. Matusky, P. (1997). *Malaysian shadow play and music—Continuity of an oral tradition*. Malaysia: The Asian Centre.

18. Tillis, S. (2001). The art of puppetry in the age of media production. In *Puppets, mask, and performing objects* (pp. 173–185). US, New York: University and Massachusetts Institute of Technology.
19. Smith, J. A., & Eatough, V. (2012). Interpretative Phenomenological Analysis. In G. M. Breakwell, J. A. Smith, & D. B. Wright (Eds.), *Research methods in psychology* (4th ed., pp. 439–460). London: SAGE Publishing.
20. Vasant, P. (2014). *Handbook of research on artificial intelligence techniques & algoriothms*. Pennsylvania: IGI Global.
21. Vasant, P. (2014). *Handbook of research on novel soft computing intelligent algorithms: theory and practical applications*. Pennsylvania: IGI Global.

Printed in the United States
By Bookmasters